转基因生物风险评估与风险管理

——生物安全国际论坛第五次会议论文集

RISK ASSESSMENT AND RISK MANAGEMENT OF THE GENETICALLY-MODIFIED ORGANISMS

—— Proceedings of the International Biosafety Forum-Workshop 5, Beijing, May 27-28, 2013

主　编　薛达元

副主编　王艳杰　CHEE Yoke Ling

中国环境出版社・北京

图书在版编目（CIP）数据

转基因生物风险评估与风险管理：生物安全国际论坛第五次会议论文集/薛达元主编．—北京：中国环境出版社，2014.5

ISBN 978-7-5111-1743-4

Ⅰ．①转…　Ⅱ．①薛…　Ⅲ．①转基因技术—生物工程—安全管理—国际学术会议—文集　Ⅳ．①Q788-53

中国版本图书馆 CIP 数据核字（2014）第 031357 号

出 版 人　王新程
责任编辑　张维平
封面设计　宋　瑞

出版发行　中国环境出版社
（100062　北京市东城区广渠门内大街 16 号）
网　　址：http://www.cesp.com.cn
电子邮箱：bjgl@cesp.com.cn
联系电话：010-67112765（编辑管理部）
010-67112738（管理图书出版中心）
发行热线：010-67125803，010-67113405（传真）

印　　刷　北京中科印刷有限公司
经　　销　各地新华书店
版　　次　2014 年 5 月第 1 版
印　　次　2014 年 5 月第 1 次印刷
开　　本　787×1092　1/16
印　　张　17.5
字　　数　410 千字
定　　价　66.00 元

前　言

据国际农业生物技术应用服务组织（ISAAA）通报，在2013年，全球27个国家有超过1 800万农民种植转基因作物，种植总面积达1.75亿 hm^2，比2012年增加3%，即增加500万 hm^2。与1996年转基因作物的种植面积170万 hm^2 相比，17年来增长超过100倍。

中国转基因作物种植面积约420万 hm^2，主要为抗虫棉，是继美国、巴西、阿根廷、加拿大、印度后，转基因作物种植面积大小为第六的国家。截至2013年，中国批准种植的转基因作物只有棉花和木瓜。虽然中国的转基因水稻已在2009年就获得安全证书，但至今尚未得到商业化生产的批准。

对于转基因农作物的安全性，国内外一直存在激烈争议，科技界也未有定论。最近几年，国内就转基因水稻的商业化问题也展开了激烈的辩论，同样没有结论。赞成者认为，经过多年的栽培实践和技术改良，大量的转基因农作物已被证明对人体和对环境安全无害；而反对者则认为，转基因农作物对人体和环境的影响具有不确定性，与传统食品相比，转基因食品存在潜在风险。2013年10月，我国61名院士联名致信，请求政府批准转基因水稻产业化，使转基因问题再次引发关注。

转基因生物对生态环境、人类健康和社会经济影响的问题已成为全球热点。2000年在《生物多样性公约》下产生了专门针对转基因生物的《卡塔赫纳生物安全议定书》，进而在2010年，又在《卡塔赫纳生物安全议定书》下产生了一项新的针对转基因生物损害的国际法——《有关赔偿责任与补救的名古屋-吉隆坡补充议定书》。另外，2012年，《卡塔赫纳生物安全议定书》第6次缔约方大会针对风险评估和社会经济影响等作出相关决定，也推动了履行议定书相关议题的工作。

为了研讨国际和国内有关转基因生物安全的最新进展，促进国家之间和专家之间、以及管理人员与研究人员之间的信息交流，经环境保护部和国家民族事务委员会正式批准，由环境保护部南京环境科学研究所与中央民族大学以及第三世界网络（TWN）主办，中国生态学学会民族生态学专业委员会、中国环境科学学会生态与自然保护分会以及中国民族地区环境资源保护研究所协办，于2013年5月27—28日在北京召开了“生物安全国际论坛第五次研讨会”。

“生物安全国际论坛”（IBF）是一个半官方论坛，始于2004年。在过去的10年中一共召开5次国际研讨会，有500多中外人员参加了论坛。该论坛的显明特点是为所有利益相关方提供了公平和自由交流的平台，包括研究人员、政府管理人员、生物技术公司、媒体、非政府社会团体及公众等，本论坛促进了中国专家与国外专家、政府与非政府机构、消费者与生物技术公司、媒体与公众之间在转基因生物安全问题上的交流。

本次研讨会的参会代表共计65名，分别来加拿大、马来西亚、美国、南非、荷兰、奥地利、拉脱维亚和中国等多个国家和相关国际机构。国内参会者包括来自中国环境保护部、外交部、农业部、国家林业局和国家质检总局等政府部门的官员和代表，高校和科研院所的专家与研究人员，第三世界网络（TWN）及绿色和平等非政府组织代表，以及对转基因生物安全感兴趣的公司、媒体和公众代表。

本次研讨会设两个主题报告和三个专题，大会进行了18个学术报告及开放式讨论。

主题报告：生物安全独立研究的最新科学进展。复旦大学生命科学学院生态与进化生物学系主任卢宝荣教授做了题为“转基因生物技术的利益及其生物安全挑战”的报告。奥地利环境保护局生物安全主管官员 Andreas Heissenberger 博士做了题为“转基因风险评估的原则、挑战及方法”的报告。

专题 1：转基因生物的风险评估进展。该专题主要根据《卡塔赫纳生物安全议定书》（以下简称《议定书》）的“风险评估与风险管理指南”而进行的风险评估国际标准的讨论，包括中国转基因生物安全的管理进展，《议定书》下风险评估标准的进展，转基因生物风险评价中的主要考虑，以及转基因水稻、转基因油菜、转基因棉花等对环境和非靶标害虫以及植物病害等方面影响的研究。

专题 2：转基因生物相关的社会经济影响。该专题主要讨论了转基因生物决策过程中考虑的社会经济因素。《议定书》允许各国在决策过程中考虑转基因生物的社会经济因素，制定相关国家生物安全法规。但是《议定书》并没有规定具体的评价标准和方法，而且调查数据不足，导致各国实施困难。该专题讨论了中国民众对转基因棉花的认知程度调查，世界各国应对转基因生物社会经济影响而采取的政策制度，南非转基因棉花的种植情况，全球和区域条件下转基因生物的社会经济影响，包括粮食安全、人类健康及经济效益等。

专题 3：转基因生物的赔偿责任与补救。该专题针对《有关赔偿责任与补救的名古屋-吉隆坡补充议定书》进行了讨论，阐述了生物安全损害、环境损害、生物安全损害赔偿、补救等概念，针对目前尚无“生物安全损害赔偿”案例的现状，分析各国有关生物安全损害及赔偿的立法、实践及面临的挑战，并提出相关建议。

为了让与会者更多的发表自己的观点，专题报告结束后，还设置了开放式讨论环节。大家针对转基因生物的损害赔偿机制、案例及转基因生物对环境、人类健康的影响，转基因生物安全的立法、监管方面等进行讨论。

本次会议得到环保部生物安全管理办公室、环保部国际合作司和中央民族大学国际合作办公室的大力支持，第三世界网络（TWN）为本次研讨会提供了资助，中央民族大学生命与环境科学学院及中国民族地区环境资源保护研究所的师生提供了大量的会务支持工作。正是大家的共同努力，才使本次国际论坛取得圆满成功，我对所有的支持和帮助表示诚挚的谢意。

我特别要感谢我的同事武建勇博士、赵富伟博士、吴力老师、成功老师和研究生王艳杰、杨京彪、刘春晖、陈晨、张家楠、王新、朴金丽、王云靓、关晴月、王程、戴蓉、杜玉欢、梁晨、胡晓燕、郑燕燕、高英等对会议组织和论文集编辑的贡献。

本论文集中的论文代表作者本人的观点，可供相关管理人员、研究人员、学校师生和公众参考。论文集编辑中难免出现错误和疏漏之处，敬请读者批评指正。

薛达元
2014 年 3 月 26 日于北京

目　录

开幕式致辞
Open Ceremony Address

主题报告
Keynote Presentations

专题 1　风险评估

Session 1　Risk Assessment of GMOs

专题 2 转基因生物相关的社会经济影响
Session 2 Socio-economic Considerations related to GMOs

会议闭幕
Closing Remarks

附 录
Appendix

开幕式致辞
Open Ceremony Address

环境保护部国家生物安全管理办公室王捷处长的致辞

Address by Mr. WANG Jie，Division Director，National Biosafety Management Office，Ministry of Environmental Protection（MEP）

各位领导、专家，女士们，先生们，大家上午好：

首先请允许我代表国家环保部生态保护司对生物安全国际论坛第五次会议的召开表示热烈的祝贺！同时，也对各位专家特别是远道而来的外国专家表示热烈的欢迎。预祝大会取得圆满成功。

转基因生物安全问题确实是国际社会所关注的焦点，现在民众越来越关注，尤其是涉及到食品安全和社会经济发展等敏感问题，2010 年国际上通过了《有关责任赔偿与补救的名古屋-吉隆坡议定书》后，《卡塔赫纳生物安全议定书》的履约行动将步入一个新的轨道。实际上，现在各个国家也都面临着如何发展生物技术和处理好生物安全这一矛盾的问题。我们人类既要发展，也要特别地关注未来的生物安全。

这已经是我第五次参加生物安全国际论坛的研讨会，前四次研讨会我都参加了，这个国际论坛与《生物多样性公约》和《卡塔赫纳生物安全议定书》的关系密切。我本人作为中国政府代表团成员，参加了《生物多样性公约》第一次缔约方大会和《卡塔赫纳生物安全议定书》第一次缔约方大会，前不久参加了在印度召开的《生物安全议定书》第六次缔约方大会。我认为转基因生物安全领域目前处在一个发展比较平稳的状态，有一些问题还没有真正搞清楚，在法律法规，尤其是国际公约这方面有了一定的进展，但是进步不明显，相对比较平稳。

中国政府为了更好地履行《生物安全议定书》国际义务，成立了涉及 24 个部委的履约协调组。一些关于生物安全的重大问题需要履约协调组进行讨论，可以利用该履约协调机制，开展一些重大的履约行动。

中国政府高度重视转基因生物安全问题：第一，制定了《农业转基因生物安全管理条例》，这是条例主要针对农业转基因生物的安全管理，中国还需要一部综合性的法规，用以管理除农业以外的多种转基因生物，环保部门早在 20 世纪 90 年代初就已着手制定生物安全综合法律，但由于部门协调困难，这项工作进展缓慢。当然，如果一个引领性的大法不能够很快出台，那么也可以考虑制定一些条例、管理办法，或者相关技术标准和规范等。第二，明确了部门的管理职责，监管机制不断完善。生物安全领域涉及多个部门，部门之间需要很好地协调，特别是关键的几个部门，像环保、农业、质检、林业、海洋等，需要相互沟通信息，相互配合，现在也正在开展这一方面的工作。第三，健全风险评估的体系，加强科技支撑。因为国际社会履行议定书的步伐相对平稳，实际上在等待一些科学研究的结果，因为任何事情都需要科学支撑，臆想是站不住脚的。所以，我们现在特别关注风险评估体系这一方面的工作，需要加强科技支撑，比如农业部门已在农业转基因生物安全领域也做了长期监测工作，环保部门也在河南安阳对转基因棉花进行了近 10 年的长期监测工作。还有其他相应的部门也都从事了若干基础性科研工作。

此外，需要积极加强信息的获取和交流，《生物多样性公约》秘书处在履行《生物安全议定书》方面花了许多精力。做出了一个特别好的信息交换机制（BCH）。中国当时也在这一方面开展了工作，但现在看来我们所做的工作并不是真正的BCH，还没有达到这个水准。所以今后要和联合国《生物多样性公约》秘书处接洽，在国内相关部门之间加强沟通。我们正在积极开展这方面工作，包括制定规划，邀请该领域有经验的国际专家来帮助建立国家 BCH 机制，目的就是为了更好地履行《生物安全议定书》。

最后是要加强公众的宣传教育。目前来看，转基因生物从科学角度来讲，存在不确定性风险。到底会对人类、生态环境、社会经济造成什么程度的影响，还在探索之中。因此，我们要坚持几个原则：①预防为主的原则，需要采取预防措施，没有风险更好，有了风险将来就无法控制了；②发展原则，在保证生物安全这个前提下，开展现代生物技术研究是必要的；③风险评估原则，要明确存在的风险，风险评估工作做不好我们心中就没有底数；④知情同意原则，有关批准转基因商业化生产等重大事件要让公众知道，并让公众共同参与生物安全管理。

最后我想表一个态度，中国政府对生物多样性和生物安全领域是高度重视的，而且我们各个部门之间也是相互的支持。我们除了发展现代生物技术，同时也会高度关注生物安全问题，没有安全就没有一切。

再次预祝研讨会取得圆满成功，也预祝各位代表在北京生活愉快，工作顺利，谢谢！

外交部条法司付长华处长的致辞

Address by Mr. FU Changhua，Division Director，Ministry of Foreign Affairs

主席先生，各位专家：

大家上午好！

这个会议很重要，所以北京特意下了今年春天的第一场雨，来欢迎大家，我很高兴参加这个研讨会，我代表外交部条法司对这个会议表示祝贺，我在的部门是负责多边环境公约谈判的部门，外交部也会同环保部和农业部等部门，参与了生物安全领域的国际谈判。

刚才王捷处长把我说的一些话已经说过了，因为中国政府历来比较重视生物安全问题，参加了《卡塔赫纳生物安全议定书》谈判的全过程，对议定书的拟定和通过做出了贡献。中国政府于 2005 年核准加入议定书，这表明我们愿意与国际社会一道加强生物安全的管理。2011 年，中国政府还同意议定书适用于香港行政区，中国还按照议定书的要求先后两次提交履约报告，中国政府也正在对是否加入《赔偿责任与补救议定书》（全称为《卡塔赫纳生物安全议定书关于赔偿责任与补救的名古屋-吉隆坡补充议定书》）进行深入的研究。

中国政府在国内也积极开展各项履约工作，已经颁布了《农业转基因生物安全管理条例》，成立了以副总理为主任的生物多样性国家委员会，统筹包括生物安全在内的《生物多样性公约》的履约问题。2012 年，中国政府通过了“联合国生物多样性十年行动计划——中国行动方案”，将以国际合作的方式，进行国内立法监督和推进生物技术的发展，解决转基因生物安全问题。

各位专家，生物安全问题涉及到生物多样性，涉及环保、农业、贸易等多个领域，我虽然不是科学家，但也了解到生物安全是一个涉及科学前沿的技术问题，还涉及立法管理，以及各个部门的协调问题，内容比较复杂。中国是最大的发展中国家，又是生物多样性最丰富的国家之一，中国还是农业大国，以及农产品进口大国。同时，中国政府也在鼓励利用现代生物技术发展农业，所以，如何在发展经济和保护环境的同时保障生物安全是我们面临的一项重要任务。中国在履行议定书方面还面临一些挑战，希望通过加强国际合作把这项工作做好。此外，中国也呼吁发达国家遵守《生物多样性公约》的有关承诺，包括为中国在内的发展中国家提供资金、技术和能力方面的帮助。

祝会议圆满成功，谢谢！

环境保护部南京环境科学研究所李维新处长的致辞

Address by Mr. LI Weixin，Division Director of Nanjing Institute of Environmental Science，MEP

各位来宾，女士们、先生们：

大家早上好！

由环境保护部南京环境科学研究所、中央民族大学、第三世界网络（TWN）共同主办，中国生态学学会民族生态学专业委员会、中国环境科学学会生态与自然保护分会、中国民族地区环境资源保护研究所协办的“国际生物安全论坛第五次研讨会”今天在北京召开。我谨代表环境保护部南京环境科学研究所对这次会议的成功举办表示热烈的祝贺，对参加此次研讨会的各位专家和朋友表示诚挚的敬意，并对参与此次会议的筹备机构和人员以及为此次会议提供资金支持的机构表示衷心的感谢。同时，借此机会感谢环境保护部，特别是国家生物安全管理办公室多年来对“国际生物安全论坛”历次研讨会的支持。

20 世纪 80 年代以来，由于基因重组及其转化技术的发展和应用，使生物技术的发展进入崭新的阶段，并且产生了巨大的经济效益。然而，转基因生物具有科学上的不确定性，这种不确定性决定了转基因生物的商业化生产和大规模应用具有一定的风险。如何评价这些风险并对风险进行管理，是目前科学管理和可持续发展的基本要求。

环境保护部南京环境科学研究是环境保护部的直属研究机构，是在国内较早开展转基因生物安全的研究机构之一。2001 年，创建了“国家环境保护生物安全重点实验室”，并于 2004 年正式成为部级重点实验室。从实验室成立以来，已承担了多项国际国内与生物安全相关的重大研究项目，开展了有关转基因生物环境安全性评价、环境影响监测以及转基因成分检测等方面的研究工作，为我国生物安全环境决策提供了有力的技术支撑，同时，也为我国履行《生物安全议定书》提供了技术支持。

环保部南京环境科学研究作为《生物多样性公约》的技术支撑单位，一直积极参与《生物安全议定书》和《赔偿责任与补救补充议定书》的相关国际谈判及其他相关事务。为了建立一个好的用于讨论转基因生物安全的交流平台，2004 年，环境保护部南京环境科学研究所经环境保护部批准，发起了“国际生物安全论坛”。论坛已于 2004 年、2005 年、2008 年和 2011 年主办过四次研讨会，历次研讨会都得到了环境保护部和国家生物安全办公室的支持，并取得了圆满成功。

女士们、先生们，转基因生物安全是国际和国内的热点问题。转基因生物安全既是科学问题，也是社会问题；转基因议题涉及食品安全和社会经济等敏感问题，受到国际社会的广泛关注。《生物安全议定书》的履行已经进入实质性阶段，但是，生物安全研究和风险评估还面临诸多挑战，许多相关问题还需要广泛研讨。因此，希望此次研讨会的举办能为相关问题的解决提供一些思路和建议。

最后，预祝本次研讨会圆满成功！谢谢！

中央民族大学国际合作处何克勇处长的致辞

Address by Prof. HE Keyong，Director of International Coorperation Office，Minzu University of China

各位，上午好：

刚才几位领导发言都提到了转基因生物安全问题，对于我们来讲，我本人作为一个消费者，尤其在中国的消费者，最直接面临的问题就是食品安全。转基因食品出现后，争议不断，中央民族大学能够为围绕转基因生物及相关产品引出的安全性问题做一些贡献，我们感到非常荣幸。

为什么说这样的贡献非常重要？因为没有学者的参加，没有学者的宣传，没有一个好的平台，我想就不会让更多的公众来关注这个问题。正因为有了广大公众的关注和参与，不管是政府还是社会各界所产生的一些影响，才会使我们围绕生物安全这个问题，研究制定一些更可靠、更有效的法规，采取一些更安全的措施。

在我的记忆当中，中央民族大学是第三次举办这个系列会议，借此我花两分钟的时间稍做介绍。中央民族大学在中国的高等学校当中比较小，全校学生只有17 000人，其中5 000多人是硕士博士研究生，还有将近 1 000 人是国际留学生，中央民族大学是一个综合性的大学，也拥有在中国高等教育当中比较熟悉的几个词，是一所 211 工程大学，也是一所 985 工程大学，得到了中央财政及其他方面支持的这么一个大学。说它是综合性但是也不全，我们提供十大门类学科的教育和研究，但是我们没有农业，没有军事，一共有 23 个学院。但是特点突出，第一是其专业和教学研究，由于叫民族大学，在全世界除了加拿大有一个民族大学之外，没有在第二个地方听说过民族大学，即便在前苏联也没有听说过，但是中国中央政府建立这样的大学，就是为了主要服务于中国的少数民族地区，以及为少数民族地区经济、社会、文化发展传承产生作用，这是它的一个使命。中央民族大学从建校开始主要是由三四所大学和科学院的人聚集起来，主要是清华、燕京和北大，开始它的起点比较高，到今天为止学校的民族学和人类学的最新排名，仍然是中国第一，有一些专业在中国学科排名位列前茅。

另外一个特点，就是我们的学生主要是少数民族，我们学生当中 60%～65%是少数民族学生，中央民族大学的校园是中国每年招生都可以看到 56 个民族的地方，由于学生的多元性，所以学生之间的相互宽容和相互尊重，是我们学校的一个特点。

另外就是我们的研究主要围绕与民族有关的一些问题，中央民族大学同时作为中央政府在民族事务方面的一个智库。还有，中央民族大学一直比较重视国际合作，到今天中央民族大学已经跟世界 120 多个大学和研究机构建立了合作关系，其中也有不少是一流大学，包括今天来的这几位外国专家所在的大学，我们都建立了关系。因此我们想借这个机会，表达我们的合作意愿。我们非常鼓励和欢迎大家跟中央民族大学的学者教授，在围绕不管是生物多样性、生物安全，还是其他领域方面进行广泛而深入的合作。

最后，预祝这次会议圆满成功，非常高兴今后大家有机会除了参加这样的会议之外，再到我们学校来。谢谢！

第三世界网络（TWN）代表 Lim Li Ching 的致辞

Address by Ms. Lim Li Ching，the representative of the Third World Network（TWN）

女士们、先生们：

谢谢各位，首先我想说非常荣幸今天能够跟大家共同举办这个会，比如说与环保部南京环境科学研究所和中央民族大学，我们已经合作了很长的时间，特别是与中央民族大学合作了多个生物安全的项目，从 2004 年至今，我们已经共同举办了五次会议。

作为观察员身份，第三世界网络参加了一系列相关会议，也参加了 CBD（生物多样性公约）及相关领域的谈判，我们认为中国完全有能力发挥重要作用，特别是推动了生物安全方面的研究工作。目前世界上相关的一些国际事务都有中国的谈判代表参加和介入，同时，我们第三世界网络也承担了一些相关义务。充分说明，有一点是非常重要的，即中国的确非常重视生物安全方面的研究和监管。

我来自马来西亚，马来西亚也非常重视生物安全，而且也承担了这方面的国际义务，同时也面临其他方面的一些压力，如商业化批准的压力。此外，我们需要掌握一些技术，并承担相关方面的责任，这的确是一种挑战。我们祝贺中国在这方面所取得的一些进展，特别是在研究方面，中国有很多科研机构介入到这一领域，我觉得中国所突出的一些重点是非常重要的，其中不乏很好的例子，对其他发展中国家也是值得学习的一个典范。

另外，其他国家也在关注中国在生物安全方面的进展，中国向其他发展中国家发出一些积极的信号，中国的确在生物安全方面起到了相当大的作用，而且在一些国际谈判中起到推动作用。我们希望本次论坛能够把一些官员，不管是来自各个部委，或者是其他相关机构，还有一些研究机构，以及一些科学家都聚集到一起，从科学和监管方面共同研究一些相关的问题，寻找到生物安全的未来发展方向，探讨在国内和国际讨论层面的发展途径。第三世界网络希望继续进行生物安全方面的工作，同时希望继续与中央民族大学和环境保护部南京环境科学研究所等机构继续合作。

我希望这不是最后一个论坛，未来还有很长的路要走。最后祝愿此次论坛取得圆满成功。谢谢！

主题报告
Keynote Presentations

转基因生物技术的利益及挑战：以科学事实增进公众对转基因生物安全了解①

Increase Public Understanding on Genetically Modified Organisms and Biosafety Issues Based on Scientific Knowledge and Facts

卢宝荣
（复旦大学生命科学学院 生物多样性与生态工程教育部重点实验室，上海 200433）
LU Bao-Rong
（Ministry of Education，Key Laboratory for Biodiversity Science and Ecological Engineering，School of Life Sciences，Fudan University，Shanghai 200433）

摘　要：转基因生物技术的飞速发展和全球转基因植物的大规模种植，引发了世界范围内的生物安全顾虑。大多数民众对生物技术、转基因产品以及生物安全的内容并不了解，因此产生了对生物技术及产品盲目的怀疑、甚至不必要的恐慌。由于全球的粮食安全问题和环境日益恶化，利用包括转基因生物技术在内的新技术来解决这些问题已成为世界的发展趋势，转基因技术产品带来的利益也已成为强手如林国际市场激烈竞争的焦点。中国作为新兴的发展中国家，不可坐失发展转基因生物技术，并占有世界一席市场的良机。为了增加民众对转基因生物技术及其产品，特别是对生物安全的了解，本文介绍了生物安全的概念及涵盖的内容，并以大量的背景知识和科学研究结果，分析和讨论了转基因技术的发展为人类带来的利益和潜在影响。旨在让公众理性和科学地看待转基因技术及其相关生物安全问题。

关键词：生物安全；遗传工程；转基因；环境影响；公众认知

Abstract：The rapid development of transgenic biotechnology and significantly increased global cultivation of genetically modified（GM）crops has aroused great concerns over the biosafety issues by public worldwide. Public has very limited knowledge on biotechnology/transgenic products and biosafety in particular，resulted in suspicion and unnecessary fear. Owning to the challenge of world food security and the deterioration of global environment，it becomes unchangeable trends to apply new technologies，including biotechnology，to solve the problem of food security. In addition，the huge economic benefits brought by biotechnology products will also become one of the completive focal points for the countries with strong biotechnological capacity. As a member

① 本文获授权转载自《植物生理学报》2013 年第 7 期，第 615-625 页，题为“以科学知识和事实增进公众对转基因生物及其安全性的了解”。

资助：国家重点基础研究发展 973 计划（2011CB100401）和转基因生物新品种培育重大专项（2013ZX08011-006）。

of the international community and an economic-emerging country, China cannot afford to lose the great opportunities to develop the technology. To increase public understanding on biotechnology, GM plants, and biosafety, this article introduced the concept and concerned areas particularly in biosafety, and discussed the benefits and potential impact associated with the technology, based on common knowledge and facts obtained from scientific research. The objective is to help public view biotechnology and biosafety in a more reasonable and scientific way.

Keywords: Biosafety, genetic engineering, transgene, environment impact, public awareness

科学技术的发展总是促进着人类社会的进步而发展，每一次科学技术和发明的成果都为人类社会的进步和人类的生活带来了巨大的利益。例如蒸汽机的发明和应用带来了工业革命的发展，伦琴射线（X-射线）的发现及其在医学诊断的广泛应用带来了医疗技术的革命，信息技术的发展对人类生活方式产生了巨大的影响（刘青峰，2006）。同样，生物技术的出现和发展及其在农业和医学领域的广泛应用，也为全球的粮食安全、生态环境改善和疾病的治疗等带来了新的机遇和巨大的利益（Hood，1988；Bennett 和 Jennings，2013）。毋庸置疑，随着人类社会的不断进步，科学技术也将不断得到发展并给人类的生活带来更多的方便和更大的利益。转基因生物技术就是现代众多具有科学技术成就的领域中对人类的生活产生最大影响的技术之一。

转基因技术在农业生产中的应用主要是对农作物品种的遗传改良。达尔文的进化思想是作物育种和品种改良的重要理论基础。按照这一理论的解释，生物体不断产生变异，而有益的变异适应于环境而被选择下来，这就产生了适应性进化，使产生了变异的生物体——新物种得以生存。作物改良的过程就是人类利用不同的方法，使栽培作物产生适应于人类需求的有利变异，并对其进行选择和保留，培养更适合于自己的需要的新型栽培农作物。转基因育种就是利用转基因生物技术在农作物品种改良过程中创造变异的有效途径之一（Suslow 等，2002）。传统的育种过程也涉及转基因，但传统育种主要通过有性杂交的方式，将优异的基因从具有该性状的供体植物（如农作物品种、野生近缘种等）转移到希望对其进行改良的目标品种（受体植物）中。只不过通过有性杂交的方式进行优异基因导入或转移，具有耗时（育种时间长）、目标性弱（难以控制希望导入的优异基因）和效率低等缺点。而且通过有性杂交的方式进行基因转移难以逾越物种之间的生殖障碍，这就大大地限制了可以利用基因资源的范围。例如，要想通过有性杂交的方法将小麦中的抗旱优异基因转移到水稻中几乎是不可能的；同时，很难通过有性杂交的方式将亲缘关系较远的野生近缘种优异基因资源转移到栽培农作物品种中，进行品种的遗传改良。

利用生物技术的方法来转移不同物种的优异基因，则是将目的基因从具有优良性状的供体生物（包括植物、动物和微生物）中克隆和分离出来，再用一个特殊的运载工具（载体），将目的基因转移到希望对其进行遗传改良的目标品种中。利用生物技术进行基因转移不仅可以达到同样的目标，而且还可以同时解决传统转基因方法中的耗时、目标性差和效率低的问题。更重要的是，利用生物技术进行转基因还可以逾越物种之间生殖隔离的障碍。从理论上讲，利用生物技术可以从任何一个供体物种中将有用的基因分离出来，而且转移到希望改良的农作物品种，这样就极大地扩展了可以利用的基因资源范围。这充分表明，转基因生物技术对作物遗传改良带来了前所未有的新机遇（Suslow 等，2002）。

事实上，自从 1994 年 5 月 18 日世界首例转基因西红柿（Flavr Savr®，CGN-89564-2）获得批准进入商品化生产以来，转基因作物的种植和应用如雨后春笋，迅速发展和扩大。据国际农业生物技术应用服务组织（ISAAA）的不完全统计，到 2012 年全球转基因作物的商品化种植的累计面积已达到了 1.7 亿 hm^2，与 1996 年转基因作物的种植面积相比，在短短的 16 年中就增长了 100 倍，带来的经济利益也极为可观（James，2012）。在农业生产中，还没有任何一项技术能够在这样短的时间内带来如此迅速的增长。这充分表明转基因技术对于农业生产和保障全球粮食安全具有重要的作用，转基因作物的种植和在农业生产中的长期商品化应用，对社会、经济以及环境方面的持续效益和影响也是巨大的。

然而，像任何一项新技术的出现和应用一样，转基因技术及其产品在对农业生产方式和生产效率带来巨大变革，以及巨大的社会经济利益的同时，也带来了全球范围内对该项技术及其应用的一系列质疑。特别是引发了全球对转基因技术及其产品应用的“生物安全”顾虑，甚至广泛的讨论和争议。一时之间，转基因的生物安全成为各大媒体频繁出现的话题，有些人甚至谈转基因色变，尽管他们可能对什么是转基因，什么是生物安全并不了解。我们提倡要科学和理性地看待转基因技术及其生物安全的相关问题，这就要求我们必须了解什么是生物安全，生物安全涵盖了哪些内容，什么是生物安全的评价和研究，如何看待转基因生物技术带来的利与弊。本文就上述一系列相关问题，特别是对环境生物安全的问题进行了介绍和讨论。

1 生物安全的概念及其涵盖内容

生物安全具有广义和狭义的概念。广义的生物安全概念泛指一切生物因素对人类健康和生态环境可能带来的负面后果及其解决方法，例如外来生物入侵、疾病的过境传输，甚至生物武器的使用等所带来的安全问题和隐患。本文所涉及的生物安全是狭义概念，特指与转基因技术相关的生物安全。狭义的生物安全定义也有诸多不同的版本，按照联合国粮食与农业组织（FAO）的解释，生物安全是指：“在生物技术的应用以及转基因植物和其他生物尤其是微生物的环境释放过程中可能对植物遗传资源、动物和植物、人类健康以及环境带来的负面影响。”通常，我们可以将生物安全理解为：“使转基因生物及其产品在研究、开发、生产、运输、销售、消费等过程中受到安全控制，防范其对生态和人类健康产生危害，以及应对转基因生物所造成的危害、损害而采取的一系列措施的总和”（卢宝荣等，2008）。也即是说，转基因生物技术及其产品在研究、开发、生产和应用的一系列过程中所可能涉及对人类健康、环境和生物多样性带来潜在的安全隐患，应该采取一系列法律法规和科学手段来对其进行评价、检测和管理，以达到避免或尽量降低这些安全隐患的目的。

转基因生物安全的内容涉及面非常广泛，包含了食品、环境、社会经济、科学研究、法律和政策等方面的内容。按照目前转基因生物安全所涉及的领域，我们可以将其大致归纳为以下几方面：

1.1 食品与饲料安全

俗话说“民以食为天”，我们最关心的应该是每天所用食用产品的安全性。由转基因作物所生产和加工而形成的食品是否存在不安全的因素，这是生物安全最关注的领域之一。例如，转基因食品可能含有外源基因（来自农作物以外的基因），是否会由于该基因而产生急性或慢性毒性、致敏性、抗营养因子以及非预期的性状等，从而对人类的健康产生影响？此外，在第一代的转基因产品中，通常包含了转基因的筛选标记，而筛选标记常常为抗生素（如卡那霉素、潮霉素等）抗性基因，这些基因是否可能对人类健康带来危害？对饲养的动物带来危害？这些都属于转基因食品生物安全的范畴（König 等，2004；Cromwell 等，2005；Marshall，2007）。

任何国家在每一个转基因作物品种进入商品化生产之前，都要对于该作物上述的食品和饲料的安全性进行长期和科学的评价，确保每一个转基因作物不存在食品和饲料方面的安全的隐患，才会对该产品颁发安全证书，允许其进入商品化生产。重要的是，按照各国的生物安全法规，现在培育的第二代转基因作物，均通过改良的生物技术而将筛选标记统统删除了，因此转基因产品中已经不再包含有抗生素等标记基因，大大地提高了转基因食品的安全性。对于用作饲料的转基因产品也需要进行与转基因食品同样严格的安全评价才能允许进行商品化生产。

1.2 环境生物安全

由于转基因作物会被释放到环境中进行大规模的商品化种植，这些作物中包含的转基因也会随之而释放到环境中。经过人工修饰的转基因是否会对生态环境产生多方面的负面影响？如何消除这些影响？这些均属于环境生物安全的范畴。国家对转基因作物的环境释放也必须进行相应的环境生物安全评价。例如，有些目标转基因（如抗虫转基因 *Bt*）含有杀虫蛋白，会杀死作物中的某一类主要害虫（称之为靶标害虫）而达到保护作物和稳定产量的目的。但这一类转基因是否能杀死农业生态系统中的一些中性昆虫、有益的天敌、甚至非昆虫的其他类型生物（称之为非靶标生物），这是人们对转基因环境生物安全的顾虑。另外，转基因是否会通过天然杂交和基因漂移而逃逸到非转基因作物品种和作物的野生近缘种，从而带来潜在的生态影响？转基因作物的大面积种植是否会影响农田生态系统中的生物多样性？转基因的大规模种植是否会对土壤生物，特别是有益的土壤微生物带来影响？这些问题均是转基因作物在进行商品化生产之前必须进行科学评价和回答的环境生物安全问题（Ellstrand，2001；Conner 等，2003；Sanvido 等，2007；Lu 和 Yang，2009）。本文后面还将以案例的方式专门对转基因的环境生物安全问题进行详细讨论。

1.3 转基因的标识和检测

为了让广大的消费者具有知情权和选择权，在我国和其他许多国家，按照转基因生物安全的法规，对于任何一种转基因产品均要求对其进行强制的标识。也即是说，生产商和销售商必须对其含有转基因成分的产品进行书面示明（金芜军等，2004；陈超和展进涛，2007；Ahmed，2002）。转基因产品的标识并不是表明该产品与其他非转基因产品有什么实质性的区别，或是存在安全隐患，而只是为消费者提供知情信息。相反，进行了转基因标识或贴有标签，正表明该产品经过了生物安全评价，应该是安全的。在有些国家如美国和加拿大，转基因产品的标识是自愿的，生产商和销售商可以不对其转基因产品进行标识，因为这一过程需要包括数量不小的花费。而我国的法律明确规定对国内生产和进口的转基因产品一律要进行标识，以满足消费者的知情权和选择权。另外，为了保证和加强对转基因产品及其加工原料标识的执行和监管，我国还成立了专业的资质机构，对销售产品是否含有转基因成分进行科学的鉴定和检测，以确保消费者的知情权得到落实。

1.4 转基因的社会、经济和伦理问题

由于转基因技术涉及巨大的经济利益，自该技术的广泛应用以来，也引发了人们对其是否会产生全球性的社会、经济和伦理影响进行了激烈的讨论（Finucane 和 Holup，2005；Einsele，2007）。例如，转基因生物技术是否会被跨国大公司所垄断，导致真正需要这些技术的人群（如小农户）和发展中国家因为付不起昂贵的费用而无法使用该技术？生物技术的垄断和转基因产品的集约化生产是否会导致两极分化的日益加剧，而使小农户变得更贫穷，甚至破产，从而导致社会的不稳定性？此外，由于不同宗教信仰的缘故，转基因生物技术也涉及一些伦理方面的问题。例如，信仰神创论的人群认为，人类不可以对上帝创造的万物进行人为的改变，包括基因改良，而素食主义者也许不愿意食用和接受含有动物基因的产品等。这些都属于转基因社会、经济和伦理问题的范畴。

1.5 转基因生物的相关管理条例和法规

全世界的转基因生物研发、种植、利用、转移、运输、贸易和管理等都是在法律的框架下进行的（Spök，2007）。生物安全的国际法规是附属于《生物多样性公约》下的《卡塔赫纳生物安全议定书》。该议定书是在世界范围内对具有活性的转基因生物进行生产、过境传输、管理以及利益分配等具有明确的规定的法律文件，全世界已有包括中国在内的100 多个国家签署同意执行该议定书。各国也建立了适合于自己国情的转基因生物安全管理法规和法律文件，便于指导和监管转基因生物的生产和商品化应用。

我国也有相对完备的转基因生物安全管理法规。2001 年 5 月 21 日，由当时的国务院总理朱镕基签署的中华人民共和国国务院令第 304 号《农业转基因生物管理条例》，就是目前我国生物安全管理的最高法律依据。此外，农业部还颁发了多个与之配套的法律文件，例如《农业转基因生物安全评价管理办法》、《农业转基因生物标识管理办法》、《农业转基因生物进口安全管理办法》、《农业转基因生物加工审批办法》等。此外，我国还成立了由多部委联合领导的转基因生物安全管理专职部门，建立了生物安全审理、评价和监管的体系。这些条例、法律和安全管理体系的建立和健全，对于我国转基因技术和转基因产品的有序发展和商品化应用提供了充分的法律保证。

1.6 转基因生物安全的公众教育和公众认知

在风险评价、风险监管和风险交流这样一个完备体系中，风险交流是非常重要的一个环节。但有关转基因生物安全的风险交流却往往不被大多数人重视。风险交流是指通过信息交流、宣传和公众教育等形式，将有关技术和产品的科学知识以及相关的安全知识传播到全社会。通过风险交流，让公众对转基因生物技术，转基因生物以及相关生物安全的一

系列科学知识和科学内容有一个正确和全面的认识，从而对有关转基因生物及其安全性的信息有正确的判断（陈桂荣，2005；许晶，2006）。这样不仅可以在政府有关部门和公众的共同努力下尽可能降低转基因产品的负面影响或潜在风险，而且还可以避免由于不具转基因生物一般知识而造成公众对信息的盲信，以及面对大量信息而不知所措，甚至不必要的恐慌。因此，应该有计划地加强公众对生物技术、转基因以及转基因生物安全知识的学习和了解，特别是从青少年的教育就开始加入有关新技术，如转基因生物技术的内容，以便提高公众对转基因技术及其生物安全的认识水平和对转基因产品的认知程度。

1.7 转基因生物安全的研究和评价体系

作为一种新技术的应用，转基因产品是否具有潜在的风险？风险有多高？是否可以通过一定的措施避免或将风险降到最低？对这些问题的回答必须依靠科学研究手段。因此，应该加强转基因生物安全的科学研究并且建立以科学为基础的生物安全评价体系（Andow 和 Zwahlen，2006；Johnson 等，2007）。转基因生物安全的评价是一个科学的过程，是以科学研究的结果和数据为依据的，大力加强相关科学领域的研究，积累大量的科学研究数据，研究有效的生物安全评价方法，对于指导转基因生物的安全评价和利用具有重要的意义。此外，在安全评价的过程中，还应该遵循与转基因生物安全评价相适应的一系列原则。例如，在食品安全评价中的“实质等同性原则”，还有其他方面评价中的如 “预防原则”、“逐步实施原则”、“个案原则”和“科学、透明原则”等。如何将各个原则进行优化而适宜地使用到转基因的生物安全评价中，达到转基因产品的可持续和安全利用以及消费者放心使用转基因产品的目的，也是生物安全的重要内容之一。

2 转基因的环境生物安全

如前所述，环境生物安全是转基因生物安全的重要领域之一。环境生物安全是指转基因生物的大规模环境释放或种植可能对环境中生物多样性、遗传结构和生态系统带来潜在影响的现象（Lu，2008）。因为环境生物安全涉及的空间范围广、时间尺度长、环境变化又受许多因素的影响，加之人类对环境变化的许多因果关系和与某一重要因子的内在联系还缺乏足够的认识，因此转基因的环境生物安全顾虑也是目前仍备受全球关注的生物安全问题之一（Lu，2012）。事实上，环境生物安全所涉及的领域也非常广泛，包括的内容也很多，但目前全球备受关注的环境生物安全问题主要包括以下几方面。

2.1 转基因对非靶标生物的影响

抗虫转基因能使受体植物（如棉花、水稻、玉米等）产生杀虫蛋白，如转 *Bt* 抗虫转基因的棉花和玉米，具有杀死棉铃虫和玉米螟虫等棉花和玉米的主要害虫的能力，从而取代杀虫农药的施用，达到防虫的效果。抗虫转基因作物的种植可以取代化学农药和杀虫剂的使用，对保护生态环境可以起到很好的作用。同时，也降低了由于化学农药的使用而带来的生产成本（购买农药和人工施药的花费），降低农产品农药残留而导致对人体健康带来的负面影响，为农业生产带来了显著的经济效益（James，2012）和生态利益（Wu 等，2008）。但是有人认为，既然 Bt 蛋白能够杀死作物的害虫，那对于农田生态系统中的其他生物，如传粉昆虫、中性的昆虫、害虫的天敌，甚至哺乳动物是否也有同样的杀除或者伤害作用呢？这就产生了对抗虫（和抗病）转基因对非靶标生物影响的顾虑。

目前，抗虫转基因作物最成功和应用最广的是 *Bt* 转基因，因此全世界针对 *Bt* 转基因的非靶标生物效应进行了多年的研究，也获得了大量的科学数据和结果。利用不同的抗虫转基因（*Bt*）作物为材料（如棉花、玉米、水稻等），在自然环境下进行对非靶标生物（如家蚕、蜘蛛等）的比较分析表明，由于 *Bt* 蛋白只是对鳞翅目的某些昆虫具有攻击的“靶点”，所以只对棉铃虫、玉米螟和水稻螟虫等鳞翅目的害虫有明显杀灭作用，而对非鳞翅目的昆虫（包括害虫）和其他门类的动物如蜘蛛、天敌昆虫和节肢动物均没有带来显著的负面影响（李保平等，2002；刘志诚等，2004；刘杰等，2006）。研究还表明，相比较于传统利用化学杀虫剂的方法来防治害虫，*Bt* 抗虫转基因对农业生态环境和生物多样性的保护和维持更加有益（刘志诚等，2004），因为绝大多数种类的化学农药和杀虫剂是没有选择杀灭作用的，往往是对所有接触到农药的生物统统都杀灭，对农业生态环境和生物多样性都有严重的负面影响。而且，在农作物中残留的化学农药还会对人类的健康和牲畜的生长发育带来不利影响。

2.2 转基因逃逸及其带来的生态环境影响

通过生物技术的方法转移到目标作物品种的外源基因，存在于转基因作物的各个器官，这些转基因会随着转基因植物营养体（如分蘖、块根和块茎）以及繁殖器官（如种子）而移动，扩散到非转基因作物或是野生近缘种的群体。通过传粉为媒介，也可以将转基因经有性杂交的方式转移到非转基因作物或野生近缘种群体。这一现象被称为花粉介导的转基因漂移，这一过程会导致转基因逃逸（Lu，2008）。有人因而担心，由于转基因植物的大规模种植，转基因会逃逸到非转基因作物品种和作物的野生近缘种，带来潜在的环境和生态风险。一些经人工修饰的转基因具有较强的自然选择优势，如果这些转基因通过基因漂移逃逸到野生近缘种群体，可能会因其受到的自然选择比未经遗传修饰的基因更加强烈而影响野生近缘种群体的进化过程，从而带来潜在的生态后果。

天然杂交和基因漂移是生物物种和物种之间经常发生的自然现象，是生物得以进化的动力。远在转基因生物出现之前，天然杂交和基因漂移就已经存在，栽培作物与野生物种的基因交流也已经自然的进行了上万年。至于转基因逃逸到作物的野生近缘种或其同种杂草群体（如杂草稻、杂草油菜）之后将带来什么样的生态和进化后果？关于这一点，我们还了解得不多（Ellstrand，2001）。在这个环境生物安全问题的驱动下，全球的科学家对作物品种之间，作物与其野生近缘种之间的基因漂移频率以及基因漂移到野生近缘种会带来什么样的生态和进化影响进行了大量的研究（Snow 等，2005；Ellstrand，2003；Lu 和 Yang，2009），也获得了很多有意义的结果（Song 等，2014；Rong 等，2007；Yang 等，2010）。关于转基因逃逸及其生态后果的内容，将在本文下一节以研究案例的方式予以详细的介绍和讨论。

2.3 转基因作物大面积种植对生物多样性的影响

由于转基因作物的许多优良的特性，受到世界上 1 亿多农户的喜爱和种植。使转基因作物在全球的种植面积从 1996 年规模化进入商品生产以来的短短 17 年中就增长了 100 倍达到了 1.7 亿 hm^2，这个面积相当于 6 个英国的国土面积。已进入商品化种植的转基因作物包括：棉花、大豆、玉米、油菜、西红柿、甜椒、茄子、马铃薯、甜菜、苜蓿、番木瓜、南瓜以及林业树木如转基因杨树等。许多重要的粮食作物如水稻、小麦等也蓄势待发，等待进入商品化生产的最佳时机。由于转基因作物迅速增加的种植面积，转基因作物的种植与快速扩展是否会影响农民对种植品种的决策，即选择大量种植转基因作物而放弃自己种植的传统农家品种，从而导致大量农家品种被少数的转基因品种所取代，使农业生态系统中的品种多样性降低。另一种看法认为，当转基因作物品种的个体通过人为混杂的方式或基因漂移，混入非转基因品种中，由于某些转基因（如抗虫、抗病或抗旱等）具有自然选择优势而被保留下来，而传统的非转基因品种可能由于不具有上述优良性状而逐渐被自然选择或农民的有意识选择淘汰，而造成传统农作物品种中不同类型基因型（遗传多样性）的丧失，最终导致某一特殊农业生态系统中生态多样性的下降。

对于上述的第一种情况导致作物品种多样性下降的担忧，其实早在 20 世纪 60 年代末，在半矮秆基因资源利用和遗传改良技术而带来的“绿色革命”过程中，就产生了许多具有跨时代意义的高产作物（如水稻、小麦和玉米）品种，这些高产品种一方面大幅度提高了作物的产量，在 20 世纪 60—70 年代解决了世界许多地区的饥饿问题，拯救了千百万人的生命。另一方面，由于这些高产品种的大面积推广，农民放弃了对传统农家作物品种的种植，从而致使许多地区的传统品种丧失，也带来了之后农作物传统品种资源保护及其保护策略的产生。

由于优良品种的大面积种植而导致农业生态系统中品种多样性下降的问题，可以通过政策以及对品种种植的合理布局来解决，而不是摈弃这些优良品种。反过来，上述事实均说明无论是“绿色革命”产生的一系列高产品种，或是“基因革命”产生的高产、优质的转基因品种具有更广泛的应用和更强的生命力。通过更适宜的政策调控，合理的种植布局，加之有意识地对特有农家品种的保护，是可以解决由于少数高产优质品种的种植而带来品种多样性降低的这一问题。而对于上述的第二种情况导致的作物品种多样性下降，则可以通过对转基因作物种植的有效管理，以及在转基因品种和非转基因品种之间设置一定的空间隔离距离来达到降低和避免基因漂移而导致的转基因混杂。欧洲许多国家曾研究和探讨一种转基因作物和非转基因作物共存（co-existence）而互助不产生影响或影响极小的种植管理办法（Devos 等，2009）。

2.4 转基因对土壤生物群落的潜在影响

转基因作物进入大规模的商品化种植，其根系的分泌物，残留在土壤中的转基因作物根系，凋落物和未被收获的作物残留部分进入土壤以后，是否会对土壤中的微生物和小型动物产生负面影响？同时这些含有转基因的残留物在土壤中进行分解的过程中，是否会影响土壤的生态性能及功能？这也是转基因环境生物安全关注的问题之一。例如有人认为，含有抗虫 *Bt* 转基因的残留物进入土壤以后，是否对土壤微生物或小型动物带来负面作用。针对这一问题，科学家也从不同的角度，并利用含 *Bt* 抗虫转基因的不同作物的残留物对土壤微生物和小型动物进行了研究。研究结果表明，含有 *Bt* 转基因残体的土壤中虽然能够检测出一定量的 *Bt* 基因或 *Bt* 蛋白的残留（Oliveira，2007；叶飞等，2010），但是这些残留物对微生物和小型动物均没有造成明显的影响。

考虑到 *Bt* 基因就是从土壤中的细菌（苏云金芽孢杆菌）中分离出来的，经过遗传修饰过的 *Bt* 基因再回到土壤中，应该不会对土壤产生额外的影响。但是，我们对这一领域的知识还相对较少，从生物安全的角度出发，可以进行更多的有关转基因对土壤生物群落的影响的深入研究（卢宝荣等，2008）。

此外，除了上述四方面的环境生物安全问题，还有专家将抗虫转基因作物（如 *Bt*）的长期和大面积种植是否会导致靶标害虫对转基因产生抗性，从而降低转基因的抗虫效果也列为环境生物安全的考虑。但是对于靶标害虫抗性产生的问题，目前广泛采用“高剂量”和“庇护所”结合的抗性治理策略来延缓靶标害虫对 *Bt* 转基因抗性的产生。由于转基因抗虫性的丧失，主要涉及抗虫基因的作用效率降低或基因资源的丧失，这不会带来任何的环境和生态方面的风险问题。因此，学者逐渐将靶标害虫或病原菌对抗虫或抗病转基因产生抗性的问题移出了转基因环境生物安全的范畴。

3 转基因逃逸及其潜在环境影响：研究案例分析

转基因逃逸是指作物中的外源转基因通过基因漂移（或称基因流）传播到环境中非转基因生物群体的现象。作为群体遗传学和进化生物学的概念，基因漂移是一个自然过程，这一过程已经在自然界的生物个体和群体之间发生了数百万年（Lu，2008）。基因漂移是生物进化的动力，是生物产生新遗传变异的源泉，可以说如果没有基因漂移发生，我们今天的大千生物世界就不会像现在这样丰富多彩。由于基因漂移与生物技术和转基因逃逸产生了联系，有人认为基因漂移可能导致转基因向非转基因农作物及其野生近缘种逃逸，进而带来潜在的社会、经济和生态环境方面的负面影响，基因漂移才再次受到了极大的关注。转基因逃逸及其潜在生态后果已经成为转基因作物商品化种植引发的全球最具有争议的环境生物安全问题之一（Ellstrand，2001；Sanvido 等，2007；Andow 和 Zwahlen，2006；Lu 和 Yang，2009）。

3.1 基因漂移的概念

从理论上讲，自然界存在两种类型的基因漂移，即基因水平转移（horizontal gene transfer）和基因漂移（vertical gene flow）。前者是指在不同生物种之间产生与有性生殖过程无关的基因转移，而后者是指具有一定亲缘关系的物种或群体之间通过有性生殖过程而产生的基因转移（Lu，2008）。基因水平转移通常发生在亲缘关系较远的物种之间，而且基因转移的概率非常低（＜0.001%），虽然基因水平转移可能在生物进化的过程中有一定的意义，但目前还没有任何研究结果显示这种类型的基因转移会带来环境生物安全方面的问题。因此，本文仅对通常意义上的基因漂移及其潜在生态风险进行讨论。

基因漂移是环境生物安全中被广泛提及的概念，通常是指由花粉漂移（传粉）介导的基因在不同的个体或群体之间的转移，通过不同个体之间的有性杂交所导致双亲基因的交换和转移。虽然通过植物的种子和无性器官（如马铃薯的块茎和水稻的分蘖）的传播，也能够导致种子或无性器官在不同空间隔离的群体之间进行扩散，但最终还是需要通过有性生殖过程，才能够产生真正意义上的基因转移（Lu，2008）。因此，花粉介导的基因漂移对讨论环境生物安全或生态风险具有重要意义。按基因漂移的对象不同，可以将其分为转基因向非转基因作物品种的漂移，即作物—作物基因漂移（例如：转基因从转基因水稻品种向非转基因水稻品种漂移），以及转基因向作物野生近缘种的漂移，即作物—野生近缘种基因漂移（例如：水稻转基因向野生稻中漂移）。由于“作物—作物”和“作物—野生近缘种”基因漂移所带来的生态后果不同，因此有必要将这两种不同类型的转基因逃逸及其导致的影响分别进行讨论。

3.2 作物—作物转基因逃逸的影响

转基因通过花粉介导的基因漂移从转基因作物向其邻近田块的非转基因作物品种逃逸比较容易产生，特别是对异花传粉植物或异交率比较高的植物（如玉米），转基因逃逸的可能性更高。我们的大量研究结果表明，水稻品种之间的基因漂移水平很低，相距 5 m 的水稻植株，基因漂移频率在 0.01 以下，因此水稻的作物—作物转基因逃逸可能性很低（Rong 等，2004；2005；2007）。转基因逃逸带来的影响主要是转基因对非转基因作物品种造成不同水平的混杂（常常被称之为转基因“污染”），从而带来食品或饲料安全的顾虑。特别是当转基因产品的生产目的并非用作食品或饲料（如用于工业原料），由作物—作物转基因逃逸产生的混杂将备受关注，并带来更大的安全问题和经济损失。有一个著名的例子可以表明转基因“污染”带来的严重问题：美国的星联转基因玉米（Startink GM Corn，转基因事件 BH-351）是作为动物饲料而被批准在美国生产，由于生产商（星联公司）未按商品化生产批准时的要求，严格实行空间隔离和不能用于食品，而在玉米一种食品中检测到了该星联玉米的转基因（*Cry9c*），造成了对所有该产品的召回和巨大的经济赔偿，最终导致星联公司的倒闭。从上述例子我们可以看出，转基因作物中的转基因向非转基因作物品种的逃逸，主要将导致“转基因污染”的问题，可能带来地区间或国家间的农产品贸易摩擦，甚至是法律方面的纠纷等（Lu，2008）。

此外，作物—作物的转基因逃逸可能还会造成传统农家品种遗传多样性的改变甚至丢失（Engel 等，2006）。这种原因导致的农作物品种遗传多样性丢失包括如下两方面内容：第一，具有某种优良性状转基因农作物的大面积种植，可能取代大量传统农家品种，造成某地区农家品种遗传资源的丧失。例如，我国推广转基因（*Bt*）抗虫棉的商品化种植仅仅10 余年，少数几个转基因抗虫棉的种植面积就超过了我国棉花种植面积的 70%（Wu 等，2008）。可以想象有多少传统的棉花品种被转基因抗虫棉所取代。同样的例子也可以在北美的转基因大豆和油菜品种中发现。第二，转基因通过基因漂移从转基因作物向非转基因农家品种逃逸，可以导致农家品种遗传完整性的降低，如果转基因具有较强的自然选择优势，逃逸的转基因可以在非转基因品种的种植和繁殖过程中不断积累和扩散，造成传统非转基因品种中丰富的基因型被取代，从而导致基因多样性的丧失。对于这方面的问题和解决方案，已在上节中进行了讨论。

3.3 作物—野生近缘种转基因逃逸的影响

已有许多生态学研究结果表明，转基因可以通过基因漂移逃逸到与农作物具有一定亲缘关系的野生近缘种（包括杂草类型）群体中（Ellstrand，2003；Lu 和 Snow，2005）。转基因逃逸到野生近缘种群体，可以改变野生近缘种的适合度，从而改变野生近缘种的生存竞争能力和入侵性，从而造成不可预测的生态环境影响（Lu，2008）。在进化生物学的概念中，适合度（fitness）是指具有特定基因型的有机生物体在特定生态环境下条件的相对繁殖成功率（Stewart 等，2003）。例如在植物中，适合度是以个体或群体的生存能力（survival）和繁殖能力（fecundity）这两方面来共同衡量的。适合度高的个体将产生大量的后代，从而在群体中逐渐取代其他的个体，反之，适合度低的个体将产生少量的后代而在群体中逐渐被其他的个体所取代。如果转基因可以通过花粉介导的基因漂移以较高的频率逃逸到野生近缘种，而同时该转基因又可以带来较高的适合度利益，那么，该转基因逃逸到野生近缘种群体将增强含有该转基因个体的生存竞争能力和入侵性，改变野生近缘种的进化潜力，从而带来负面的生态环境影响。因此，对转基因的逃逸概率以及转基因所带来的适合度利益或成本效应的分析，已成为转基因逃逸及其生态影响的核心问题（Stewart 等，2003；Lu 和 Snow，2005；Lu 和 Yang，2009）。

通常，转基因向野生近缘种群体逃逸可能带来以下几方面的生态影响：首先，转基因逃逸到野生近缘种群体可能导致新类型杂草的产生。杂草是广泛分布于全球农田生态系统的特殊植物类型，杂草对农田的入侵，不仅导致农作物的不同程度的减产，还常常由于其有害性状如毒性和过敏性等，造成作物产品的品质下降。作物向野生近缘种的转基因逃逸可能将有明显自然选择优势的性状（如抗虫，抗除草剂和抗旱等）转移到作物的野生近缘种或同种杂草类型，从而改变这些物种群体的生存竞争能力和入侵性，形成难以控制的恶性杂草，对农业生产带来杂草控制和管理的困难，特别是当逃逸到野生近缘种群体的转基因具有较高的适合度优势时，可能带来的杂草问题将更为严重，导致较大的生态环境影响。

其次，转基因逃逸到农作物的野生近缘种群体，可能会由于转基因对野生近缘种的“遗传污染”而导致野生近缘种群体的遗传多样性降低，当转基因具有显著的适合度劣势，而同时又有大量转基因不断地转移和渐渗到野生近缘种群体中，这将会导致遗传同化（genetic assimilation）效应，从而导致野生近缘种群体的遗传单一化，甚至群体遗传多样性的丧失。另一方面，具有很强自然选择优势并能够增强适合度的转基因整合到野生近缘种的基因组中，可能会由于选择性剔除（selective sweep）效应而致使野生近缘种群体遗传多样性的降低甚至丢失，造成野生近缘种遗传资源的丧失（卢宝荣等，2009；Lu，2013）。栽培作物的野生近缘种是作物进行遗传改良的重要种质资源，保证这些资源的长期生存和可持续利用对作物育种和粮食安全十分重要。因此，对外源转基因逃逸到农作物的野生近缘种群体将对种质资源产生什么影响还有待于进一步的研究（Lu，2103）。

4 转基因生物技术及其应用的利与弊

毋庸置疑，转基因作物商品化应用和种植所带来的巨大利益，已经被全球近一亿农民应用该项技术的热情，以及全球转基因作物的种植面积的迅速扩展所证明（James，2012）。而大量的生物安全研究和已经获得的科学数据表明，转基因生物应用所带来的生物安全问题远比人们想象的要小得多。就转基因的环境生物安全而言，到目前为止，所有的研究结果还没有证实已在生产上广泛应用的转基因（如 *Bt* 抗虫基因）产生了明显和负面的生态环境影响。相反，研究证明长期种植抗虫转基因（*Bt*）棉花带了显著的正面生态效应（Wu 等，2008）。这些转基因还是指具有较强自然选择优势（如抗虫、抗病）的基因。其实还有大量的转基因不具有任何的自然选择优势或与环境的影响没有直接关系，例如按人类需求对作物的品质、口感或是营养成分进行改善的基因，它们对环境应该不会产生任何影响。

任何新生事物的出现和新技术的应用都不会一帆风顺，因为这些新生事物和新技术在其刚刚出现之时并不一定非常完美，可能会存在一些不足。但是在科学和技术的发展过程中，新技术某些不足可以逐渐得以改善的。例如，最早蒸汽机汽车的使用，常常会产生事故甚至有爆炸的危险，但是现代的汽车已经被改善得具有很高的安全性。又如，第一代的转基因产品都含有抗生素筛选标记基因，但是随着转基因技术的发展，所有标记基因都可以被删除，因而新一代的转基因产品已经不再含有筛选标记基因。这反映了人类从“必然王国”向“自由王国”的发展过程。随着人类对自然规律的深入认识和科学技术的进一步发展，总是可以将对人类有用的技术或产品改善到利益最大而风险最小的状态。

值得一提的是，世界上并不存在“零风险”的事物，例如对人类非常有用的医药、手机、电脑、交通工具等，在使用时就没有风险吗？事实是人类目前非常高兴地使用着这些技术的成果。人类对科学技术及其产品的利用是本着“两利相权取其重，两害相权取其轻”的原则，尽量利用能带来巨大利益的技术，尽管该技术可能存在一定风险。就如现代交通工具的使用，尽管每年仅在中国就会有远大于 10 万次的严重交通事故和伤亡发生，但是人们使用这一快捷而有利交通工具的热情并没有丝毫下降的趋势，也没有任何希望终止使用现代交通工具的迹象。又如在棉花的生产中，由于已经有大量害虫的发生，严重威胁着棉花产量，为了控制虫害保住棉花的产量，人类只有两种选择：一是大量喷洒化学农药杀虫剂，二是使用含 *Bt* 转基因的抗虫棉。前者会杀死棉田中的所有生物包括对人类产生伤害，同时在土壤和棉花产品中均有大量农药的残留；而后者则是有“靶向性”地杀死主要害虫棉铃虫。两种方法均可以控制虫害、保证产量，利弊关系一目了然，人们将如何做出选择，也应该是显而易见。

5 展望

如前所述，从 1996 年转基因作物有一定规模的商品化生产到现在（James，2012），全球转基因作物的种植面积呈快速的增长势态，全球有一亿多农户种植了转基因作物，这充分证明转基因生物技术所带来的巨大社会经济利益及其强大的生命力。这是一项能产生很大经济效益的技术，在当今国际竞争强手如林的形势下，各个国家都希望在这个市场中占有一席之地和分到一杯羹。中国不去占领这个市场，其他国家将毫不客气地占领这一市场。

我国的转基因技术研发和转基因产品应用的形势不容乐观，据 2012 年资料的统计，我国的转基因作物（包括林木）的种植面积仅为 400 万 hm^2，以转基因作物种植面积的排序，中国排在第 6 位（James 2012）。全球排在第 1～5 位的国家分别是美国（6 950 万 hm^2）、巴西（3 660 万 hm^2）、阿根廷（2 390 万 hm^2）、加拿大（1 160 万 hm^2）和印度（1 080 万 hm^2）（James，2012）。从上述数据看，中国转基因作物的种植面积还不到排在全球第 5 位印度的一半，就更不用和与我国经济发展水平的相似的巴西相比了（比我国多 9.11 倍）。可以想象，我国在国际转基因市场所占的份额还不足 4%。在未来全球资源趋于紧缺，粮食安全面临日益严重挑战和各国的经济利益严重冲突的形势下，不发展生物技术，不用我国的优势技术去占领国际市场，我国就会处于被动的局面。“逆水行舟不进则退”，在 2003 年我国的转基因作物种植面积排名第 4（210 万 hm^2），与当时排名第 7 的印度（小于 100 万 hm^2）相比较，还有很大的优势。10 年过去，我们的转基因作物种植的面积被西方和印度远远抛在后面。我国的农民难道不需要新技术么？这其中的原因值得我们去深思。

世界在发展，科技在进步，无论我们喜欢与否，转基因生物技术一定会在全球范围内不断发展。由世界科技进步而给各国带来的发展机遇是一样的，只有紧紧抓住发展的机遇，才能让科学技术的成果为国家的发展和人民生活水平的提高做出应有的贡献。和其他高新技术一样，转基因生物技术在世界的经济发展中占有重要的地位。但是转基因生物技术还不是一个完美的技术，还有改善和发展的空间。因此，目前还存在食品安全和环境安全的顾虑。但是，本着“两害相权取其轻，两利相权取其重”的原则，我们是因为该技术还存在一定的缺陷就完全摒弃它，还是让它通过不断改良和完善来造福人类、造福中国人民？全世界很多国家都对这一技术的发展持积极支持的态度，为了保证这一技术可持续和健康发展，各国都在发展自己的转基因生物安全评价方法和技术。相信随着生物安全问题的逐步解决，转基因技术及产品将为人类带来更加安全和环境友好的产品。

参考文献

[1] 陈超，展进涛，2007. 国外转基因标识政策的比较及其对中国转基因标识政策制定的思考. 世界农业，（11）：21-24.

[2] 陈桂荣，2005. 公众对转基因食品的了解和接受程度. 昆明理工大学学报，5：14-17.

[3] 金芜军，贾士荣，彭于发，2004. 不同国家和地区转基因产品标识管理政策的比较.农业生物技术学报，12（1）：1-7.

[4] 李保平，孟玲，万方浩，2002. 转基因抗虫植物对天敌昆虫的影响. 中国生物防治，18：97-105.

[5] 刘杰，陈建，李明，2006. 转 *Bt* 棉花对蜘蛛生长发育及捕食行为的影响. 生态学报，26：945-949.

[6] 刘青峰，2006. 让科学的光芒照耀自己：近代科学为什么没有在中国产生. 北京：新星出版社.

[7] 刘志诚，叶恭银，胡萃，2004. 抗虫转基因水稻和化学杀虫剂对稻田节肢动物群落的影响. 应用生态学报，15：2309-2314.

[8] 卢宝荣，傅强，沈志成，2008. 我国转基因水稻商品化应用的潜在环境生物安全问题. 生物多样性，16：426-436.

[9] 卢宝荣，夏辉，杨箫，等，2009. 杂交—渐渗进化理论在转基因逃逸及其环境风险评价和研究中的意义. 生物多样性，17：362-377.

[10] 许晶，2006. 转基因技术在美国和欧洲的认知及传播. 前沿，2：204-208.

[11] 叶飞，宋存江，陶剑，李长林，2010. 转基因棉花种植对根际土壤微生物群落功能多样性的影响. 应用生态学报，21：386-390.

[12] Ahmed FE，2002. Detection of genetically modified organisms in foods. Trends Biotechnol，20：215-223.

[13] Andow DA，Zwahlen C，2006. Assessing environmental risks of transgenic plants. Ecol Lett，9：196-214.

[14] Bennett DJ，Jennings RC，2013. Successful Agricultural Innovation in Emerging Economies-New Genetic Technologies for Global Food Production. Cambridge：Cambridge University Press.

[15] Conner AJ，Glare TR，Nap JP，2003. The release of genetically modified crops into the environment - Part

II. Overview of ecological risk assessment. Plant J，33：19-46.

[16] Cromwell GL，Henry BJ，Scott AL，Gerngross MF，Dusek DL and Fletcher DW，2005. Glufosinate herbicide-tolerant（LibertyLink®）rice vs. conventional rice in diets for growing-finishing swine. J Anim Sci，83：1068-1074.

[17] Devos Y，Demont M，Dillen K，Reheul D，Kaiser M，Sanvido O，2009. Coexistence of genetically modified（GM）and non-GM crops in the European Union. A review. Agron Sustain Dev，29：11-30.

[18] Einsele A，2007. The gap between science and perception：The case of plant biotechnology in Europe. In：Green Gene Technology. Research in an Area of Social Conflict，Fiechter A & Sautter C（eds）. Heidelberg：Series-Advances in Biochemical Engineering/Biotechnology，107：1-11.

[19] Ellstrand NC，2001. When transgenes wander，should we worry？ Plant Physiol，125：1543-1545.

[20] Ellstrand NC，2003. Current knowledge of gene flow in plants：Implications for transgene flow. Phil Trans R Soc B，358：1163-1170.

[21] Engels JMM，Ebert AW，Thormann I，De Vicente MC，2006. Centres of crop diversity and/or origin，genetically modified crops and implications for plant genetic resources conservation. Genet Resour Crop Evol，53：1675-1688.

[22] Finucane ML，Holup JL，2005. Psychosocial and cultural factors affecting the perceived risk of genetically modified food：An overview of the literature. Soc Sci Med，60：1603-1612.

[23] Hood L，1988. Biotechnology and Medicine of the Future. J Am Med，259：1837-1844.

[24] James C，2012. Global status of commercialized biotech/GM crops：2012. ISAAA Brief No. 44. New York：ISAAA.

[25] Johnson KL，Raybould AF，Hudson MD，Poppy GM，2007. How does scientific risk assessment of GM crops fit within the wider risk analysis？ Trends Plant Sci，12：1-5.

[26] König A，Cockburn A，Crevel R，Debruyne E，Grafstroem R，Hammerling U，Kimber I，Knudsen I，Kuiper H，Peijnenburg A et al.，2004. Assessment of the safety of foods derived from genetically modified（GM）crops. Food Chem Toxicol，42：1047-1088.

[27] Lu B-R，2008. Transgene escape from GM crops and potential biosafety consequences：an environmental perspective. Trieste：International Centre for Genetic Engineering and Biotechnology（ICGEB），Collection of Biosafety Reviews，4：66-141.

[28] Lu B-R，2012. China：earlier experiences and future prospects. In：DJ Bennett，RC Jennings（Eds）：Successful Agricultural Innovation in Emerging Economies - New Genetic Technologies for Global Food Production. Chapter 9. Cambridge：Cambridge University Press.

[29] Lu B-R，2013. Introgression of transgenic crop alleles：Its evolutionary impacts on conserving genetic diversity of crop wild relatives. J System Evol，51：245-262.

[30] Lu B-R，Snow AA，2005. Gene flow from genetically modified rice and its environmental consequences. BioScience，55：669-678.

[31] Lu B-R，Yang C，2009. Gene flow from genetically modified rice to its wild relatives：assessing potential ecological consequences. Biotechnol Adv，27：1083-1091.

[32] Marshall A，2007. GM soybeans and health safety - a controversy reexamined. Nat Biotechnol，25：981-987.

[33] Oliveira AR，Castro TR，Capalbo DMF，Delalibera I，2007. Toxicological evaluation of genetically modified cotton（Bollgard®）and Dipel® WP on the non-target soil mite *Scheloribates praeincisus*（Acari：Oribatida）. Exp Appl Acarol，41：191-201.

[34] Rong J，Xia H，Zhu YY，Wang YY，Lu B-R，2004. Asymmetric gene flow between traditional and hybrid rice varieties（*Oryza sativa*）estimated by nuclear SSRs and its implication in germplasm conservation. New Phytol，163：439-445.

[35] Rong J，Song ZP，Su J，Xia H，Lu B-R，Wang F，2005. Low frequency of transgene flow from Bt/CpTI rice to its nontransgenic counterparts planted at close spacing. New Phytol，168：559-566.

[36] Rong J，Lu B-R，Song ZP，Su J，Snow AA，Zhang XS，Sun SG，Chen R，Wang F，2007. Dramatic reduction of crop-to-crop gene flow within a short distance from transgenic rice fields. New Phytol，173：346-353.

[37] Sanvido O，Romeis J，Bigler F，2007. Ecological impacts of genetically modified crops：Ten years of field research and commercial cultivation. In：Green Gene Technology. Research in an Area of Social Conflict，Fiechter A，Sautter C（eds）. Heidelberg：Series- Advances in Biochemical Engineering/Biotechnology，107：235-278.

[38] Snow AA，Andow DA，Gepts P，Hallerman EM，Power A，Tiedje JM，Wolfenbarger LL，2005. Genetically modified organisms and the environment：Current status and recommendations. Ecol Appl，15：377-404.

[39] Spök A，2007. Molecular farming on the rise - GMO regulators still walking a tightrope. Trends Biotechnol，25：74-82.

[40] Stewart CN，Halfhill MD，Warwick SI，2003. Transgene introgression from genetically modified crops to their wild relatives. Nat Rev Genet，4：806-817.

[41] Suslow T，Thomas B，Bradford K，2002. Biotechnology Provides New Tools for Plant Breeding. Agricultural Biotechnology in California Series，Publication 8043，USA： UCANR Publications（http://www.plantsciences.ucdavis.edu/bradford/8043.pdf）.

[42] Vogel G，2006. Genetically modified crops - Tracing the transatlantic spread of GM rice. Science，313：1714-1714.

[43] Wu KM，Lu YH，Feng HQ，Jiang YY，Zhao JZ，2008. Suppression of cotton bollworm in multiple crops in China in areas with Bt toxin-containing cotton. Science，321：1676-1678.

Risk Assessment of GMOs：Key principles，Challenges and Ways Forward

转基因生物的风险评估：关键原则、挑战与未来方向

Dr. Andreas Heissenberger

（Umweltbundesamt – Environment Agency Austria 奥地利环境保护局）

ENVIRONMENT AGENCY AUSTRIA umweltbundesamt®

Risk assessment of GMOs

Key principles, challenges and ways forward

Andreas Heissenberger

1

ENVIRONMENT AGENCY AUSTRIA umweltbundesamt®

Content

- Introduction
- Definitions
- Risk assessment protocols (OECD, CODEX, CPB)
- Principles
- Challenges
- Way forward

2

ENVIRONMENT AGENCY AUSTRIA umweltbundesamt®

Why risk assessment of GMOs?

- GMOs/LMOs are different!
  - Transfer of genes across species barriers
  - Different to conventional breeding
- New characteristics – new possibilities – new risks!
- Long (scientific) debate about the risks, still ongoing!
- OECD: "Blue Book", Recombinant DNA Safety Consideration, 1986
- Risk assessment is a requirement in many national, legislative frameworks and regional and international agreements and often a prerequisite for the authorization of GMOs

ENVIRONMENT AGENCY AUSTRIA umweltbundesamt®

Definition of Risk

- Cartagena Protocol - Training Manual on Risk assessment:
  - Risk is often described as the combined evaluation of hazard and exposure
  - "Hazard" is defined as the potential of a stressor to cause harm to a biological system (e.g. a species) (UNEP/IPCS, 1994).
  - "Exposure" means the contact between an agent and a receptor. Contact takes place at an exposure surface over an exposure period (WHO, 2004)
  - Risk = the combination of the magnitude of the consequences of a hazard, if it occurs, and the likelihood that the consequences will occur

4

umweltbundesamt

Protocols/guidance

- OECD
  - www.oecd.org
  - 38 Consensus documents published (compiled in 4 Volumes – available also on Google books)
  - Biology of crops, trees, and micro-organisms; traits; molecular characterization; unique identifier
- Codex Alimentarius Commission
  - www.codexalimentarius.net
  - collection of standards, codes of practice, guidelines and other recommendations
  - "foods derived from biotechnolgy" (no environmental RA)
  - Documents on general principles, plants, animals, micro-organisms

5

umweltbundesamt

Protocols/guidance

- European Food Safety Authority
  - www.efsa.europa.eu
  - Responsible institution for evaluation of risk assessments in the EU
  - Several guidance documents, e.g. food and feed, environmental risk assessment, statistics, stacked events, comparators
- Cartagena Protocol on Biosafety
  - www.cbd.int/biosafety
  - AHTEG developed guidance documents
  - In 2012 COPMOP6 recognized work, decided to test documents for their practicability
  - Currently ongoing
  - General document ("roadmap") plus mosquitos, abiotic stress, stacked genes, trees, monitoring

6

umweltbundesamt

Protocols – Cartagena Protocol

- Key document is the "Roadmap"
- Identifies overarching issues
  - Criteria for relevance of data
  - Scientific robustness and quality of data
  - Transparency, reproducibility (reporting, ...)
  - Identification and consideration of uncertainty
- Context and scoping of risk assessment
- Stepwise risk assessment procedure

7

umweltbundesamt

Scoping

- Scoping: define the extent and the limits of the risk assessment process
- consideration of protection goals, and other factors (use of GMO, agricultural practice, scale, ...)
  - National level, regional, international context
  - Different for different GMOs
- Points to consider
  - selection of relevant assessment endpoints or representative species on which to assess potential adverse effects
  - establishing baseline information
  - if possible, establishing the appropriate comparator(s)

8

umweltbundesamt

Principles – Cartagena Protocol

- Step 1: "An identification of any novel genotypic and phenotypic characteristics associated with the living modified organism that may have adverse effects on biological diversity in the likely potential receiving environment, taking also into account risks to human health."
  - Points to consider regarding the characterization of the LMO
  - Points to consider regarding the receiving environment
  - Points to consider regarding the potential adverse effects resulting from the interaction between the LMO and the receiving environment

9

umweltbundesamt

Proinciples – Cartagena Protocol

- Step 2: "An evaluation of the likelihood of adverse effects being realized, taking into account the level and kind of exposure of the likely potential receiving environment to the living modified organism."
- Step 3: "An evaluation of the consequences should these adverse effects be realized."
- Step 4: "An estimation of the overall risk posed by the living modified organism based on the evaluation of the likelihood and consequences of the identified adverse effects being realized."
- Step 5: "A recommendation as to whether or not the risks are acceptable or manageable, including, where necessary, identification of strategies to manage these risks"

10

umweltbundesamt

Risk assessment process (Cartagena)

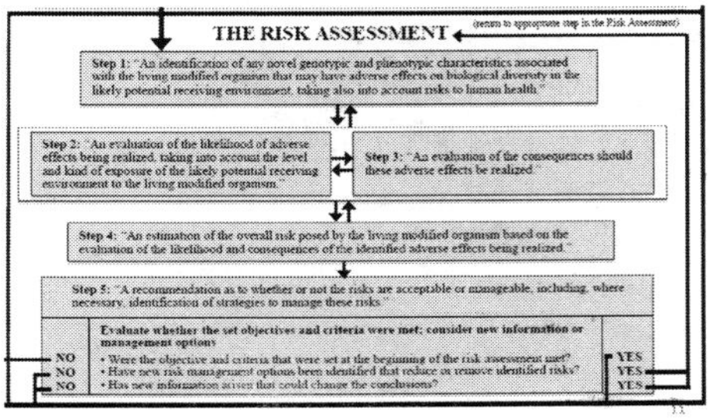
THE RISK ASSESSMENT
(return to appropriate step in the Risk Assessment)
Step 1: "An identification of any novel genotypic and phenotypic characteristics associated with the living modified organism that may have adverse effects on biological diversity in the likely potential receiving environment, taking also into account risks to human health."
Step 2: "An evaluation of the likelihood of adverse effects being realized, taking into account the level and kind of exposure of the likely potential receiving environment to the living modified organism."
Step 3: "An evaluation of the consequences should these adverse effects be realized."
Step 4: "An estimation of the overall risk posed by the living modified organism based on the evaluation of the likelihood and consequences of the identified adverse effects being realized."
Step 5: "A recommendation as to whether or not the risks are acceptable or manageable, including, where necessary, identification of strategies to manage these risks."
Evaluate whether the set objectives and criteria were met; consider new information or management options
• Were the objective and criteria that were set at the beginning of the risk assessment met?
• Have new risk management options been identified that reduce or remove identified risks?
• Has new information arisen that could change the conclusions?
NO
NO
NO
YES
YES
YES

11

umweltbundesamt

Challenges

- **General problems**
  - Direct testing is scarce - focus on indirect evidence and assumption based reasoning
  - Lack of testing according to GLP (toxicology)
  - Possible secondary effects disregarded in many cases
  - Definition of comparators (baseline)
- **Regional data often missing**
  - field trials in US or South America only
  - Results extrapolated to other environments without scientific justification
  - African, Asian, South American countries: special ecosystems and megadiverse countries
  - Crop specificities (Rice, Cassava etc.)

umweltbundesamt

Challenges

- **Evaluation of field trial results not clear and of poor statistical quality**
  - Bad experimental setup (power analysis, problem formulation)
  - Pooling of data from different field trials and different years
  - Pooling of data from different geographic regions (e.g. USA, Chile)
- **Research**
  - Biosafety research lacks funding
  - Access to GMO material controlled by companies

umweltbundesamt

Challenges

- **Data interpretation**
  - Significant differences explained by "biological variation" or classified as "biologically not relevant"
- **Guidance documents**
  - Available?
  - Suitable?
  - Binding?
  - Followed?
- **Organizational Challenges**
  - Participation to working groups (Cartagena Protocol, OECD, etc.)
  - Qualification of staff, training

umweltbundesamt

Decision making

- Decision should be based on outcome of risk assessment
  - Risk identified?
  - Risk manageable?
  - Measures needed proportionate?
  - Monitoring/surveillance necessary?

15

umweltbundesamt

Decision making

- In practice decision also based on other reasons
  - General policy (pro or con)
  - Public opinion (media)
  - Stakeholder pressure (industry, NGOs)
  - Economic considerations
  - Fear (WTO)

16

umweltbundesamt

Way forward

- Regulatory issues
  - Development of recognized framework (Cartagena Protocol)
  - Development on detailed guidance for developers/applicants and risk assessors
  - Involvement of regulators and scientists
  - Decisions based on scientific grounds
  - Upcoming issue: socio-economic aspects

17

umweltbundesamt

Way forward

- Scientific issues
  - Independent studies
  - Standardized test protocols (feeding studies)
  - Long term effects – monitoring, modeling
  - Regional aspects – problem formulation, selection of test organisms and comparators
  - Start discussions on "new" traits, e.g. stress tolerance, new ingredients
  - Stick to case-by-case approach: data derived from testing the relevant GMO
  - Follow good scientific practice (statistics, data interpretation, ...)

18

umweltbundesamt

Conclusions

- Principles of risk assessment generally recognized
- Protocols and guidance for risk appraisal and assessment are available
- Protocols/guidance do not contain detailed information on methodology to be used or on the interpretation of data/results – different viewpoints, many problems
- Discussions led already to an improvement over the past years, but still many things to do

19

umweltbundesamt

Contact & Information

Dr. Andreas Heissenberger
andreas.heissenberger@umweltbundesamt.at

Umweltbundesamt
www.umweltbundesamt.at

20

Andreas Heissenberger 先生报告的要点：

奥地利环保局在转基因生物风险评估方面也做了很多工作，我们这个部门也希望加强这方面的监管工作。从欧盟的角度来讲，我们也是一个比较重要的机构，欧盟在水、土壤、废物、能源、气候变化以及生物安全方面做了很多规定，我们也做了很多关于履行《卡塔赫纳生物安全议定书》的研究工作，成立了专门的执行机构，我本人代表奥地利官方参与了转基因生物安全相关的谈判工作。我是一个生物化学家，也是生物学家，已经从事这方面工作 15 年了。开始我们搞一些转基因宣教方面的工作，后来从事风险评估方面的工作。另外，社会经济层面我研究得多一点，监管方面也在介入。首先给大家介绍这个议定书是怎么回事，以及我们现在面临的挑战、风险评估的方法和原则，同时介绍未来的走向。

1．前面第一个发言人已经讲到了，目前转基因和非转基因的情况是不一样的，这就是为什么很多国家已经开始进行评估，特别是在投放市场之前。由于转基因食品已到了我们的餐桌，从基因转移的角度来讲，是物种方面的相互交叉；从传统的培植角度来说，转基因生物不可能用传统培植方法实现。转基因生物不管是从动物还是植物的角度来讲，都是有新的进展，虽然可能会引起其他风险，包括食品中的微生物和相关环境风险，但同时也给我们带来一些新的可能性，包括一些防虫方面的好处。过去一段时间内，大家对这个问题进行长时间讨论，而且，现在这种辩论也在进行当中，从全球来讲，国际监管方面的文件从 1986 年开始发布，如 OECD 组织发布的“关于重组 DNA 安全考虑”的蓝皮书。评估是审批程序中最重要的前提，也就是说一个转基因产品，特别是转基因食品，在很多国家要投放市场之前都要进行风险评估，中国、欧盟各国都是这样，但是，也有一些国家例外，像美国不需要评估，直接报告政府批准就可以了。

2.《生物安全议定书》中的“风险”从根本上说是一种综合因素，可是有实际的危害，也可能是潜在的危害，如对一些生物物种或人类健康产生一些危害。所谓“暴露”是指受到转基因生物的影响而产生的结果，是空间和时间方面的综合因素，同时我们看到这种风险是一种可能性的。所以我们在做风险评估的时候，一定要记住这个定义，最基本的要素在很多文件中已经表述，我简单讲几个，例如国际上的一些指南性文件，基本上都遵循这些方面的原则和定义。

（1）经合组织（OECD）的文件：是第一个比较宽泛的文件。该组织发表相关文件后，38 个共识性文件也相继发表，可以从谷歌图书库获得 4 卷。主要是一些对最基本问题的分析，如植物生物学问题，并没有谈到比较具体的转基因问题。而且，这些分析对风险评估也是有用的，此外，还涉及一些分子方面的特征描述，独特标识等，特别是个别国家政府的相关立法已经出来了。

（2）食品法典委员会（www.codexalimentarius.net）：从国际层面讲，已经有 190 多个国家加入《生物多样性公约》，他们也在讨论生物安全问题，包括食品安全。其他国家也有一些转基因方面的文件，包括国际食品法典也有相关的具体规定，但是，关于环境或风险评估方面，这一块相对来说发表得比较少。

（3）欧盟食品监管机构：这些监管机构发表了很多指南性和比较具体的文件，目前，欧盟拥有转基因方面最详细的文件。欧盟食品安全监管机构主要负责转基因生物的申请和科学评估。此外，有很多的文件每天都在定时和不定时地更新，而且这些文件从科学角度来说的确是最新的。

（4）卡塔赫纳生物安全议定书：从国际层面来讲最重要的一点就是大家达成了《卡塔赫纳生物安全议定书》，另外还有一些指南性的文件，及一些专门的特设工作组。的确产生了一些好文件，比如说对于风险评估路线图、一些携带基因方面的动植物或者是一些抗性的东西，已经有相关的文件。包括一些不同的转基因植物或者动物，他们有可能相互交叉。《卡塔赫纳生物安全议定书》是各个国家都要遵循的最基本文件，其主要原则都是相同的，所以，它是唯一一个具有国际层面高度的文件，最为重要。

风险评估路线图作为关键文件，需要明确几个主要问题：进行风险评估需要给相关数据设定标准，确保数据的科学性、准确性和透明性，包括汇报和报表。此外，对于不确定的风险评估，一定要充分考虑才能做出较好的结论。当然科学工作是我们永远也做不完的，不管怎么说，随着一些新技术的产生，我们可以利用这些新技术，使风险评估工作做得更好。关于风险评估的内容和范围界定，风险评估的步骤都是逐渐完善的内容。

3．什么是范围界定？实际上范围界定是要设定一个相关的极限，比如说我们需要做什么？我们怎么去做？做什么和怎么去做是要分开的。在实际工作中，我们要考虑几个关键点，如进行 GMO（转基因生物）的环境风险评估时我们要考虑多样性问题，包括一些濒临灭绝的物种，GMO 如何应用，包括工业行业及食品方面等。其中一个主要目标是要保证人类健康，不能危害人类健康。农业种植方式一旦发生变化，转基因对这种变化到底好不好，对生物多样性有什么影响等，都要充分考虑其中的潜在危险，我想到目前科学方面的一些实验，大部分科学实验规模比较小，有的时候就几亩几公顷，面积太小了，一旦谈到商业化生产的时候，就是一个截然不同的问题。

另外从政治和环境的综合角度来看这几个问题，有国家层面的问题，有地区的问题，东南亚的环境跟欧洲和北美环境完全不一样，跟中国的环境也完全不一样，这几个方面要充分考虑进来。当然我们要意识到这些问题的范围界定，都涉及到一些 GMO 问题。在很多情况下我们根本搞不清楚从环境保护角度希望我们做什么样的工作，特别是转基因生物一旦释放到环境当中所产生的反应。有些已经释放到环境中的转基因生物，实际上已经发生变化了，从环境科学的角度来说，为时已晚，所以我们应该设定一个尺度或指标，我们应该做什么样的工作，例如找出 GMO 的相关特点，及可能产生的负面影响，不光是对环境，还可能对人类健康产生不利影响，但我们一定要给它分类，同时还要充分考虑其他受体的环境，不同国家的环境不一样，不同地区的环境也不一样。

4．我们要考虑进行一些风险评估，这些风险有可能发生，一发生就会相互起作用，所以我们不愿意看到风险的发生。因此，第一步我们要搞清楚在 GMO 相互作用之前，对人类健康产生什么样的影响，对环境有什么样的影响，GMO 出来之后，新的物种也会出现。第二步设想风险的可能性。第三步设想风险的结果。第四步还需要估计一下，通过第三步的评估，看对其他环境产生什么外移作用。可以接受的，这是一种风险。我们看到这几个步骤，实际上是相互影响的，比如说从一开始的时候，从管理角度来讲，要搞一个策略的东西，一开始我们要有防范的意识，进行一些相关研究。

5．主要挑战和问题——我们当然总有问题，总有挑战，特别是在应用的时候：

（1）从宽泛角度来讲，我们应该意识到目前没有太多直接的一些证据，关于 GMO 直接的测试，大部分工作是一些非直接的，一些建模模式，或者是一些假设的证据。我们对杂交研究还是很充分的，如果没有风险，或者是有风险，有的时候考虑得不是很充分，有的时候只考虑到了一些相互作用，或者是（致敏性）。我们要看转基因的毒性，这一方面的研究实际工作做得不太好，只限于实验室方面的要求，现在只是针对一些很小的动物在做一些实验，这个是不够的。

（2）关于不利影响或次生影响，我们考虑得不是很充分，比如说在印度，我们看到有一些实验，利用转基因棉花防治害虫，但用了转基因棉以后就产生了另外一个害虫，而且在印度目前没有对此进行科学的研究，一旦有了这种新的能抗转基因棉花的害虫，对产量将产生什么样的影响，对环境又将产生什么样的影响等，这些工作还没有做。还有，关于一些指标如何定义也的确是很困难。搞了十五六年的培植工作，这些指标跟非转基因指标是非常不同的，不同的话就没有可比性。所以我们要进行一些转基因方面的指标研究，看看哪些指标比较合适。

（3）关于区域性方面的工作：从区域来讲问题也不少，我们也考虑得不是很充分，大部分数据来自转基因国家，也就是 27 个国家，可能还有 5～7 个国家，他们照样也是大量生产转基因作物，但是没有被纳入考虑范围。如果我们仔细研究这些国家的情况，会发现有五六个欧盟国家，他们也种转基因作物，可能也就是几公顷，量不是很大，因此我们的数据非常有限，特别是对风险评估来讲是不够的，这些数据对于区域的风险评估是非常不充分的。关于环境方面的问题，不管是中国，还是在欧洲，这些数据好像现在大部分来自于美国。关于环境，我们的科学依据也比较少，好像只有一个科学依据，即阿根廷的情况，但是，阿根廷的科学依据又不适合欧洲的情况，数据的确是有的，但是具有科学依据的数据并不充分。

（4）国际层面一些具体问题我们也要充分考虑，包括一些亚洲、非洲生态平衡方面的问题，还有南美，我们有一些超级生物多样性大国，这些国家生物多样性极为丰富，你就没有更仔细去考虑他们。我们要去想一下一些不同作物的具体情况，例如，亚洲的水稻地位非常重要，有些地区可能是木薯非常重要，但是对于木薯方面的研究又不是很充分，关于转基因方面的大部分数据都与玉米、大豆有关系，因此，需要做一些更加充分的研究工作和数据统计以及试验数据等，这都是我自己做的一些工作。只是，这一方面的实验数据和样本都比较少。只有大量汇聚不同情况、不同时间、地点的数据，达到一定的数据量后，才有利于我们进一步讨论和改善研究工作。

（5）关于致敏性方面的研究，我们的分析研究力度还是比较大的。从科学角度来讲，我想大家都同意我这样一个观点，如果有一些不同的意见，我们应该找出这些问题的不同意见在哪里，问题在哪里，科学方面的一些差异在哪里，科学的差距又来自于哪些地区，原因又是什么，这些工作目前还没有做。大家通常会发现，我们的确有一些差异，可是这些差异又与生物方面的研究没有太多的关系，有的时候没有差异，也可能是研究不太深入。如果与生物方面不是特别相关的，我就想问大家这么一个问题，你干吗要做这个实验，既然不相关你还花了很多钱，那就没有意义了。

我们对转基因的监管方面应该更加严格，指南性的文件并不是方方面面都写得很详细，比如说文字方面，卡塔赫纳议定书中就没有涉及这一问题，我们要看一些指南文件对我们的工作是否适宜？如果适宜的话，如欧盟有一些指南文件可能与中国的一些情况类似，可以拿其中的一部分，但是要小心，可不是欧洲所有的情况都适合中国。而且有些还具有限制性，这些更麻烦，因此要慎重。此外，如何纳入更多的机构，需要工作组制定更好的指南性文件。风险评估是一个挑战，发展中国家特别面临没有资金，没有相关人员介入，以及人员素质等问题，人员队伍变化很大，一会走一会又来，不是太稳定。

6. 决策：这个决策跟风险评估是有关系的，我们也要搞清楚这两者之间的关系，风险评估的确是为了 GMO 的监管和决策服务的。从原则角度来说是一种协议上的安排，从全世界的角度来讲也是一种协议的安排，也就是说主管方面的决定，要基于风险方评估，的确找到了一种风险，但是这个风险是可控的，或者虽然没有找到，但是这个风险还是可控的，我们是否应该采取一些措施适度控制这个 GMO，如果有监管和监控，这些监管和监控是不是长期支持这些监督方面的工作。同时，从决策来说，好像是基于不同的资源，不同的渠道，有时候不同的国家有不同的政策，美国比较开放，觉得搞转基因没有事，他们欢迎，审批方面没有“前提知情同意”要求，门槛也比较低。在欧盟又是另外一种情况，欧盟对 GMO 相关方面的要求非常严格，在欧盟批准 GMO 商业化生产是非常困难的。奥地利政府又有政治方面决策，禁止大面积种植 GMO，保证 GMO 的安全性，如果不安全就不能种，我们对环境的风险评估是非常严格的，而且相关的研究工作也非常充分。

此外，关于媒体和公众的意见，政客所面临的压力，以及相关人员遇到的情况等，都是要考虑的。特别是发展中国家常面临这些方面的问题，如非洲或东欧的发展中国家，要考虑经济因素，考虑是否还需要转基因，转基因对食品安全有好处吗？是种好，还是不种好？……我们奥地利就不允许种植转基因作物，奥地利是欧洲唯一一个不种转基因作物的国家，我们人口也就是北京人口这么多，的确是一个很小的国家，但是从经济方面来讲我们考虑得很充分。这些方面对于决策都是需要充分考虑的方面。另外，有些国家特别是一些小国家和穷国的恐惧心理也要考虑到，因为在加入 WTO 时，加拿大、美国、阿根廷就害怕欧盟一些相关的法律，对这些小国家来说，你要不用 GMO，经济成本又比较高，尤其是这些小国必须要遵循 WTO 的国际义务。上述几个方面从决策来说一定要充分考虑，当然还有其他的因素，也都是非常重要的。

7. 未来走向：最后，我们今后怎么做，特别是如何完善 GMO 的风险预测？两个方面，一个是监管方面，另一个是科学方面。先说监管方面，我们的确有这样一个必要性，我们要意识到从全球的层面搞一个框架性的东西，《卡塔赫纳生物安全议定书》相关的专家一开始就做了很好的工作，但是由于政治方面的原因，一些相关的文件并没有被所有的国家签署，所以我们还需要另外一轮的测试。但是这个文件如果被所有的国家都认可签署，我想它是一个最好的监管基础，相关指南性的文件还要更加充分一点。我们不是说光搞一个框架和路线图，我们需要非常细节的方法、详细的数据，还有监管人员和科学家应该坐在一起，有的时候科学家比监管部门更重要，有的时候监管部门更重要，这两者都应该共同介入，所以对于监管部门的决策来说，他们应该基于科学方面的依据，最起码现在要比以前做得更好。

从监管方面来说，有一点是非常重要的，就是风险评估怎么和社会经济层面更有机地联合起来，这一点我想有一些新的问题已经出现了，可能明天大家要进一步进行讨论。从科学角度来看，我想最重要的一点就是搞科学工作一定要独立，我们要搞一些独立性的研究工作。GMO 到底给我们带来什么好处，我们已经有一些规范化的议定书。我们要搞一些长期的数据采集和监控，特别是关于科学数据方面的监测，需要更加科学的研究。同时需要一些更好的建模，还有应该强调区域方面的问题。另外我们要看一下前瞻方向性问题，有一些新特征也出来了，还有一些不同的人有反对的声音，不喜欢这种转基因生物，还有其他行业的声音，这些方面我们就没有经验了，而且没有一些充分的文件和议定书，我们的经验也不充分，目前全面的研究工作还不能完全展开。现在我们只是根据个案来进行科学方面的研究，如果说你遵循某个个案，那你就遵循这个个案，从 GMO 的家度来讲，我们一定要进行一些个案或者基于事实方面的数据，特别是我们谈论的一些逐步分析的方法，一定要遵循科学依据，进行数据分析。

这些都是比较重要的方面，我们怎么样去改善风险评估，我们如何遵循相关风险评估的方法，这些方法实际上大家已经认可了，目前讨论得又不是很多。这些方面的内容议定书已经有了，需要方法和数据的更加细化。这些指南性的文件，还有整个对转基因方面进行讨论已经有了很多的改善，在过去 10 年中通过观察，我们也发现已经有质量方面的改善，特别是数据的质量，数据分析的质量也提高了，但是我们要走的路还有很长，我们需要完成的事情还有很多，有很多要求没有达到，这是我个人的观点，我们也希望未来朝好的方向发展。

谢谢！

专题 1　风险评估
Session 1　Risk Assessment of GMOs

中国农业转基因生物安全管理进展

The Progress on Biosafety Regulation of Agri-GMOs in China

付仲文
（中国农业部科技发展中心 副处长、高级农艺师）
FU Zhong-wen，
（Deputy Division Director，Center of Science and Technology Development，Ministry of Agriculture of China）

我国转基因生物安全管理进展

付仲文
农业部科技发展中心
2013年5月27日，北京湖北大厦

内容提要

一、管理框架

二、基本政策及其变化

三、如何看待农业转基因

一、管理框架

◆法规体系

◆管理体系

◆技术支撑体系

◆ 法规体系

国务院
- 农业转基因生物安全管理条例

农业部
- 农业转基因生物安全评价办法
- 农业转基因生物标识管理办法
- 农业转基因生物进口安全管理办法
- 农业转基因生物加工审批办法
- 相关公告、技术指南、标准和规范

质检总局
- 进出境转基因产品检验检疫管理办法

◆ 管理体系

部际联席会议

职责：研究、协商农业转基因生物安全管理的重大问题

组成：农业部牵头，农业、科技、卫生、商务、环境保护、检验检疫等部门组成。目前有11个

农业部

职责：负责安全评价、监督管理、体系建设、标准制定、进口审批和进口标识管理

机构：农业部农业转基因生物安全管理领导小组及农业部农业转基因生物安全管理办公室

◆ 管理体系

县级以上农业行政主管部门

职责：负责本区域监督管理，生产、加工和标识许可

机构：各省农业转基因生物安全管理办公室（挂靠在农业厅科教处）

质检总局

负责进出境转基因检验检疫

◆ 技术支撑体系

□ 农业转基因生物安全委员会

➢《条例》、《安评办法》：设立国家农业转基因生物安全委员会，负责农业转基因生物的安全评价工作。由从事农业转基因生物研究、生产、加工、检验检疫、卫生、环境保护等方面专家组成，每届任期三年

➢ 历届情况：

✓2002—2004年，第一届，58人

✓2005—2008年，第二届，74人

✓2009—2012年，第三届，60人

✓2013—2015年，第四届，64人

◆ 技术支撑体系

□ 全国农业转基因生物安全管理标准化技术委员会

➢ 2004年11月30日，《关于成立全国农业转基因生物安全管理标准化技术委员会（SAC/TC276）的复函》

➢ 负责转基因植物、动物、微生物及其产品的研究、试验、生产、加工、经营、进出口及与安全管理方面相关的国家标准制修订工作

➢ 委员41名，秘书处设在农业部科技发展中心

➢ 发布实施标准102项，已报批待发布标准15项，已立项在研标准超过40项

◆ 技术支撑体系

□ 转基因检测机构

2005年9月，转基因生物安全监督检验测试机构列入农业部第五批部级质检中心筹建计划。截至2013年4月，已有39个机构通过了“2+1”认证，涵盖了“综合性、区域性、专业性”三个层次、“转基因植物、动物、微生物”三个领域和“产品成分、环境安全、食用安全”三个类别，初步形成了功能完善、管理规范的农业转基因生物安全检测体系，为《条例》及配套规章实施提供了重要技术保障。

二、基本制度及政策变化

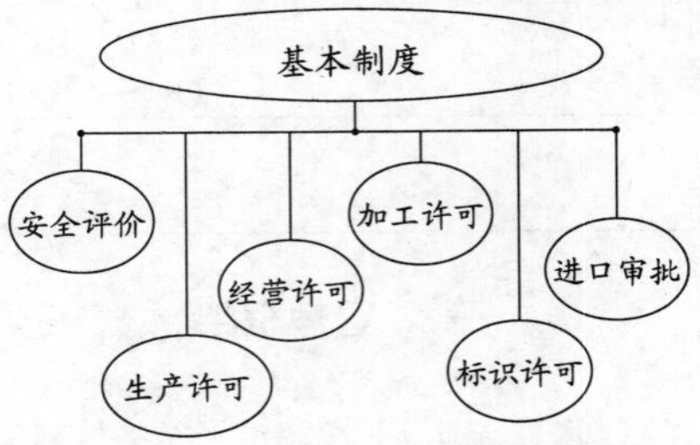

法规管理对象

利用基因工程技术改变基因组构成，用于农业生产或者农产品加工的动物、植物、微生物及其产品

□ 转基因种子、种畜禽、水产苗种和微生物

□ 转基因产品

□ 直接加工品

□ 含有转基因成分的产品

法规管理范围

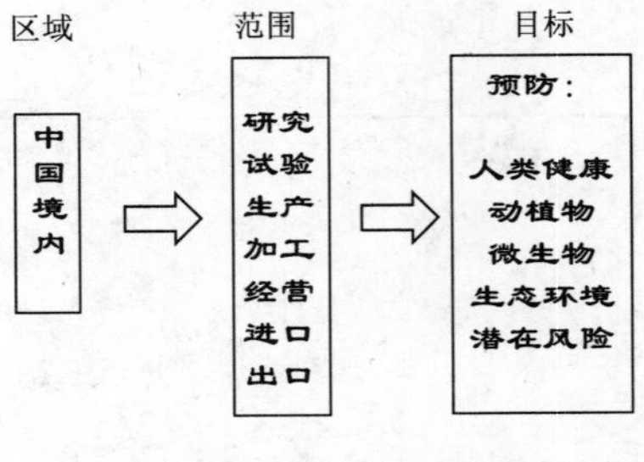

安全评价制度

3类评价对象 ⇨ 动物  植物  微生物 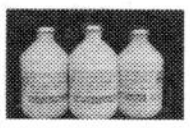

4个安全等级

5个评价阶段

2种评价方式

安全评价制度

3类评价对象

4个安全等级 ⇨

- I：尚不存在危险
- II：具有低度危险
- III：具有中度危险
- IV：具有高度危险

5个评价阶段

2种评价方式

安全评价制度

3类评价对象

4个安全等级

5个评价阶段 ⇨

1. 实验研究
2. 中间试验
3. 环境释放
4. 生产性试验
5. 安全证书

2种评价方式

安全评价制度

3类评价对象

4个安全等级

5个评价阶段

2种评价方式 ⇨ 报告制 / 审批制

◆ 安全评价—报告制

(1) 本单位生物安全小组审查
(2) 直接报农业部
(3) 农业部科技发展中心审查
(4) 不上安委会，有问题咨询
(5) 书面反馈意见，不收费

◆ 安全评价—审批制

(1) 本单位生物安全小组审查
(2) 试验所在省级主管部门审查
(3) 科技发展中心初审
(4) 安委会科学技术审查
(5) 批复文件为审批书或不批准文件、安全证书，环境释放及以上缴安全评价费

◆ 在国内研发申报程序

➢ 中外合作、合资、独资：从实验研究报，所有等级均审批

➢ 除中外合作、合资、独资外：
—实验研究I、II，本单位管理
—实验研究III、IV和所有等级中间试验报告农业部
—环境释放、生产性试验和申请安全证书审批

◆评价内容与方法

经济合作与发展组织（OECD）

➢ 1986年《重组DNA安全性考虑：用于工业、农业和环境的重组DNA生物安全性》

➢ 1992年《生物技术安全性考虑》

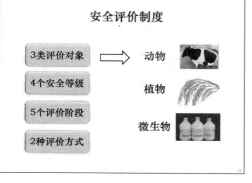

➢ 1995年以来，研制出了一系列与转基因风险分析相关的基础共识文件70多个

国际食品法典委员会（CAC）

- 2000年FAO/WHO：关于重组DNA植物食品的健康安全问题
- CAC于2000年成立生物技术食品政府间特别工作组
  - 《现代生物技术食品风险分析原则》
  - 《重组DNA植物食品安全评价指南》
    - ✓ 附件1：潜在致敏性评价
    - ✓ 附件2：营养品质改良重组DNA植物食品安全评价
    - ✓ 附件3：含少量重组DNA植物材料食品安全评价
  - 《重组DNA动物食品安全评价指南》
  - 《重组DNA微生物食品安全评价指南》

◆ 中国：以转基因植物为例，在评价内容与方法上与OECD和CAC标准接轨

- ✓ 分子特征：DNA—RNA—蛋白质—性状
- ✓ 遗传稳定性
- ✓ 环境安全：生存竞争能力，抗性效率，基因漂移，对非靶标和生物多样性影响
- ✓ 食用安全：营养成分和抗营养因子分析，外源表达产物的毒性和过敏性

- ➢ 比较法：转基因与非转基因比较，重点考察两者差别及差异可能带来的影响（好、坏、没影响）
- ➢ 应考虑预期效应和非预期效应（可测、不可测）

■ 安全评价实施成效

◆ 我国农业转基因生物安全证书[生产应用]发放情况

截至目前，共批准发放7种作物安全证书

1997年 耐储存番茄（已过有效期）

抗虫棉花（续）

1999年 改变花色矮牵牛

抗病辣椒（已过有效期）

2006年 抗病番木瓜（续）

2009年 抗虫水稻转植酸酶玉米

■ 安全评价实施成效

◆ 我国进口用作加工原料农产品安全证书[进口]发放情况

截至目前共批准发放境外研发商5种作物32个转化体安全证书

- ✓ 棉花6个（抗虫、抗除草剂）
- ✓ 甜菜1个（抗除草剂）
- ✓ 油菜7个（抗除草剂）
- ✓ 大豆5个（抗除草剂、品质改良）
- ✓ 玉米13个（抗虫、抗除草剂）

◆ 《条例》进口安全管理范围

- ➢ 国内单位引进用于研究与试验的所有转基因生物
- ➢ 境外公司向中国出口用于生产的所有转基因生物
- ➢ 关于进口用作加工原料的转基因产品（2012年4月）
  - ✓ 国家质检总局对所有含有转基因成分的产品进行检测，贸易商要获得安全证书。农业部对已列入标识目录的产品仍按原办法执行；对未列入标识目录的产品，采取只审批安全证书，不办理标识批准文件

进口审批制度

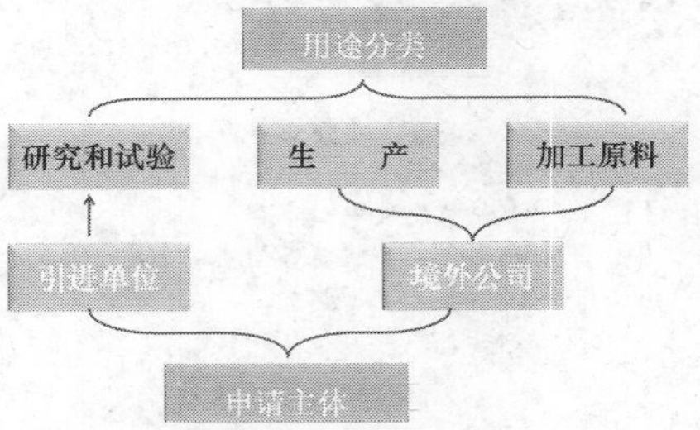

◆ 进口管理程序——审批制

- ➢ 用于研究和试验，引进单位
  - ——用于研究，从实验研究报
  - ——用于试验，从中间试验报
- ➢ 用于生产，境外公司，从中间试验报
- ➢ 用作加工原料或直接消费品，境外公司，直接申请安全证书，中国境内检测

◆ 进口用作加工原料管理程序

① 境外研发商提交进口用作加工原料安全证书申请

② 农业部受理，提交安委会审查

③ 如符合要求，发放入境材料批件

④ 试验材料进口

⑤ 开展身份验证、环境安全及食用安全检测

⑥ 申报书及综合评价报告提交安委会审查

⑦ 如符合要求，发放境外研发商进口用作加工原料安全证书

⑧ 境外贸易商凭研发商安全证书，办理每批次进口安全证书

转基因标识制度

国际标识制度选项汇总	中国
➢ 标识目的 ✓ 消费者知情权和选择权 ✓ 安全因素（可追溯）	➢ 标识目的 ✓ 消费者知情权和选择权
➢ 适用范围 ✓ 标识目录 ✓ 标识的豁免	➢ 适用范围 ✓ 标识目录
➢ 标识方式 ✓ 强制性或自愿性 ✓ 定性或定量 ✓ 标识阈值	➢ 标识方式 ✓ 强制性 ✓ 定性
➢ 标签表述方法 ✓ 转基因或只标出差异部分	➢ 标签表述方法 ✓ 转基因
➢ 阴性标识 ✓ 非转基因标识	➢ 阴性标识 ✓ 未规定

✓ 我国制度特点：标识目录、强制标识、定性标识

✓ 标识目录由农业部商国务院有关部门制定、调整和公布

■ 第一批标识目录（2002年发布实施）

- 大豆种子、大豆、大豆粉、大豆油、豆粕
- 玉米种子、玉米、玉米油、玉米粉
- 油菜种子、油菜籽、油菜籽油、油菜籽粕
- 棉花种子
- 番茄种子、鲜番茄、番茄酱

生产许可制度

生产转基因植物种子、种畜禽、水产苗种，应当取得农业部颁发的生产许可证

申请条件

- □取得安全证书并通过品种审定
- □在指定的区域种植或者养殖
- □有相应的安全管理、防范措施
- □农业部规定的其他条件

加工许可制度

在中国境内从事具有活性的转基因生物为原料生产加工活动的单位，应当取得省级人民政府农业行政主管部门颁发的《农业转基因生物加工许可证》

经营许可制度

经营转基因植物种子、种畜禽、水产苗种，应当取得农业部颁发的经营许可证

申请条件

——有专门的管理人员和经营档案

——有相应的安全管理、防范措施

——农业部规定的其他条件

与其他法规衔接

➢ 《条例》第十七条 转基因植物种子、种畜禽、水产苗种，利用农业转基因生物生产的或者含有农业转基因生物成份的种子、种畜禽、水产苗种、农药、兽药、肥料和添加剂等，在依照有关法律、行政法规的规定进行审定、登记或者评价、审批前，应当依照本条例第十六条的规定取得农业转基因生物安全证书。

➢ 《安办评法》附录1：转基因植物在取得农业转基因生物安全证书后方可作为种质资源利用。用取得农业转基因生物安全证书的转基因植物作为亲本与常规品种杂交得到的杂交后代，应当从生产性试验阶段开始申报安全性评价。

转基因作物管理总体流程

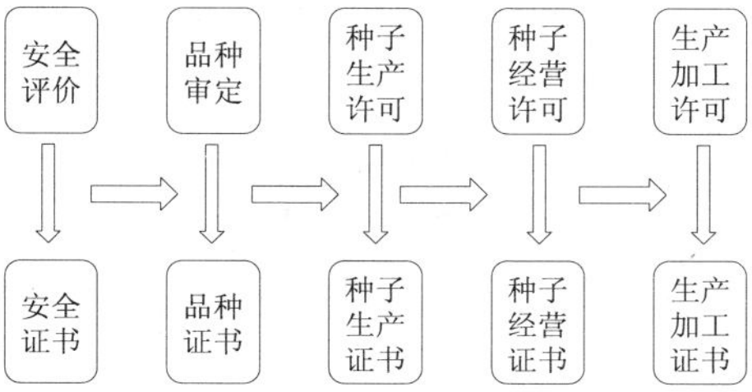

■ 相关政策变化

◆抗虫棉政策演变

◆其他政策变化

◆ 抗虫棉政策：2004年9月

➢ 2004年9月，农业部公告410号
- ✓ 区域扩大：一个省一批简化为按生态区审批
- ✓ 阶段简化：生产性试验改为直接申请安全证书
- ✓ 育种单位可直接或与基因专利单位联合申请
- ✓ 三种类型：扩区；跨区；新申请（衍生品系）
- ✓ 首次指定了5家检测机构
- ✓ 首次提出检测和监测报告指标要求：

（1）靶标害虫抗性效率监测报告；
（2）杀虫蛋白表达含量的检测报告；
（3）抗虫稳定性和纯合度的生测报告。

◆ 抗虫棉政策：2008年2月

➢ 2008年2月，农业部公告989号
- ✓ 分两种类型：扩区加跨区；新申请（衍生品系）
- ✓ 5家检测机构不变
- ✓ 类型1：不需要检测报告
- ✓ 类型2：检测报告2个

（1）杀虫蛋白表达量的检测报告
（2）对靶标害虫抗虫性的生测报告

◆ 抗虫棉政策：2011年12月

➢ 2011年12月，农业部公告1693号
- ✓ 分两种类型：不变。
- ✓ 检测机构：不再是原来指定的5家，通过双认证的检测机构
- ✓ 检测报告：不变（1）杀虫蛋白表达量的检测报告；（2）对靶标害虫抗虫性的生测报告。
- ✓ 盲样检测：由农业部科技发展中心组织

◆ 抗虫棉政策：2012年5月

➢ 2012年5月，根据国务院8号文件清理整顿种业的文件精神，规定：
- ✓ 关于第三方盲样检测
  - • 所有抗虫棉申请，包括续申请、扩区、新申请等，全部要求开展**第三方盲样检测**
- ✓ 关于检测报告：不变
  - • 杀虫蛋白表达量的检测报告
  - • 对靶标害虫抗虫性的生测报告
- ✓ 关于检测结果
  - • 今后对申请生产应用安全证书的项目，安委会只对通过评审的项目进行排序，我部根据生产需要决定发放生产应用安全证书的数量

◆ 抗虫棉政策：2013年2月

➢ 2013年2月，《关于开展转基因抗虫棉转基因成分验证检测的通知》（农科（执法）函［2013］36号）
- ✓ 在提交抗虫棉检测样品时，增加200g样品用于验证检测，并按要求提交承诺书。

◆ 有关政策：2013年3月2日

➢ 关于转基因植物品种命名
- ✓ 要符合《农业植物品种命名规定》（农业部令2012年第2号）

➢ 关于生产应用综合评价报告
- ✓ 首次申请生产应用安全证书，需提交《农业转基因生物品种（或品系）生产应用综合评价报告》，包括对我国生产、贸易、社会等方面影响。

➢ 关于转基因生物活性样品
- ✓ 首次申请生产性试验和安全证书的，申请截止日30个工作日前提供所申报转基因生物活性样品和技术资料。

三、如何看待农业转基因

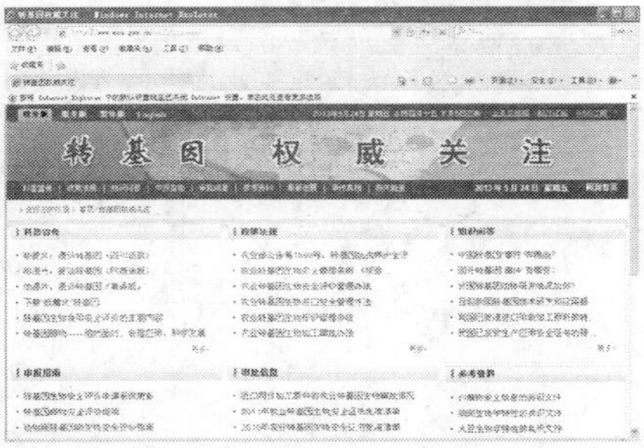

■ 正式回答一些社会关注的问题

■ 转基因产品发生过安全事故吗？

✓ 转基因技术及其产品的安全研究近40年

✓ 转基因农产品安全食用15年

✓ 迄今为止，批准上市的转基因农产品没有发生一起经过科学反复论证证实的生物安全问题

■ 转基因食品的安全性有无定论？(1)

✓ **安全食品：**“食品无毒无害，符合应当有的营养要求，对人体健康不造成任何急性、亚急性或慢性危害”——《食品安全法》

✓ **不安全食品：**“食品中有毒有害物质对人体健康有危害的公共卫生问题”——世界卫生组织

✓ **转基因食品：**不属于食品安全问题——世界卫生组织

✓ **转基因食品安全问题谁说了算：**国际食品法典委员会（制定国际食品标准，170多个国家组成，WTO承认的国际仲裁标准）。类似电、飞机能否应用

■ 转基因食品的安全性有无定论？(2)

有定论：

✓ 通过安全评价获得生物安全证书的转基因食品是安全的，可以放心食用

✓ 转基因食品安全评价遵循的是个案分析原则，不能笼统地谈转基因食品的安全性问题

✓ 迄今为止，转基因食品没有发生一起经过证实的食用安全问题

■ 如何对待转基因技术？

一发展仍是硬道理

一我国转基因技术研发的“十六”字方针：

✓ 加快研究、推进应用、规范管理、科学发展

一强化生物安全管理理念，克服两种极端观点：

✓ 认为转基因技术与常规技术一样的观点

✓ 认为农业转基因技术风险不可控不能发展的观点

结 束 语

- 研发转基因技术是我国必然的战略选择
- 我国转基因安全管理是符合国际组织技术规范，在国际上也是最严格的国家之一
- 转基因风险不能与已发生的食品安全事件混为一谈
- 通过安全评价并获得安全证书的转基因食品是安全的，未发生一起安全事件，与非转基因食品具有同等安全性，可以放心食用
- 转基因技术与传统技术既一脉相承又有区别，必须依法坚强监管。认为两者等同不需专门管理或者认为风险不可控进而阻挠发展的观点都是不正确的

谢 谢 ！

Advancing Risk Assessment under the Cartagena Protocol on Biosafety

推进《卡塔赫纳生物安全议定书》下的风险评估

Lim Li Ching
（Senior Researcher，*Third World Network* 第三世界网络生物安全专家）

Abstract: Parties to the Cartagena Protocol on Biosafety established an Ad Hoc Technical Expert Group on Risk Assessment and Risk Management（AHTEG）in 2008 to develop further guidance on specific aspects of risk assessment and risk management. The outcome is the 'Guidance on Risk Assessment of Living Modified Organisms', which comprises three parts: the 'Roadmap' for risk assessment of living modified organisms（LMOs）; guidance on conducting risk assessments on LM abiotic stress-tolerant plants，stacked plants，LM trees and LM mosquitoes; and guidance on monitoring LMOs released into the environment. With the development of the Guidance，Parties now have a framework which they can use and adapt to their conditions，needs and obligations，consistent with the Protocol and in particular its Annex III on risk assessment. The Guidance has produced a number of innovations and improvements on appraising risks. At the same time，it contains a few general weaknesses in its approach. Nonetheless，the overall outcomes of the AHTEG activities，on the balance，provide good guidance for advanced understanding of risk appraisal that will be useful for the implementation of risk assessment frameworks to strengthen national biosafety legislation. Despite the challenges ahead，putting the Guidance to use will be the best measure of its ability to uphold the core objectives of the Cartagena Protocol. In addition，following their 6th meeting in 2012，Parties have continued the work on risk assessment by continuing online forum discussions and establishing a new AHTEG. These activities and their outcomes will continue to be important for shaping and advancing our understanding of risk assessment.
Keywords: Risk assessment，living modified organisms（LMOs），Cartagena Protocol on Biosafety，guidance

Introduction

The Cartagena Protocol on Biosafety is a protocol under the Convention on Biological Diversity（CBD）. It was adopted on 29 January 2000 in Montreal，and entered into force on 11 September 2003. As of August 2013，there are 166 Parties，most of which are developing countries.

The Cartagena Protocol is the only international legally-binding instrument that deals exclusively with living modified organisms (LMOs), or as they are more commonly known, genetically modified organisms (GMOs). The objective of the Protocol is to contribute to ensuring an adequate level of protection in the field of the safe transfer, handling and use of LMOs resulting from modern biotechnology that may have adverse effects on the conservation and sustainable use of biological diversity, taking into account risks to human heath, and specifically focusing on transboundary movements.

Therefore, for the first time in international law there is recognition that LMOs are different from other naturally occurring organisms and may carry special risks and hazards and thus need to be regulated internationally. The Protocol recognizes that LMOs may have biodiversity, human health, and socio-economic impacts; and that these impacts must be risk assessed or taken into account when making decisions.

Risk assessment is therefore the central obligation of Parties to the Cartagena Protocol, as elaborated in Article 15. Risk assessment is to be conducted in a "scientifically sound manner", in accordance with the annex on risk assessment (Annex III of the Protocol), and taking into account recognized risk assessment techniques. Furthermore, precaution is the basis for decision-making and risk assessment under the Protocol, and the precautionary principle is put into operation in the Protocol's provisions.

1 Risk Assessment Work under the Cartagena Protocol

While Annex III of the Cartagena Protocol provides general guidance on risk assessment, it has been recognized that this is not enough. In 2008, at the Fourth Conference of the Parties serving as the Meeting of the Parties (COP-MOP4), Parties requested specific guidance on risk assessment. To this end, they established an Ad Hoc Technical Expert Group (AHTEG) on Risk Assessment and Risk Management.

The AHTEG was given the mandate to develop a 'roadmap' on the necessary steps to conduct a risk assessment, and provide examples of relevant guidance documents, and to develop further guidance on specific aspects on risk assessment. The AHTEG reported back to COP-MOP5, and its mandate was further extended for another two years to produce a revised version of the guidance it had developed (following scientific review and testing), as well as provide further guidance on new specific topics of risk assessment.

The final outcome of the ATHEG's work is a 'Guidance on Risk Assessment of LMOs', which comprises a 'roadmap' for risk assessment of LMOs; specific guidance on living modified plants with stacked genes or traits, living modified plants tolerant to abiotic stress, living modified trees and living modified mosquitoes, as well as guidance on monitoring of LMOs released into the environment.

2 Significance of the Guidance on Risk Assessment of LMOs

The Guidance is a significant development in the Protocol's history. Parties to the Protocol now have a risk assessment framework that can be used and adapted to their conditions, needs and obligations. The objective of the Guidance is to provide a reference that may assist Parties and other Governments in implementing the provisions of the Protocol with regards to risk assessment, in particular its Annex III.

Hence the Guidance was drafted to be consistent with the Protocol and its Annex III on risk assessment.

While Annex III is general in nature, the Guidance is more specific and further elaborates on the risk assessment process generally as well as in relation to specific LMOs and traits. It is applicable to different potential receiving environments and societies, but at the same time is based on consistent standards. An assurance of its quality is provided by the fact that the Guidance was subjected to numerous rounds of feedback and peer review. It is also meant to be a 'living' document that can be updated and strengthened as experience is gained with its use.

The Guidance has several innovations and strengths that make it an especially valuable contribution to the risk assessment literature. It recognizes that an environmental assessment of risks does not happen in a vacuum, but that the criteria and needs of the risk assessment may be informed by other actors or aspects. For example, it recognizes that there may be a need to allow for stakeholder participation prior to conducting the actual risk assessment, to help identify protection goals, assessment endpoints and risk thresholds relevant to the assessment. The Guidance also shows how the conduct of the risk assessment relates to other stages in the risk appraisal process (including the pre-risk assessment scoping process and post-risk assessment evaluation of its outcomes).

On the issue of the quality and relevance of information, the Guidance provides clear requirements that the information provided should be of high scientific quality. This means that such information needs to be transparent, reproducible, and verifiable, if necessary, by access to research material and raw data. This is not necessarily the case in current practice, and so is a good step forward.

Importantly, the Guidance recognizes the need for describing the nature and sources of uncertainty in the risk assessment, including the impact on the estimated level of risk and on the conclusions and recommendations of the risk assessment. Uncertainties should also be communicated to decision makers. Such considerations of uncertainty strengthen the scientific validity of a risk assessment. In cases where the nature of the uncertainty implies that it cannot be addressed through the provision of more data, the Guidance recognizes that this may be dealt with by risk management and/or monitoring.

The Guidance also recognizes that risk assessment is a process that may involve a stepwise and iterative advancement, where risks at smaller spatial or time scales of release may be assessed before larger releases are permitted to take place. In addition, as results are gathered at each step of the risk assessment process and new information arises, certain steps may need to be revisited. Analyzing risks within the context of alternative options is also considered as important in the Guidance, particularly when determining the acceptability of the risk.

A significant contribution is that the Guidance now includes guidance on monitoring LMOs released into the environment. This is viewed as important for risk assessment and risk management because no specific guidance is available from the Protocol, or internationally. The Guidance provides a robust, comprehensive approach to develop a monitoring plan: what to monitor, how to monitor, where to monitor, how long to monitor for, and how to communicate monitoring results. It recognizes two general categories of monitoring: case-specific monitoring, to address uncertainties identified in the risk assessment, and general monitoring, to address uncertainties that were not identified in the risk assessment, including long-term effects.

While the Guidance is an important step in advancing our understanding of biosafety under the Protocol, nonetheless, there remain some weaknesses. One is its overemphasis on comparative risk assessment, for example allowing the use of very broad comparators, which would tend to underestimate differences and overestimate similarity or equivalence between comparators. The Guidance also fails to provide a link to other aspects of risk that may be taken into account in decision-making, such as socio-economic, legal and ethical issues, all of which are critical for biosafety decision-making.

On balance, however, the overall outcomes of the AHTEG activities provide good guidance for advanced understanding of risk appraisal that will be useful for the implementation of risk assessment frameworks to strengthen national biosafety legislation.

3 Further Work on Risk Assessment

While the Guidance was welcomed by COP-MOP5 and its progress commended by COP-MOP6, some Parties insisted on its further testing. As such, the Guidance will be tested nationally and regionally for further improvement in actual cases of risk assessment and in the context of the Cartagena Protocol on Biosafety, and the results of the testing will be reported back to COPMOP7 in 2014.

COP-MOP6 thus mandated the continuation of work on risk assessment. It extended the open ended online forum and established a new AHTEG to provide input to structure and focus the process of testing the Guidance, and to help analyze the results gathered; to coordinate development of a package that aligns the Guidance with the training manual 'Risk Assessment of Living Modified Organisms', developed by the Secretariat of the Convention on Biological Diversity; and to consider development of guidance on new topics.

Other elements of the the COP-MOP6 decision on risk assessment include a process and mechanism for testing of the Guidance; a mechanism for updating the background documents to the Guidance; the alignment of the Guidance and training manual; and training courses and workshops on risk assessment and risk management.

It is clear that discussions on risk assessment under the Cartagena Protocol will continue to evolve and progress. At the national level，it would be important to put the Guidance to use in actual risk assessments. This will be the best measure of its ability to uphold the Protocol's objectives. Parties can also test the Guidance according to the questionnaire that has been developed，and possibly through workshops or focus group discussions. Active participation in the online forum and AHTEG also provides an opportunity for Parties to shape and advance the risk assessment discussions under the Protocol.

References

[1] Guidance on Risk Assessment of LMOs，http://bch.cbd.int/onlineconferences/guidance_ra.shtml.

[2] Tappeser，B. and Quist，D. 2012. Advancing risk assessment under the Cartagena Protocol on Biosafety. Paper presented at Scientific Conference 2012 Advancing the Understanding of Biosafety：GMO Risk Assessment，Independent Biosafety Research and Holistic Analysis，28-29 September 2012，Hyderabad，India.

[3] Heinemann，J.A. and Quist，D. 2012. The AHTEG Guidance on Risk Assessment of LMOs. TWN Briefings for COP-MOP6 #3，Third World Network：Penang.

遗传修饰生物体环境风险评价中的主要考虑[①]

Main Considerations for Environmental Risk Assessment of GMOs

魏　伟
（研究员，中国科学院植物研究所，北京　100093）
Wei Wei
（*Institute of Botany*，*Chinese Academy of Sciences*，Beijing 100093）

摘　要：一般地，风险评价被定义为一个科学过程，因此应该具有科学性。生物多样性公约下的生物安全议定书是遗传修饰生物体（GMOs）生物安全管理的唯一国际性法律文件，风险评价是生物安全议定书国际谈判的焦点之一。从科学角度来看，GMOs 对生物多样性的影响主要包括但并不限于以下方面：①从 GMOs 到其野生近缘种或其他生物的基因流动；②通过食物链或耕作措施的改变对其他非靶标生物的影响；③对生物地球化学过程和营养循环的影响；④竞争优势以及同本土物种间相互作用；⑤害虫的抗性进化。由于 GMOs 的使用用途、释放的时空规模以及释放的环境不同，风险评价所需要考虑的因素可能会有不同。GMOs 的风险评价遵循个案原则，并需要考虑所有必要的信息。不确定性是科学研究中的特点，需要在风险分析中考虑到科学分析和科学数据的不确定性。在风险评价时，需要选择一个合适的参照物对比分析 GMOs 的风险，非遗传修饰生物通常拿来作对照。在这里，实质等同性原则的可行性有限，已经不能满足基于科学事实的风险评价。在生物安全议定书框架下，已经建立了分五步走的风险评价路线图，以及五个针对不同议题的风险评价指南文件。基于风险评价的风险管理建议是风险评价的重要结果和产出，应该是风险评价有机整体的一部分。GMOs 环境释放以后，风险监测就成为保证生物安全的重要手段。除了已经建立的指南文件，其他还有些议题也急需建立风险评价指南，例如遗传修饰作物在其相应受体植物起源中心的风险评价。遗传修饰作物和同种的传统受体作物的共存也是风险评价中需要考虑的问题，要考虑前者对后者的可能影响及其和谐共存。GMOs 风险评价也需要考虑其他一些因素，如社会和经济因素、公众意识与参与以及责任与补偿等。遗传修饰作物的发展迅猛，应用广泛，因此本文以遗传修饰作物为例，来讨论生物安全风险评价与管理中需要考虑的问题。

关键词：生物多样性，生物安全议定书，共存，遗传修饰作物，风险评价，不确定性，监测

① 自然科学基金面上项目（No. 31270578）资助。Supported by the National Natural Science Foundation of China（NSFC No. 31270578）。

Abstract: In general，risk assessment is defined as a science-based process，thus should be scientifically sound. In scientific aspect，the effects on biodiversity which is focused in Biosafety Protocol and may include but not limit to the following: ①Gene flow from GMOs to their wild relatives and other organisms; ②Effects on nontarget organisms through food chain or new/changed practices; ③Effects on biogeochemical processes and nutrient cycles; ④Competition advantages and interaction with native species; ⑤Resistance evolution of insect pests. Points of consideration for risk assessment may vary according to the intended use，the release scale and the receiving environment etc. Risk assessment for GMO should be carried out on a case-by-case basis and take into consideration all necessary information. Uncertainty is an inherent and integral element of scientific analysis and should be identified and considered in risk assessment，which could have critical effect on the results and conclusion of risk assessment and strategies for management. Appropriate comparators need to be chosen and applied in the assessment. Past experience and practices of the non-modified recipient organism should also be considered，but the principle of substantial equivalence has limited use and may not meet the request of scientific evidence-based analysis. Guidance on risk assessment of GMOs had been developed under the framework of Biosafety Protocol，which includes the five-step roadmap of risk assessment and five specific guidance documents on five specific topics，respectively. Risk management recommendation is one final result and output and should be one part of risk assessment. After the environmental release of GMOs，risk monitoring is a crucial measure to ensure biosafety regarding of uncertainty in risk assessment. In addition to the current specific topics，for which risk assessment guidance had been developed，other topics should also be considered and included in the risk assessment from a scientific view，such as the risk assessment of GM crop in the origin center of the non-GM recipient plant. Co-existence is an important issue that should be considered among others. Social and economic considerations，Public Awareness and Participation as well as Liability and Redress should also be taken into account in risk assessment.

Keywords: biodiversity，biosafety protocol，co-existence，genetically modified crops，risk assessment，uncertainty，monitoring

遗传修饰生物体（genetically modified organisms，GMOs）是利用现代生物技术获得的具有新异遗传组成和特征的生物体，所使用的生物技术能够打破物种类群间的生殖隔离，有别于传统育种技术。在国内，普遍使用转基因生物（transgenic organisms），主要指将外源基因转化进入受体基因组形成的生物，但是相对于现代生物技术来说，这个名词有其一定的局限性，比如利用双链 RNA 技术实现基因沉默或使用基因敲除技术而形成的生物体就很难用转基因生物这个名词来全部包含。而目前国内的生物安全评价和管理的法规及条例均针对转基因生物，因此要满足遗传修饰生物体安全评价与管理的要求，国内的生物安全法规还有很大的提升和完善空间。

目前来讲，在遗传修饰生物体中，发展比较完善和应用比较广泛的是遗传修饰作物（GM crops），本文将主要以遗传修饰作物为例，来讨论生物安全风险评价与管理中需要考虑的问题。随着技术的进步和利益的驱动，从 1996 年商业化释放以后，遗传修饰作物的面积逐年上升，到 2012 年，全球遗传修饰作物面积达到 1.7 亿 hm^2，而且在发展中国家中种植的遗传修饰作物的面积首次超过发达国家（James，2013）。然而，这里要注意的是，除了中国等极个别国家，其他发展中国家种植的遗传修饰作物均是发达国家的产品，因此，简单地认为遗传修饰技术在发展中国家的发展已经超过发达国家是不正确的。在目前全球商业化释放的遗传作物中，主要是抗虫、抗除草剂、抗病毒或兼性抗除草剂和抗虫的作物，因而引起了人们对其安全性的担心和争论，因此世界范围内，各国政府均对遗传修饰生物可能造成的危害持谨慎态度，并开展了遗传修饰生物的风险评价和风险管理。

1 遗传修饰生物体的风险评价

1.1 概念与原则

普遍认为，遗传修饰生物包括作物的风险是实际上发生危险与其可能发生概率的综合作用。联合国环境署于 1995 年在其发布的国际生物技术安全的技术指南中提出，风险评价是一个过程，是基于遗传修饰生物体可能引起的不利影响以及这些不利影响发生的可能性及后果，估计其风险的过程。因而，生物安全也可定义为管制、管理或控制遗传修饰生物体使用和释放可能带来不利的环境和健康影响的风险。生物安全涉及多个科学领域，包括生物学、生态学、微生物学、分子生物学、动植物病理学、昆虫学、农业、医学以及法律、社会经济影响和公众意识等（生物多样性公约秘书处，2000）。因此遗传修饰生物安全评价主要包括两方面的内容，一为人类健康风险，二为环境风险。此外，国际上还有伦理、法律以及社会经济方面影响的讨论与争论。

在遗传修饰生物体包括遗传修饰作物的风险评价中，一般有三个主要的原则需要遵循，一是预防原则，即在风险发生以前，要进行评价，并采取措施预防风险的发生；二是个案原则，即每个遗传修饰生物体的性状不同，用途不同，其安全影响也不同，就要采取不同的评价方法和过程；三是科学原则，即在风险评价中要注重科学性，以事实为依据，以科学为准绳开展评价，不能混入伪科学的概念，不能进行模糊评判。

1.2 风险概要

遗传修饰生物包括作物的环境风险评价需要考虑以下几个方面的风险：①从遗传修饰生物到其野生近缘种或其他生物的基因流动，包括通过具有亲缘关系的亲本在不同代别之间的流动（垂直基因流），也包括不同类群间的流动，如从植物到微生物（水平基因转移）。②通过食物链或耕作措施的改变对其他非靶标生物的影响，由于遗传修饰生物体在遗传上发生了变化，不管是基因沉默、基因敲除，还是携带了新的基因，都会影响到生物体的基因表达，生物体基因的表达或其产物的变化，能够通过食物链影响到其他营养级水平上的生物。而且对于作物来说，由于种植了具有新异特性的遗传修饰作物，耕作措施就会有所变化，比如抗除草剂大豆的引入，必然会增加某一特定除草剂的用量，从而田间杂草就有可能进化出对该除草剂的抗性。③对生物地球化学过程和营养循环的影响，新的基因或修饰了的表达产品以及改变了的次生代谢物质可通过植物根系分泌物进入土壤，对土壤地球化学循环过程，如碳、氮循环等产生影响。④竞争优势以及同本土物种间相互作用，遗传修饰生物具有了新异的特性，有时具有选择优势，成为外来入侵物种。⑤害虫的抗性进化，长期种植抗虫的遗传修饰作物，会使害虫产生耐受性，从而使该遗传修饰作物降低，甚至失去对害虫的控制。

中国大面积释放转基因抗虫棉始于 1997 年，2002 年，薛达元主笔撰写了《关于 Bt 棉在中国对环境影响的调查概要》①，明确地提出了 Bt 棉可能带来的风险，并提出警告：①随着抗虫棉的大面积释放，次要害虫可能代替棉铃虫，上升为重要害虫；②在中国耕作制度下，害虫有可能进化出对抗虫棉的抗性。该报告一经面世即产生巨大争议，受到生物技术狂热支持者的强烈批判。然而，该报告的警告在若干年后不幸成为现实。

研究人员根据 1997 年到 2008 年的监测数据（Lu et al.，2010），发现随着抗虫棉种植比例的增加，一种次要害虫盲椿蟓种群数量在不断增加，而且针对盲椿蟓的农药施用量也在增加，作者认为其原因是随着抗虫棉种植面积的增长，针对棉铃虫的杀虫剂的用量在降低，因而减弱对其他害虫的控制。从生态学上来讲，棉铃虫受到 Bt 棉的控制，其所在的生态位可能就产生了空隙，其他具有类似功能的物种就有可能填补进来，盲椿蟓可能恰巧受益于棉铃虫的退出，占领了棉铃虫所在的生态位。而次生害虫的增加，可能也与遗传修饰作物本身有关。有研究表明，由于外源基因的插入，可能会导致受体基因组的表达，比如降低了次生代谢防御物质的含量，导致相关害虫的数量如蚜虫在遗传修饰作物上的增加（Hagenbucher et al.，2013）。

在害虫对抗虫遗传修饰作物的抗性进化方面，美国学者比较了抗虫棉大面积释放前 1992—1993 年与 2002—2004 年美洲棉铃虫田间害虫种群对抗虫棉的耐受性情况，发现大面积释放 5～7 年后田间种群对抗虫棉进化出了抗性（Tabashnik et al.，2008）。有学者认为，可用复合抗性来应对害虫的抗性进化，然而，最近的研究发现，同时表达 Cry1Ac 和 Cry2Ab 蛋白的复合抗虫棉也存在潜在问题，对 Cry1Ac 蛋白有耐受性的田间抗性种群对该复合抗性植株也有一定的耐受性，即该复合抗性植株不能完全控制害虫（Brévault et al.，2013）。

遗传修饰作物释放以后，会通过种子或花粉为媒介，把外源基因或修饰的性状转移进入其亲缘种中，导致对非遗传修饰作物的污染，或对野生遗传资源的污染。如果逃逸的性状具有选择优势，则可能会提高受体野生植物的生态适合度，导致一定的生态学后果（魏伟等，1999）。例如，栽培油菜收获遗漏的种子很容易在田间形成自播植物，而油菜的生命力旺盛，一株植物能够结实上万粒种子，因此遗传修饰油菜基因扩散的风险较高。随着大面积抗除草剂的遗传修饰油菜的释放，2000 年，加拿大田间报道了能够抗三种除草剂的自播油菜，作者认为是花粉流的结果（Hall et al.，2000）。这种花粉介导的外源基因逃逸也能导致野生资源的污染。Warwick 等（2008）在抗除草剂油菜田附近收集其野生亲缘种——野生白菜，发现了对除草剂有抗性的野生植株。更为有意义的是，作者发现了一株具有抗性的野生白菜，其染色体倍性与野生亲本完全一致，为二倍体，且发生抗性分离的后代染色体倍性也是二倍体。由于栽培油菜的染色体为四倍体，野生白菜为二倍体，如果是杂交种，则倍性应该处于两亲本之间。但发现的抗性野生白菜的染色体为二倍体，所以预示着抗除草剂的外源基因已经在野生种群中整合和固定，这种固定在生态学上有重要意义，有进一步扩散的可能。

油菜的异花授粉比率较高，而大豆是严格自交的植物。一般认为栽培大豆与野生大豆

① 薛达元. 2002. 关于 Bt 棉在中国对环境影响的调查概要。

是不可能发生基因交流的。然而，中国农业科学院的研究人员（Wang et al.，2010）在采集的栽培大豆和野生大豆种子中，发现了性状的分离，在蔓生的野生大豆植株中出现了直立的半野生大豆，收集的栽培大豆和野生大豆的种子萌发结实后，出现了类似半野生大豆形态的种子，说明栽培大豆与野生大豆之间曾经有过基因交流，更重要的是说明了栽培大豆与野生大豆能够进行基因交流。中国是大豆的起源中心，在野外具有丰富的野生大豆资源。遗传修饰的大豆，比如携带外源抗除草剂基因大豆的大面积释放，可能会导致外源基因对野生大豆资源的污染和危害。

2 风险评价中的两个重要问题

如前所说，遗传修饰作物发展迅速，2012 年全球商业化种植面积，是刚开始种植时的万倍以上。应当说，这些遗传修饰作物在释放前都进行了严格的风险评价，大多都是确认安全的情况下才批准商业化种植的。然而，认为那些风险评价过程中认定为安全的遗传修饰作物永远是安全的观点是不正确的。因为科学有一定的不确定性，局限于当时科学认识，界定为安全的东西，可能随着科学研究的深入，由于一些新证据的发现或环境的变化，要对原来认为安全的遗传修饰作物重新评价。由于不确定性的存在，在商业化释放遗传修饰生物体以后，要有跟踪监测，发现问题要进行及时的评价和管理。

2.1 不确定性

首先，不确定性来源于已有知识的限制，对于世界的认识，人类是存在局限性的。一些关键信息的缺失会阻碍做出正确的风险判断。人类探知自然的好奇心永不停止，对自然的认识是不断更新的，随着认识的深入，会对目前遗传修饰作物的安全性有更深更完善的认识。原先认定为安全的东西，在新认识面前也许就不安全了，而原先不安全的事物，在新知识的帮助下能够变得安全。原来不确定的认识，可能也会在新信息的支持下会得到确认。例如，2001 年加州大学伯克利分校的学者 Quist 和 Chapela（2001）发现墨西哥本地的玉米遭到遗传修饰玉米的污染，其研究结果在著名杂志《自然》发表后，招致多方指责，《自然》编辑部在有关方面的压力下不得已宣布撤稿。然而，随后有关墨西哥本土玉米遭到污染的证据不断出现，相关研究报道不断发表，《自然》最终承认墨西哥本土玉米受到生物技术玉米的污染已成为事实（Gilbert，2013）。由于墨西哥是玉米的起源中心，最近墨西哥政府也公开表示要禁止遗传修饰玉米的进口及其在墨西哥的任何种植活动①。

其次，遗传修饰生物的安全性也会受到所接收环境的变化而受到影响，例如，全球气候变暖条件下，遗传修饰生物与环境的互作关系可能由于其本身基因表达的变化以及环境水温等条件的变化而发生变化（魏伟等，2009），其对环境的不利影响可能减弱也可能会加强。

再次，遗传修饰生物体终端使用目的的不同也会导致不同的风险，例如，用做食品的与用做饲料的遗传修饰生物体的风险评价策略与方法可能有所不同，在实际生活中

① http://ecowatch.com/2013/10/16/mexico-bans-gmo-corn-effective-immediately.

发生风险的类型与危害的程度也会有所区别。又如，不同环境下，严格隔离下的应用与开放空间的环境释放，遗传修饰生物体的风险大小以及评价与管理对策可能会有不同的要求。

最后，遗传修饰生物体释放的规模和时间尺度也会为风险评价带来不同的影响，大规模或小规模释放，长期或短期释放，均是影响风险评价的不确定因素。

鉴于不确定性的存在，遗传修饰生物的生物安全评价中要充分考虑不确定性的影响，在管理中要留有预案，一旦确认不确定性会带来一定程度的风险，就需要跟上合适的管理措施，避免风险的发生或尽量减少损失。

2.2 风险评价过程中参照物的选择

遗传修饰生物体的风险评价要有参照物做参考，这个参照物也就是我们通常所说的对照。一般地，我们选择非遗传修饰的近等基因系作为对照，评价遗传修饰生物体的安全性。然而，随着大面积遗传修饰作物的种植及其带来的经济利益，面临商品社会冲击的农民种植传统作物的积极性一再降低，有可能加剧作物种植的单一化窘况，传统作物在不断地丧失。例如，随着国内大面积种植 *Bt* 抗虫棉，现在已经很难找到合适的不携带 *Bt* 外源基因的传统棉花品种作为风险评价的对照了。

1993 年，针对来自遗传修饰生物体的食品的风险评价，世界经济合作与发展组织（OECD）提出了实质等同性原则，即如果一种遗传修饰食品与现存的传统同类食品相比较，其特性、化学成分、营养成分、所含毒素以及人和动物食用和饲用情况是类似的，那么它们就具有实质等同性。很显然，该原则仅适用于食品安全的讨论，并且该定义主观性比较强，没有数量化的概念，在实际操作中很难把握，有很大的局限性（魏伟等，2001）。1999 年《自然》杂志专门刊文评论认为“一种遗传修饰食品与其传统同类食品在化学成分上相似并不证明其食用安全性”（Millstone et al.，1999）。因此，Millstone 等（1999）认为，在食品安全评价中要摈弃实质等同性这个伪科学命题。同样，在环境安全评价中也要避免这个所谓实质等同性原则的干扰。

3 风险评价与风险管理的国际条约

《生物多样性公约》（以下简称公约）是目前唯一旨在保护地球生物资源的国家性公约。1992 年 6 月 1 日由联合国环境规划署发起的政府间谈判委员会第七次会议在内罗毕通过，同年 6 月 5 日，由签约国在巴西里约热内卢举行的联合国环境与发展大会上签署，于 1993 年 12 月 29 日正式生效，目前有 193 个缔约国，只有美国等几个为数不多的国家不是缔约方。中国于 1992 年 6 月 11 日签署该公约，1992 年 11 月 7 日批准，1993 年 1 月 5 日交存加入书，正式加入。遗传修饰生物体的生物安全评价与管理是《公约》的一个热点，《公约》在第 8 条的原生境保护条款中有相关规定：

“8（g）. 制定或采取办法以酌情管制、管理或控制生物技术改性活生物体在使用和释放时可能产生的风险，即可能对环境产生不利影响，从而影响到生物多样性的保护和持久使用，也要考虑到对人类健康的风险。”

这里生物技术“改性活生物体”是公约里的用语，相当于我们说的遗传修饰生物体。

公约注意到了进行生物安全评价与管理的重要性，在其 19 条第 3 款里规定：

"缔约方应考虑对一项议定书的需求及其形式，以规定适当程序，特别包括事先知情协议，管理可能对生物多样性的保护和持久使用产生不利影响的改性活生物体的安全转移、处理和使用。"

经过各缔约国的努力和谈判以及各方讨价还价，旨在管理生物技术改性活生物体（遗传修饰生物体）越境转移、处理和使用的生物安全议定书于 2000 年 1 月底在加拿大蒙特利尔召开的生物多样性公约缔约国大会上通过。由于原来预定在哥伦比亚的卡特赫纳会议上达成最终文本，因而最终达成的议定书定名为《卡特赫纳生物安全议定书》（以下简称《生物安全议定书》）。《生物安全议定书》2003 年 9 月 11 日生效，目前 166 个缔约方。中国于 2000 年 8 月 8 日签署该议定书，2005 年 4 月 27 日核准该议定书，2005 年 9 月 6 日正式成为缔约方。《生物多样性公约》和《卡特赫纳生物安全议定书》在中国的国家联络点是国家环保部，规定了国家环保部在中国遗传修饰生物体生物安全管理中的职责。

生物安全议定书非常关注风险评价与风险管理，在其附录 3 中列出了风险评价与管理的步骤，并将改性活生物体及其产品同时纳入风险评价与风险管理的范畴。2008 年《公约》秘书处为了更好地履行生物安全议定书，邀请有关缔约方、非缔约方及相关组织的代表，成立了风险评价与风险管理特设技术专家组，专门就评价与管理制定相关指南文件①，包括一个风险评价与管理路线图和五个分对象文件：复合抗性作物的风险评价、抗非生物因子作物的风险评价、改性蚊子风险评价、改性树木风险评价和生物安全监管和监测。同时该专家组也提出一些需要制定指南文件的备选主题：①生产药物和工业产品的改性活生物体的风险评价，包括对脊椎动物和无脊椎害虫的影响；②改性动物的风险评价，包括鱼类；③改性微生物和病毒的风险评价；④改性活体在多样性起源中心释放的风险评价；⑤将人类健康整合进入环境风险评价的指南；⑥营养成分改变的改性植物风险评价；⑦利用 dsRNA 技术制造的改性活生物体的风险评价；⑧合成生物学技术产生的改性活生物体的风险评价；⑨在小规模耕作制度下改性活生物体与非改性活体间的"共存"；⑩环境风险评价的社会—经济因素；⑪改性藻类的风险评价。2012 年，在印度海德拉巴召开的第六届生物安全议定书缔约国大会上，有些国家由于担心这个专家组会制定过多的文件，阻碍其本国生物技术的发展，在生物技术工业组织的支持下，主张停止该专家组的工作。经过各方妥协，最终同意延长该专家组的工作至 2014 年的第七届缔约国大会再讨论确定。

改性活生物体的越境转移、释放和使用也涉及其他方面的因素，如社会经济因素的考虑、改性与非改性生物体共存问题、公众意识与参与以及责任与赔偿等问题。经过多年的博弈，《卡特赫纳生物安全议定书》之"责任赔偿和补救的补充议定书"于 2010 年在第五届生物安全议定书缔约国大会上通过，以解决一旦源于越境转移的改性活生物体可能给生物多样性的保护和可持续利用造成损害时采取的应对措施。该补充议定书为改性活生物体的损害设定了赔偿责任及补救的最低国际标准，然而，遗憾的是该补充议定书未将改性活生物体的产品列入管辖范围，为今后的有效实施埋下隐患。该议定书的通过是发展中国家做出许多让步的结果，从 2010 年缔约国大会通过至今尚未

① http：//bch.cbd.int/onlineconferences/guidance_ra.shtml.

达到签署国的法定生效数量（需要至少 40 个生物安全议定书缔约方签署），该补充议定书还没有生效。

4 风险评价的公众意识与参与

遗传修饰作物包括大豆、玉米、油菜、木瓜等已经进入了公众的生活，目前公众多关注食品安全，因为食品安全与其自身密切相关。公众似乎忽视了一个更重要的问题，即遗传修饰生物体对环境的影响及其风险。因为即使来自遗传修饰生物体的食品是安全的，生产食品也可能需要遗传修饰作物的环境释放，如果不能很好地管理和控制风险，将造成对环境的危害，例如污染传统作物的种子或野生遗传资源。与食品安全问题相比，环境安全问题造成的危害可能更大，环境污染的治理与恢复可能长期的努力。所以在与公众交流生物技术风险的时候，需要明确和介绍其所带来的潜在环境危害，倡导公众对环境安全的重视。

《卡特赫纳生物安全议定书》在公众意识与参与方面有明确规定，《生物安全议定书》第 23 条第 2 款规定：

“各缔约方应按照其各自的法律和规章，在关于改性活生物体的决策过程中征求公众的意见，并在不违反关于机密资料的第 21 条的情况下，向公众通报此种决定的结果。”

然而，国内转基因生物安全相关的法规与条例却没有做到这一点，相关主粮如转基因水稻的安全证书只是颁发后再向公众通报，遗传修饰玉米、大豆等的进口决定都是事后公布，没有在决策过程中征求公众的意见，很显然是违反生物安全议定书相关规定的。即使是机密材料（比如具体的基因序列等）不便于公布，也需要事先告知公众这一过程，并说明不公开的理由。因此，目前国内的生物安全管理办法、法规、条例及其执行上存在有需要弥补的漏洞，亟须出台更完善的法案。

鉴于遗传修饰生物体本身及其释放存在不确定性，甚至安全隐患，因此要谨慎对待其安全问题。来自遗传修饰生物体的食品和商品要有标识，给予公众知情权和选择权。欧盟国家及俄罗斯、沙特、澳大利亚、印度、日本和韩国等国家以及中国都强制要求标识来自遗传修饰生物体的食品（也称为遗传修饰食品 GM foods，因为现阶段商业化种植的均是转入外源基因的遗传修饰作物，也可统称为转基因食品），而在美国是自愿标识。然而，最近情况有了变化，越来越多的美国人要求标识遗传修饰食品，美国康涅狄格州和缅因州已经通过法案，要求强制标识所有遗传修饰食品，以便消费者做知情选择，其他大约 40 个州也在考虑强制标识的法案[①]。美国的现代生物技术走在世界前列，大豆、玉米、油菜等遗传修饰作物的种植面积比例巨大，公众尤其是中国公众特别关注身为生物技术巨人的美国的国民每年吃了多少来自遗传修饰生物体的食品。与公众交流信息时要提供正确的知识，例如美国人每年都摄入大量遗传修饰食品的说法是没有依据的。

根据美国农业部网站的资料[②]，2013 年美国遗传修饰玉米和大豆占其国内玉米和大豆

① http：//www.huffingtonpost.com/maria-rodale/the-power-of-labeling-pre_b_4098539.html.

② http：//www.ers.usda.gov/data-products/adoption-of-genetically-engineered-crops-in-the-us.aspx.

总种植面积的 90%和 93%，2009 年这两种作物遗传修饰品种种植比例分别是 85%和 91%，2007 年这两个比例分别为 73%和 91%。这就是说，在最近这几年里，全美分别有 10%～27%的玉米和 7%～9%的大豆种植的是传统的非遗传修饰品种。联合国粮食与农业组织（FAO）有各国粮食贸易的统计数据①，但最新数据只到 2009 年。美国是玉米和大豆的净出口国，在 2009 年美国玉米总产约 33 255 万 t，直接用于食品的是 391 万 t，约占总产的 1.17%，其余的用于出口、饲料、加工（如生物燃料等），这个比例与 2007 年基本持平。2009 年美国大豆总产 9 141.7 万 t，出口 4 051.3 万 t，占总产 44%多，其余的主要用作加工原料，直接用于食用的只有区区 1 万 t，占总产的 0.011%，2007 年这一比例是 0.018%。由此看出，美国直接用于食品的玉米和大豆的比例远远低于其传统品种的产量，所以在美国食品超市里经常能够看到标签为来自非遗传修饰作物的产品。因而，美国人每年人均摄入的转基因食品的量非常有限。而中国的情况却不同，且不说水稻和小麦是主粮，即使是大豆，直接用于食用的大豆占 2009 年大豆国内总消费量（国内总产+进口=5 967.7 万 t）的 8.4%，是美国直接用于食品比例的 800 多倍。所以说，遗传修饰大豆一旦在中国全面放开，中国人摄入的转基因成分的量远远大过美国人。

5 结语

遗传修饰生物技术本来是一门中性的科学技术，其最初的应用是在实验室里验证基因的功能。后来天才的科学家们发现了其潜在的应用价值，尝试着用其造福人类社会。然而，这门具有风险性的技术现在已经越来越多地用来谋取利润，在商业利益的驱动下，尤其是在美国得到了很大的发展。当然，这门技术有其积极的一面，大规模应用时，能够在一定程度上节约劳动力和生产成本。但是，任何技术不能滥用。目前尤其在中国，凡是人们能够想到的植物和动物都被施以遗传修饰的技术进行改造，亟须采取措施防止遗传修饰技术的滥用。任何有风险的技术都要施以合适高效的管理才能走上正确的轨道。另一方面，中国的生物安全管理效率比较低下，例如，转基因水稻在尚未获得商业化批准以前，就已经频频在市场上出现，至今没有得到有效解决和控制，说明中国的生物安全评价与管理任重道远。一旦以盈利为唯一目标、轻视人类与环境健康的生物技术私营公司在中国发展壮大，形成影响力，中国的生物安全管理和监管将面临巨大挑战，所以一定要尽快完善相关立法。不能以发展的名义而置安全于不顾，不能借口还有其他更严重的环境问题而忽视遗传修饰生物体的风险问题。对于任何一个国家来讲，现代生物技术作为一项高科技无疑是需要掌握和发展的，但我们要重视技术探索，谨慎使用。在国家巨款支持、国际巨头频频叩门的情况下，要保持清醒头脑，正视中国国情，沿着正确的轨道发展生物技术。同时任何安全的考虑都要考虑到剂量的因素，考虑到中国人对主粮和遗传修饰食品的消耗量，还要考虑到在中国大量分布的宝贵遗传资源不能受到污染和危害。在保证人类健康和环境安全的条件下，有序地发展生物技术。

① http：//faostat.fao.org/site/368/DesktopDefault.aspx？PageID=368#ancor.

参考文献

[1] 联合国，1992. 生物多样性公约.

[2] 生物多样性公约秘书处，2000. 卡特赫纳生物安全议定书，蒙特利尔.

[3] 魏伟，钱迎倩，马克平，1999. 转基因作物与其野生亲缘种间的基因流.植物学报，41：343-348.

[4] 魏伟，钱迎倩，马克平，等，2001. 转基因食品安全性评价的研究进展.自然资源学报，16(2)：184-190.

[5] 魏伟，2009. 转基因油菜生态学风险评价. //薛达元（主编）转基因生物风险评估与安全管理——生物安全国际论坛第三次会议论文集. 中国环境出版社，68-79.

[6] Brévault T，Heuberger S，Zhang M，Ellers-Kirk C，Ni X-Z，Masson L，Li X-C，Tabashnik BE，Carrière Y，2013. Potential shortfall of pyramided transgenic cotton for insect resistance management. Proceedings of the National Academy of Sciences of USA doi：10.1073/pnas.1216719110.

[7] Hagenbucher S Wäckers FL，Wettstein FE，Olson DM，Ruberson JR，Romeis J. 2013. Pest trade-offs in technology：reduced damage by caterpillars in Bt cotton benefits aphids Proceedings of the Royal Society B：Biological Sciences 280：20130042.

[8] Hall L，Topinka K，Huffman J，Davis L，Good A. 2000. Pollen flow between herbicide-resistant *Brassica napus* is the cause of multiple-resistant *B. napus* volunteers. Weed Science 48：688-694.

[9] James C. 2012. Global Status of Commercialized Biotech/GM Crops：2012. ISAAA Brief No. 44. ISAAA：Ithaca，NY.

[10] Lu Y-H，Wu K-M，Jiang Y-Y，Xia B，Li P，Feng H-Q，Wyckhuys KAG，Guo Y-Y. 2010. Mirid bug outbreaks in multiple crops correlated with wide-scale adoption of Bt cotton in China. Science 328：1151-1154.

[11] Millstone E，Brunner E，Mayer S. Beyond ‘substantial equivalence’. Nature，401：525-526.

[12] Gilbert N. 2013. A hard look at GM crops. Nature 497：24-26.

[13] Quist D，Chapela I. Transgenic DNA introgressed into traditional maize landraces in Oaxaca，Mexico.Nature 414：541-543.

[14] Tabashnik BE，Gassmann AJ，Crowder DW，Carrière Y，2008. Insect resistance to Bt crops：evidence versus theory. *Nature Biotechnology* 26：199-202.

[15] Wang K-J，Li X-H. 2010. Interspecific gene flow and the origin of semi-wild soybean revealed by capturing the natural occurrence of introgression between wild and cultivated soybean populations. Plant Breeding 130：117-127.

[16] Warwick SI，LÉGÈRe A，M.-J. Simard M-J，James T. 2008. Do escaped transgenes persist in nature? The case of an herbicide resistance transgene in a weedy Brassica rapa population. Molecular Ecology 17：1387-1395.

转 *Bt* 基因水稻及常规 SY63 水稻叶片结构对臭氧浓度升高的响应

Leaf Morphology and Ultrastructure Responses to Elevated O_3 in Transgenic *Bt*（Cry1Ab/Cry1Ac）Rice and Conventional Rice under Fully Open-air Field Conditions

刘　标

（环境保护部南京环境科学研究所，南京 210042）

LIU Biao

（Nanjing Institute of Environmental Science，Ministry of Environmental Protection，Nanjing　210042）

摘　要：高浓度臭氧（O_3）对植物的形态结构、生理生化、生长繁殖等会造成不同程度的损伤。本文利用中国农田开放式 O_3 浓度升高（O_3-FACE）研究平台，在 O_3 浓度升高和田间自然条件（Ambient）下盆栽种植转 *Bt* 基因水稻 Bt 汕优 63（Bt-SY63）及其亲本常规汕优 63（SY63），分别于 64-DAS（Days after seeding 育苗后）、85-DAS 和 102-DAS 调查了两种水稻的叶片形态结构和超微结构指标。研究结果表明，与 Ambient 下相比，O_3 浓度升高没有使 Bt-SY63 和 SY63 水稻叶片的长度、宽度、面积、气孔长度发生显著变化，但是使 Bt-SY63 叶片气孔密度在 64-DAS 时显著降低；O_3 胁迫导致两种水稻叶片都出现了类囊体肿胀时间提前，嗜锇颗粒占细胞面积比例（PCRA）增大，淀粉粒占细胞面积（SCRA）减小，均加速了两基因型水稻叶片衰老。与 SY63 相比，O_3 胁迫下 Bt-SY63 类囊体肿胀的时间更早，叶绿体降解的程度更大，叶肉细胞壁厚度、叶绿体占叶肉细胞面积比例（CCRA）、PCRA 和 SCRA 等指标波动更大，持续时间更长，表明 O_3 胁迫对 Bt-SY63 叶片超微结构造成的损伤比 SY63 更大。

Background: Elevated tropospheric ozone severely affects not only yield but also the morphology，structure and physiological functions of plants. Because of concerns regarding the potential environmental risk of transgenic crops，it is important to monitor changes in transgenic insect-resistant rice under the projected high tropospheric ozone before its commercial release.

Methodology/Principal Findings: Using a free-air concentration enrichment（FACE）system，we investigated the changes in leaf morphology and leaf ultrastructure of two rice varieties grown in plastic pots，transgenic Bt Shanyou 63（Bt-SY63，carrying a fusion gene of Cry1Ab and Cry1Ac）and its non-transgenic counterpart（SY63），in elevated O_3（E-O_3）versus ambient O_3（A-O_3）after 64-DAS（Days after seeding），85-DAS and 102-DAS. Our results indicated that E-O_3 had no significant effects on leaf length，leaf width，leaf area and stomatal length for both Bt-SY63 and SY63 but significant effects on leaf thickness and stomatal density for

Bt-SY63. O_3 stress caused early swelling of the thylakoids of chloroplasts, a significant increase in the proportion of total plastoglobule area in the entire cell area（PCAP）and a significant decrease in the proportion of total starch grain area in the entire cell area（SCAP）, suggesting that E-O_3 accelerated the leaf senescence of the two rice genotypes. Compared with SY63, E-O_3 caused early swelling of the thylakoids of chloroplasts and more substantial breakdown of chloroplasts in Bt-SY63.

Conclusions/Significance: Our results suggest that the incorporation of Cry1Ab/Ac into SY63 could induce unintentional changes in some parts of plant morphology and that O_3 stress results in greater leaf damage to Bt-SY63 than to SY63, with the former coupled with volatility in CCAP(the proportions of total chloroplast area in the entire cell area), PCAP and SCAP. This study provides valuable baseline information for the prospective commercial release of transgenic crops under the projected future climate.

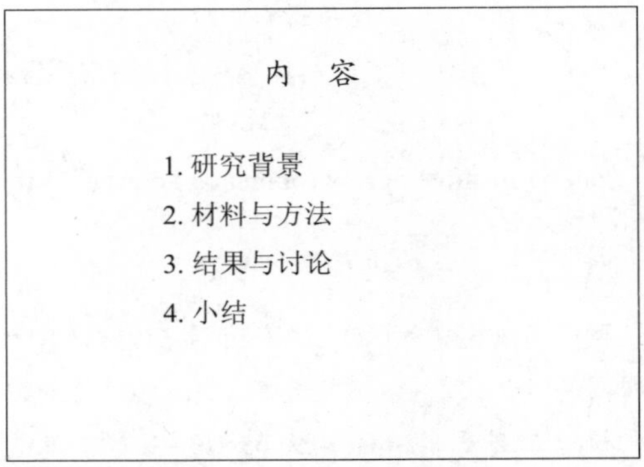

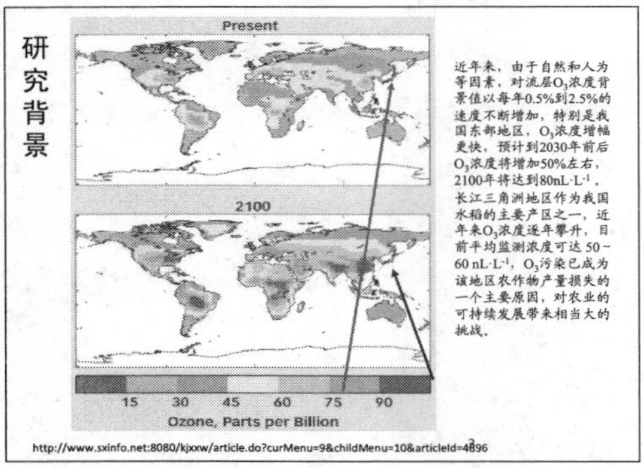

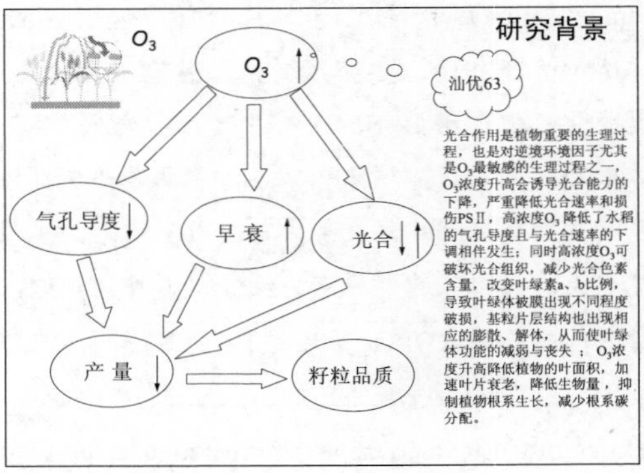

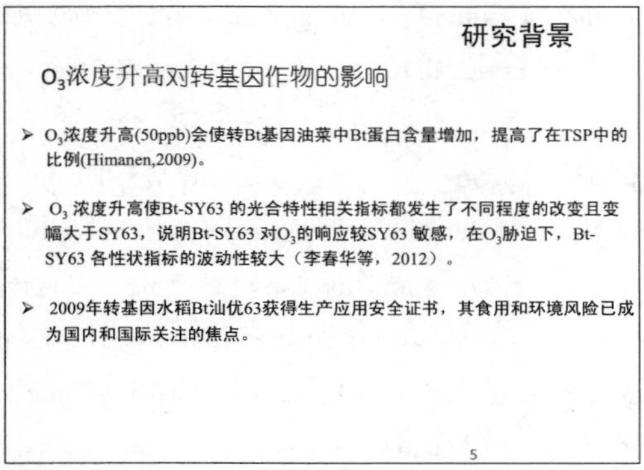

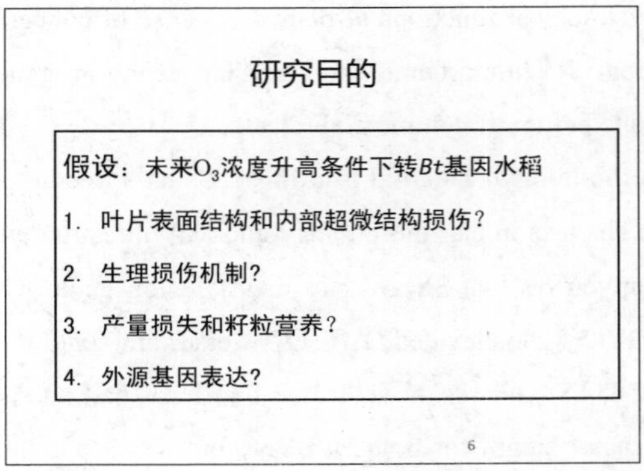

2. 材料与方法

2.1 试验地概况

试验地位于江苏省江都市小纪镇良种场的中国O_3-FACE系统研究平台，该地耕作方式为水稻—冬小麦复种，是典型的稻麦复种农田生态系统 。

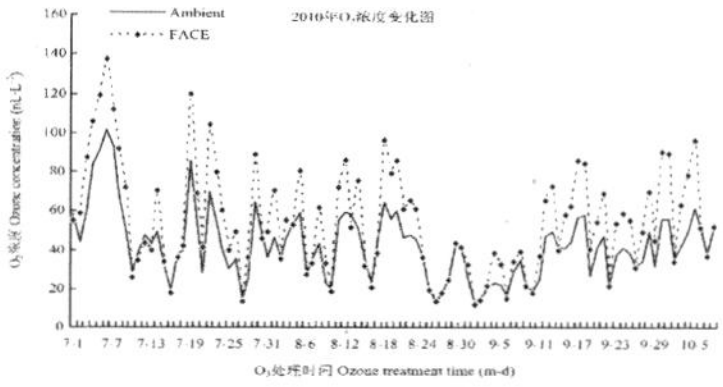

2. 材料与方法

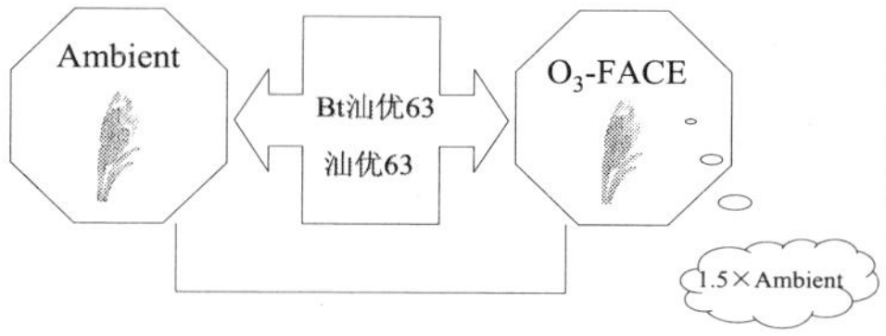

供试水稻品种汕优63（SY63）和转Cry1Ab/Ac基因汕优63（Bt-SY63）由华中农业大学植物科技学院提供
。

2. 材料与方法

2.2 测定内容与方法

2.2.1 叶面积、叶片长度、叶厚度

采用叶面积仪(AM100， ADC， UK)，分别于抽穗期（8月16日，处理35 d），灌浆期（9月6日，处理51 d）和成熟期（9月23日，处理67 d）每圈每一品种各选取5～8片长势一致的倒二叶在原位条件下进行叶面积和长度的测定。

采用叶厚度传感测量仪测定叶厚度。

2.2.2 气孔密度和大小

取叶中部约2cm×2cm，通过导电胶固定样品，采用日立TM1000Mini-电子扫描显微镜观察，放大300倍统计气孔密度，放大1000倍测量气孔长度。

2. 材料与方法

2. 2 测定内容与方法

2. 2. 3 超微结构观察

通过取样、固定 、浸透、包埋、超薄切片与染色、透射电镜观察，采用专业图像分析软件Image-pro-plus 6.0对超微结构下叶肉细胞壁厚度、叶肉细胞面积、叶绿体面积、嗜锇颗粒面积以及淀粉粒面积进行测量。

2.2.4 *Bt*基因的定量PCR和Bt蛋白含量的测定

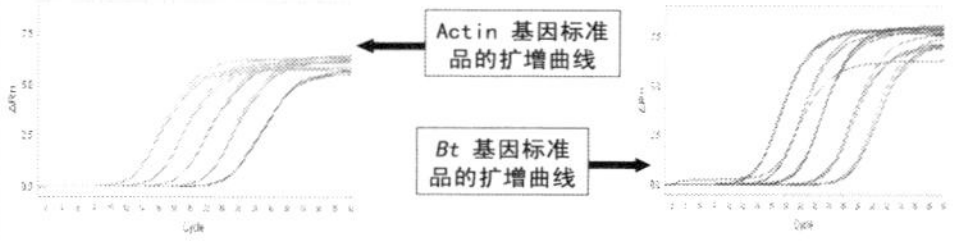

3. 结果与讨论

3. 1 开放式臭氧浓度升高对Bt-SY63和SY63叶宽、叶长及叶面积的影响

品种	处理	处理时间					
		长度(cm)			宽度(cm)		
		35d	51d	67d	35d	51d	67d
Bt-SY63	A-O_3	48.27±1.38b	47.43±2.29b	46.95±2.71a	1.62±0.29a	1.84±0.09a	1.84±0.03a
	F-O_3	52.11±3.61ab	46.60±1.72ab	49.37±3.15a	1.84±0.15a	1.89±0.13a	1.87±0.08a
SY63	A-O_3	55.23±4.24a	54.89±1.21a	46.47±4.09a	1.89±0.08a	1.96±0.04a	1.81±0.08a
	F-O_3	54.17±4.02ab	51.19±6.87ab	49.07±3.40a	1.84±0.14a	1.88±0.11a	1.83±0.10a

Ambient（田间自然条件）下35 d 和51d 时SY63叶长度显著大于Bt-SY63叶长度。

3. 结果与讨论

3. 1 开放式臭氧浓度升高对Bt-SY63和SY63叶宽、叶长及叶面积的影响

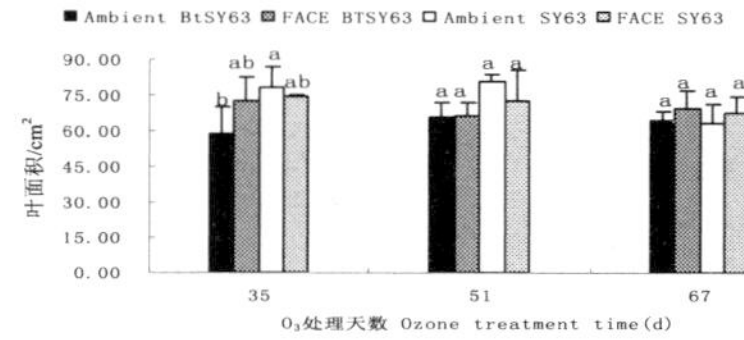

Ambient（田间自然条件）下35d 时 SY63 叶面积比Bt-SY63大22.4%（$P<0.05$），其他比较无显著性差异。

3. 结果与讨论

3. 2 开放式臭氧浓度升高条件下Bt-SY63和SY63叶的气孔大小和气孔密度变化

表2 不同处理时间Bt-SY63和SY63叶片气孔长度比较（μm）

品种	处理	处理时间 (d)		
		35d	51d	67d
Bt-SY63	A-O_3	21.3±2.0a	22.0±1.6a	22.7±1.1a
	F-O_3	22.7±1.3a	23.5±1.2a	23.2±1.8a
SY63	A-O_3	24.9±1.2a	23.9±1.8a	22.8±1.3a
	F-O_3	23.9±1.4a	23.7±1.5 a	22.7±2.6a

表3 不同处理时间Bt-SY63和SY63叶片气孔密度比较（个/cm^2）

品种	处理	处理时间		
		35d	51d	67d
Bt-SY63	A-O_3	4.7±0.3a	4.5±0.3a	3.6±0.6a
	F-O_3	4.1±0.3b	4.3±0.2a	3.6±0.8a
SY63	A-O_3	4.5±0.6 ab	4.0±0.2 a	4.0±0.8a
	F-O_3	4.3±0.7 ab	4.2±0.4 a	4.0±0.7a

3. 结果与讨论

3. 3 开放式臭氧浓度升高对Bt-SY63和SY63叶厚度的影响

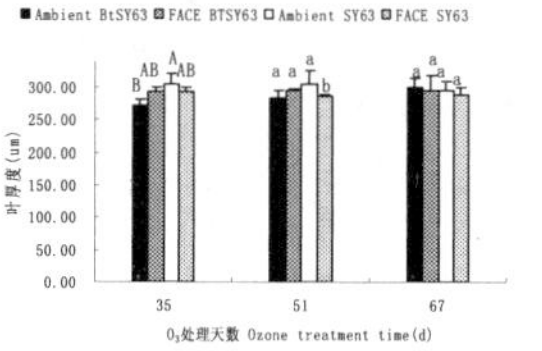

- Ambient下，SY63的叶厚度在35d和51d时显著高于Bt-SY63，67d时无显著差异。
- O_3处理下，Bt-SY63的叶厚度与Ambient下相比在35d和51d时分别增加6.9%和4.3%，67d接近成熟期有所减小。
- O_3处理下三个时期的SY63叶厚度均比Ambient小。

O_3处理下，Bt-SY63叶厚度在35d抽穗期时略小于SY63，51d灌浆期时显著地大于SY63（$P<0.05$），67d成熟期时大于SY63

- Ambient下SY63的叶厚度在35d和51d时比Bt-SY63大11.3%（$P<0.01$），67d时两品种叶厚度相当。O_3处理下，Bt-SY63的叶厚度与Ambient下相比在35d和51d时分别增加6.9%和4.3%，67d接近成熟期有所减小。而O_3处理下三个时期的SY63叶厚度均比Ambient下小。O_3处理下，Bt-SY63叶厚度在35d抽穗期时略小于SY63，51d灌浆期时显著地大于SY63（$P<0.05$），67d成熟期时大于SY63。
- 本试验中田间自然状态下35d抽穗期时Bt-SY63的叶厚度显著地小于SY63，到了67d灌浆期，臭氧处理的Bt-SY63水稻叶厚度反而显著地大于臭氧处理的SY63叶厚度。也就是说，臭氧使Bt-SY63水稻的叶厚度增加，使SY63的叶厚度减少，暗示着Bt-SY63和SY63在叶厚度对臭氧的响应上截然相反，从另一个侧面也说明，在叶片厚度上，臭氧浓度升高以相反的方向拉大了Bt-SY63和SY63之间的差异。

3. 结果与讨论

3.4 开放式臭氧浓度升高条件下Bt-SY63和SY63的叶片超微结构变化

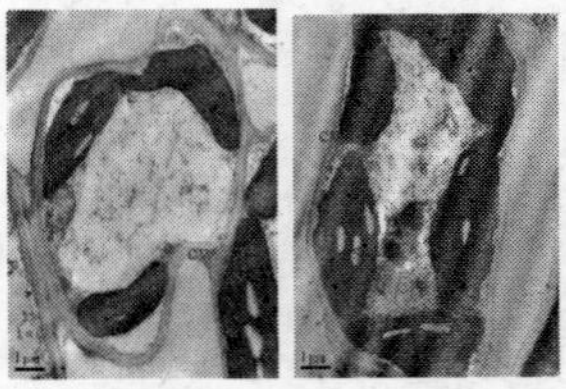

ch-叶绿体
cw-细胞壁
pl-嗜锇颗粒
s-淀粉粒
ts-类囊体
g-基粒
n-细胞核

分蘖期 BtSY63-A　　分蘖期 SY63-A

图 3 Ambient 下分蘖期 BtSY63 和 SY63 叶绿体超微结构

- 大多数叶绿体为凸透镜状的长方形，SY63比Bt-SY63更凸些，均被密集分布的基粒（基质电子密度大导致基粒不清楚）和基质类囊体占满，基质类囊体紧密互相平行。
- 叶绿体紧贴着细胞壁靠细胞边缘分布，能看到淀粉粒。细胞器结构清晰。SY63比Bt-SY63细胞质更丰富。

3. 结果与讨论

3.4 开放式臭氧浓度升高条件下Bt-SY63和SY63的叶片超微结构变化

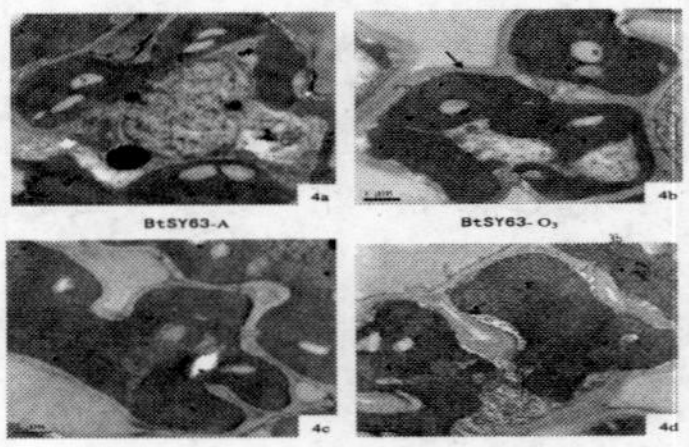

ch-叶绿体
cw-细胞壁
pl-嗜锇颗粒
s-淀粉粒
ts-类囊体
g-基粒
n-细胞核

BtSY63-A　　BtSY63-O_3

SY63-A　　SY63-O_3

图 4 35d 时 BtSY63 和 SY63 叶绿体超微结构

处理时间35d后观察（左图），Ambient下的Bt-SY63和SY63叶绿体变成稍园的凸透镜形和不规则形状，Ambient下的SY63叶绿体面积占叶肉细胞面积比(CCRA)和淀粉粒面积占叶肉细胞面积(SCRA)分别比Bt-SY63大28.65%（$p<0.05$）和2.32%（$p<0.05$）。

处理时间35d后,O_3处理下的Bt-SY63（图4b）与Ambient下（图4a）相比，类囊体开始肿胀，叶绿体整体开始模糊，叶肉细胞壁厚度显著增加（$p<0.05$）且变得粗糙（图4b箭头所示），嗜锇颗粒面积占叶肉细胞面积（PCRA）显著增加（$p<0.05$），SCRA显著减少（$p<0.05$）。

O_3处理下的SY63（图4d）与Ambient下相比，叶绿体整体仍很清晰，PCRA极显著增加（$p<0.01$）。O_3处理下的Bt-SY63与SY63相比，Bt-SY63叶绿体整体比SY63模糊，PCRA显著少于SY63（$p<0.05$）。

3. 结果与讨论

3.4 开放式臭氧浓度升高条件下Bt-SY63和SY63的叶片超微结构变化

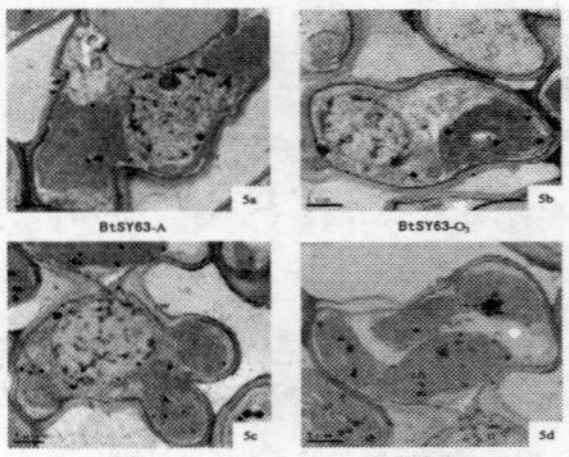

ch-叶绿体
cw-细胞壁
pl-嗜锇颗粒
s-淀粉粒
ts-类囊体
g-基粒
n-细胞核

BtSY63-A　　BtSY63-O_3

SY63-A　　SY63-O_3

图 5 51d 时 BtSY63 和 SY63 叶绿体超微结构

- 51d 时的超微结果显示（图5），Ambient下的Bt-SY63和SY63相比，SY63的淀粉粒面积占叶肉细胞面积(SCRA)显著少于Bt-SY63。O_3处理下的Bt-SY63（图4b）与Ambient下相比，核膜开始破损，叶绿体面积占叶肉细胞面积比例（CCRA）极显著增加（$p<0.01$），嗜锇颗粒面积占叶肉细胞面积比例（PCRA）显著增加（$p<0.05$）。O_3处理下的SY63（图4d）与Ambient下相比，核膜也开始破损，叶绿体整体开始模糊。
- O_3处理下，Bt-SY63的 CCRA显著地大于SY63（$p<0.05$），叶绿体也比SY63模糊。O_3处理前后两品种均能辨别叶绿体基粒的存在。

3. 结果与讨论

3.4 开放式臭氧浓度升高条件下Bt-SY63和SY63的叶片超微结构变化

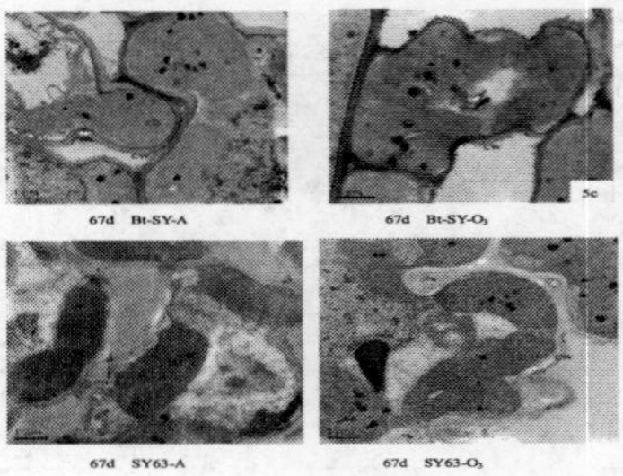

67d Bt-SY-A　　67d Bt-SY-O_3

67d SY63-A　　67d SY63-O_3

ch-叶绿体
cw-细胞壁
pl-嗜锇颗粒
s-淀粉粒
ts-类囊体
g-基粒
n-细胞核

图 6 67d 时 BtSY63 和 SY63 叶绿体超微结构

◈ 67d时是接近收获期的时候，此期超微结构（见图6）显示，Ambient下的Bt-SY63和SY63叶肉细胞与51d相比，细胞和叶绿体都开始有不同程度的降解，类囊体都开始肿胀，Bt-SY63的叶绿体被膜和核膜破损，开始出现降解，细胞器严重降解，SY63的核膜出现破损。Ambient下的Bt-SY63和SY63相比，Bt-SY63的叶绿体被膜破损，而SY63叶绿体被膜基本完整，Bt-SY63的细胞器降解比SY63严重。O_3处理下的Bt-SY63与Ambient下相比，最明显的变化是严重解体的叶绿体，几乎占满了整个细胞，已不能辨别基粒的存在，叶绿体基质颗粒状情况恶化，细胞器降解严重。O_3处理下的SY63与Ambient下相比，类囊体肿胀变形严重，叶绿体基质也是颗粒状，亦不能辨别基粒的存在，但其叶绿体仍具有一定的形状。细胞器部分降解。O_3处理下Bt-SY63和SY63相比，SY63的叶绿体仍具有一定的形状，叶绿体降解和细胞器降解程度均比Bt-SY63轻，损伤轻。

◈ 67d时是接近收获期的时候，此期超微结构（见图6）显示，Ambient下的Bt-SY63和SY63叶肉细胞与51d相比，细胞和叶绿体都开始有不同程度的降解，类囊体都开始肿胀，Bt-SY63的叶绿体被膜和核膜破损，开始出现降解，细胞器严重降解，SY63的核膜出现破损。 67d时，Ambient下的Bt-SY63和SY63相比，Bt-SY63的叶绿体被膜破损，而SY63叶绿体被膜基本完整，Bt-SY63的细胞器降解比SY63严重。O_3处理下的Bt-SY63与Ambient下相比，最明显的变化是严重解体的叶绿体，几乎占满了整个细胞，已不能辨别基粒的存在，叶绿体基质颗粒状情况恶化，细胞器降解严重。O_3处理下的SY63与Ambient下相比，类囊体肿胀变形严重，叶绿体基质也是颗粒状，亦不能辨别基粒的存在，但其叶绿体仍具有一定的形状。细胞器部分降解。O_3处理下Bt-SY63和SY63相比，SY63的叶绿体仍具有一定的形状，叶绿体降解和细胞器降解程度均比Bt-SY63轻，损伤轻。

3. 结果与讨论

3.4 开放式臭氧浓度升高条件下Bt-SY63和SY63的叶片超微结构变化

表4 不同处理时间Bt-SY63和SY63叶片超微结构

细胞特征	处理时间 (d)	Bt-SY63		SY63	
		$A\text{-}O_3$	$F\text{-}O_3$	$A\text{-}O_3$	$F\text{-}O_3$
叶绿体被膜完整性	35d	完整	完整	完整	完整
	51d	较完整	较完整	较完整	较完整
	67d	破损	破损	基本完整	基本完整
类囊体肿胀程度	35d	正常	较肿胀	正常	正常
	51d	正常	较肿胀	正常	较肿胀
	67d	较肿胀	严重肿胀	较肿胀	严重肿胀

臭氧浓度升高使得Bt-SY63和SY63发生最明显的变化有两个，一个是类囊体肿胀变形占满整个叶绿体，已不以基粒的形式存在，这个现象也是植物衰老过程中的现象，所以也有人认为类囊体肿胀变形更像是臭氧胁迫诱导的衰老的结果.本试验田间条件下后期Bt-SY63和SY63也发现叶绿体中类囊体肿胀现象，但本试验中臭氧浓度升高条件下67d接近收获期时，Bt-SY63和SY63叶绿体中已辨别不出基粒的存在，而Ambient下能看到基粒，说明叶绿体基粒结构的严重破坏确实是臭氧胁迫引起的。

3. 结果与讨论

3.5 开放式臭氧浓度升高条件下Bt-SY63叶片Bt蛋白的表达量

表5 各采样期Bt-SY63水稻和SY63水稻叶中Bt蛋白含量变化　　单位：μg/g

		2010年			2011年		
品种	处理	分蘖期	抽穗期	成熟期	分蘖期	抽穗期	成熟期
Bt-SY63	$A\text{-}O_3$	8.54±2.59 a	9.20±2.06a	3.86±1.26a	9.65±2.49 a	11.70±3.39 a	9.69±1.55a
	$E\text{-}O_3$	8.54±0.53 a	9.98±2.64a	5.26±1.30a	8.08±1.63 a	10.50±1.95 a	10.18±3.52a
SY63	$A\text{-}O_3$	–	–	–	–	–	–
	$E\text{-}O_3$	–	–	–	–	–	–

3. 结果与讨论

3.6 开放式臭氧浓度升高条件下Bt-SY63叶片Bt蛋白的转录

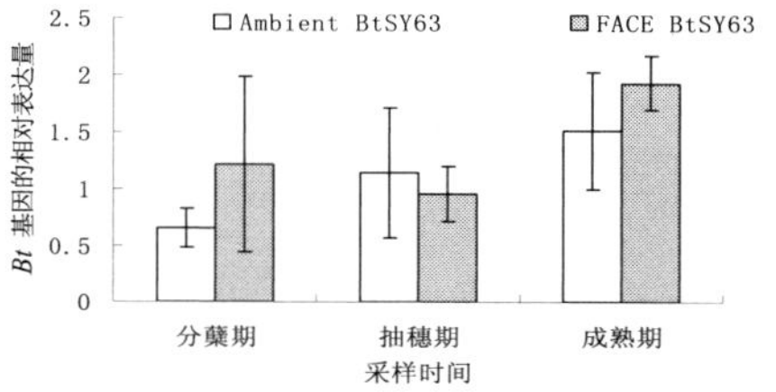

◈ 臭氧浓度升高使Bt-SY63叶肉细胞壁厚度在生长还很旺盛的抽穗前后即显著增加，这可能也是本试验中臭氧升高条件下Bt-SY63叶厚度比Ambient下增加的部分原因，而SY63叶肉细胞壁厚度在臭氧处理前后无明显变化，说明在叶肉细胞壁对臭氧浓度升高的反应上看，Bt-SY63比SY63更敏感。

◈ 类囊体肿胀变形直接导致叶绿体肿胀，使得臭氧处理条件下Bt-SY63叶绿体面积占叶肉细胞面积比例（CCRA）在51d和67d时持续显著增加。实际上，CCRA的此时“增加”可能并不是有效光合膜系统的增加，相反正是光合膜系统的膨胀损伤，而SY63的CCRA仅在51d时增加但不显著，另外，67d时Bt-SY63叶绿体已严重降解，而SY63叶绿体仍具有一定的形状，这些似乎说明Bt-SY63比SY63损伤更严重。

4. 小结

- 臭氧浓度升高对Bt-SY63和SY63的气孔密度和叶厚度等表面形态结构，细胞和叶绿体等的超微结构产生了不同程度的影响，O_3处理后导致两品种叶绿体类囊体肿胀，同时伴随着嗜锇颗粒显著增加，淀粉粒数量下降，叶绿体基质颗粒状，最终叶绿体被膜破损，直至基粒消失，核膜破损，细胞器不同程度的降解；
- 与SY63相比，Bt-SY63叶绿体类囊体肿胀的时间更早，叶绿体降解的程度更大，叶绿体面积占叶肉细胞面积比例（CCRA）、嗜锇颗粒面积占叶肉细胞面积比例（PCRA）、淀粉粒面积占叶肉细胞面积(SCRA)等指标波动大，持续时间长，Bt-SY63比SY63对于臭氧胁迫更敏感，胁迫损伤更大。
- 臭氧处理对外源Bt基因的转录和蛋白表达没有显著影响。

转基因油菜的风险评估：油菜与其近缘野生种之间的基因流研究

Risk Assessment of GM Oilseed Rape: Gene Flow between Oilseed Rape and its Wild Relatives

刘勇波[①]，李俊生，黄海，张细桃

（中国环境科学研究院环境基准与风险评估国家重点实验室，100012，北京）

LIU Yong-Bo[①]，LI Jun-Sheng，HUANG Hai，ZHANG Xi-Tao

（State Key Laboratory of Environmental Criteria and Risk Assessment，Chinese Research Academy of Environmental Sciences，Beijing 100012，China）

摘　要：随着转基因作物大面积的商业化种植，转基因作物的生物安全问题受到普遍关注，其中转基因作物与近缘野生种间基因流一直是人们关注和探讨的焦点。转基因油菜因其容易与近缘野生种发生杂交而成为研究转基因作物基因流风险的模式作物之一。本文从以下几个方面评估转基因作物基因流发生的风险：①转基因油菜与其近缘野生种的杂交难易程度：研究发现油菜(*Brassica napus*)能与 *B. rapa*，*B.juncea* 等在自然条件下自由杂交。②转基因油菜花粉漂移的距离：最远距离能达到 3 ~ 5 km。③种子在土壤中存活的年限：转基因油菜果荚容易裂开使得种子掉落在土壤中，形成种子库；研究发现最长年限能存活近 10 年；种子发芽长成自播植物，进一步扩散转基因。④相对适合度和竞争能力：转基因油菜自播植物或杂交后代在自然界的生存取决于他们的相对适合度和竞争能力。⑤发生基因渐渗的后果：基因渐渗可影响植株的形态特征与种群动态变化。

关键词：油菜；转基因；基因流；基因渐渗；近缘野生种

Abstract: With the commercial release of transgenic crops (GM) conferring novel traits，biosafety assessment of GM is concerned，and consequences of gene flow and introgression are still one main concern. Numerous studies have focused on the probability of occurrence of gene flow between transgenic crops and their wild relatives and the likelihood of transgene escape，which should be assessed before the commercial release of transgenic crops. This report focuses on this issue for GM oilseed rape (*Brassica napus*), a species that can cross easily with its wild relatives. We discuss the risk assessment of GM oilseed rape with four aspects followed. ①The possibility between oilseed rape and its wild relatives is reviewed, and studies found that *B. napus*, *B. juncea* and *B. rapa* could hybridize naturally in fields. ②Dispersal distance of pollen of oilseed rape is far，and there was

① Correspondence author：liuyb@craes.org.cn.

基金项目：国家自然科学基金（31200288）；教育部留学回国人员科研启动基金（教外司留 2013-693）.

study found that the maximum distance achieved 3 ~ 5 kilometers. ③Oilseed rape seeds fell in soil and buried for many years, and the seeds were viable after ten years in soil. Volunteers from these seeds are the source of further gene flow. ④The survive of GM plants depends on the relative fitness and competitive capacity of GM oilseed rape, volunteers or progeny compared to non-GM plants. ⑤The introgression of transgenes to wild relatives might have impacts on plant morphology and population dynamics.

Key words: Oilseed rape; transgenes; gene flow; introgression; wild relatives

引言

由于转基因作物涉及许多重要的种质资源物种，如大豆、玉米、棉花、油菜、甜菜、南瓜、木瓜和苜蓿等，这些作物在全世界广泛种植 [1]，而且这些转基因作物中，绝大多数都能和它们的近缘野生种发生杂交 [2,3]。所以，转基因作物和近缘野生种间的基因流和基因渐渗成为转基因作物生物安全管理最主要的关注问题之一。

物种间的杂交是基因流的前提，基因流是基因渐渗的前提。然而，不是所有的种间杂交都能导致基因流的发生。只有满足某些合适的条件时，才能成功获得杂交种，比如，作物与野生种间的性亲和、花期的重叠、花粉和种子的扩散、形成可育的后代。同样，也不是所有的基因流都能导致基因渐渗的实现。转基因渐渗到接收的物种或种群中，主要依赖于发生杂交和渐渗的各种障碍，种间杂交后代的进化和命运，个体适合度，转基因在后代表达的潜在代价和利益等。

通过种间杂交的种间基因流和随后的基因渐渗，在通过自然种群获得遗传适应性的过程中有着重要的生物学作用。人们意识到，基因渐渗能够频繁发生且能繁殖出可育的甚至高度适应的后代[4,5]；其实许多进化科学家们早已关注基因渐渗在许多植物类型进化中的作用[6,7,8]。

本文将综述油菜和野生种间基因流和基因渐渗的发生过程和可能后果。在释放转基因作物前忽略了这一主题，但是到目前为止仍有许多学者对其进行了一系列的实验研究，并获得了一些研究成果。

1 油菜（*Brassica napus*）与近缘野生种间的杂交

为了更好地了解油菜与野生种间发生基因流的风险，首先要明白十字花科植物各种间的相互关系。十字花科是一个大科，包含有 370 多个属，超过 3 000 个种。其中三个物种，油菜（*B. napus*，AACC，$2n = 4x = 38$）、野芥菜（*B. juncea*，AABB，$2n = 36$）和埃塞俄比亚芥（*B. carinata*，BBCC，$2n = 34$）都是四倍体，起源于三个二倍体物种：黑芥（*B. nigra*，AA，$2n = 16$）、芸薹（*B. rapa*，CC，$2n = 20$）和甘蓝（*B. oleracea*，BB，$2n = 18$）；图 1 显示了它们之间著名的“U”型三角关系。

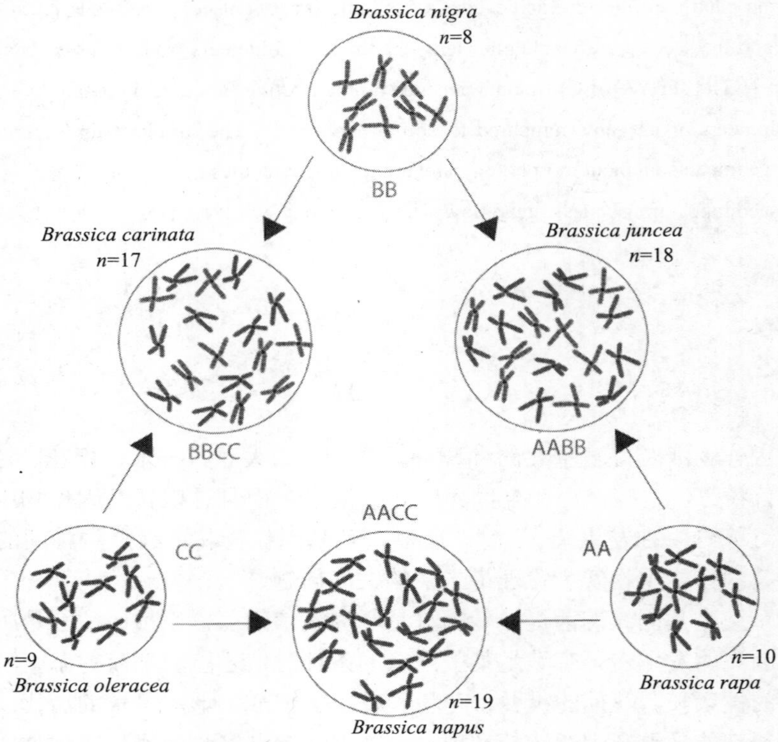

图 1　芸薹属物种间的"U"型三角染色体遗传关系图

"U-triangle" representing the genomic relationships between Brassica species（re-drawn from U 1935）

由于它们之间的亲密关系，油菜能和其他 5 个物种发生杂交，但是和每个物种杂交的难易程度不一样。而且，种间杂交还受很多其他外来因素影响，包括风速和风向、虫媒的移动[9]、植株在田野里的位置以及花朵在植株上的具体位置等[10]。*B. napus*，*B. rapa* 和 *B. juncea* 这三个物种间能够在自然条件下频繁地发生杂交[11-13]，所以，从油菜到 *B. rapa* 或到 *B. juncea* 的转基因流也许在自然授粉下就能发生[12-14]。尽管油菜和甘蓝（*B. oleracea*）在自然界发生杂交的成功率比较低，但是通过人工授粉能够成功获得杂交种。在英国，通过流式细胞仪和微卫星标记（SSR），发现了油菜和甘蓝之间的自然杂交种[15]。虽然还没有研究报道这三个四倍体物种和黑芥之间成功地发生自然杂交[11]，但是当把雄性不育油菜种植在黑芥地里的时候，观察到了从黑芥到油菜的自然杂交[16]，可是这并不能说明它们之间就能相互杂交。通过胚珠培养，获得了芸薹和黑芥之间的杂交种[17]。通过人工授粉能实现油菜和埃塞俄比亚芥菜，以及野芥菜和埃塞俄比亚芥菜之间的杂交[18]。黑芥和埃塞俄比亚芥菜之间的杂交种只有当埃塞俄比亚芥菜作为母本的时候才能获得[19]。

这三个四倍体物种不能和白芥（*Sinapis arvensis*）发生自然杂交[11,20]，意味着转基因不可能从油菜（*B. napus*）逃离到白芥[21]。但是，油菜和野萝卜（*Raphanus raphanistrum*）之间却能在自然界中能发生杂交，在加拿大和丹麦，有研究获得了以雄性不育油菜为母本的杂交种[22-24]；在澳大利亚和法国，有报道称发现一些以野萝卜为母本的杂交种[25-27]，但是杂交率都比较低；而且杂交率在一定程度上依赖于环境条件，如油菜品种、野萝卜种群

以及种植密度等 [22-28]。另外，合子合成前的花粉萌发和胚珠授精的遗传多态性也是妨碍种间杂交的抑制因素[29]。

关于油菜和其近缘野生种以及相关物种间更详细的杂交情况，有几篇综述已经进行了一些论述，可以参考，如：Rieger et al. [30]，FitzJohn et al. [31]，Andersson and de Vicente 等。

2 转基因流

各种不同类型油菜种类之间发生大量转基因流，结果导致在田野中出现许多自播植物，例如在加拿大，栽培一些抗某一种油菜后，在农田中发现有些油菜植株能同时抵抗三种不同类型的除草剂 [33-35]。转基因流不仅能在作物种植田地及其附近区域发生（同域性），而且还能够通过花粉和种子进行长距离扩散，从而把转基因从转基因油菜传播到更远处的近缘野生种群中。

2.1 种子传播

种子大多具备休眠特性，所以转基因作物的种子能够在土壤中存活多年，这可能导致种子落粒后在田野中建立转基因种子库。作物收割时候的种子落粒、作物浅耕、土壤翻耕时间等被认为是阻止种子进入土壤种子库的一些关键因素[36]。油菜作物（*B. napus*）同样具有这些特征，例如果荚容易开裂的落粒特性和种子容易被诱导二次休眠等，从而导致油菜种子库在土壤中的长期存在。有报道发现了油菜地里自播植物的更新生长，进一步说明油菜种子能够在翻耕和非翻耕地里休眠[37]。除了转基因作物的生产和耕种地等区域的本地落粒外，种子也能被携带到更远处，人类是一个转基因流发生的媒介，尤其是在转基因种子的运输和贸易过程中的种子传播。例如，在日本一些主要的贸易港口和运输公路等地，曾发现几株转基因油菜植株；而这些转基因种子在日本是不允许种植的，因此极大可能是由于进口转基因油菜种子在运输和交易过程中落下的[38,39]。同样，在韩国一个海港口附近的路边，发现了一株转基因野生玉米植株（*Zea mays*）[40]。这些转基因种子库的种子能够在接下来的数年里陆续萌发长成自播植物，然后自播植物产生的种子能够继续补充土壤种子库，成为进一步发生基因流的源头或者桥梁。

2.2 花粉扩散

油菜是自花授粉植物（自交亲和），但也能发生异花授粉，异交率大约为 12%到 47% [10]。油菜花粉传播的媒介主要是风和昆虫[41]。花粉调节的基因流受许多因素影响，包括开花时间、基因型、风向和风速、花粉源种群和接收花粉种群之间的距离等。油菜花粉的扩散距离可以达到 1.5 km，而且距离花粉源 1.5 km 处的花粉数量足以产生种子（平均每立方米有 22 颗花粉粒）[42]。转基因基因流可以在距离油菜花粉源几公里外的地方发生，如表 1 所示，从几百米到几千米以外都有不同程度的基因流发生 [43,44]。而且，对单棵植株而言，大约有一半的花粉分布在植株周围三米以内的地方，所以，如果用整个实验区域代替单棵植株来计算，将低估转基因花粉的比例[45]。

然而，大多数的转基因杂交事件发生在距离花粉源 10 m 以内的范围，50 m 以外杂交率显著降低[46]。一般情况下，基因流发生的程度和频率随着花粉源和接收种群间距离的增加而逐渐降低[34]。但是也有一些研究发现，远距离的花粉扩散是随机的。例如，有研究发现从 33 m 到 2 000 m 的远距离花粉扩散并不是一个逐渐减少的过程，风显著影响了花粉的扩散方向和距离[44]。同样，Rieger et al. [43]发现了隔离授粉事件的随机分布，这可以由两点解释：油菜的多种授粉媒介（风和昆虫）以及大尺度范围的花粉源。另外，Cresswell[47]发现传粉媒介调节的基因流与植株空间异质性无关，植物聚集对基因流的影响将取决于斑块间距离的空间尺度范围。

表 1 转基因油菜因花粉扩散发生基因流的最远距离

Pollen-mediated gene flow（GF）occurred at the maximum distance from pollen resources of GM oilseed rape（*B. napus*）

参考文献	研究的最远距离/m	发生的最远距离/m	基因流发生概率/%	转基因性状 GM
Scheffler et al. 1995	400	400	0.004	抗草铵膦 glufosinate
Beckie et al. 2003	800	400	0.04～0.05	抗草铵膦，草甘膦 glyphosate，glufosinate
Cai et al. 2008	2 000	2 000	<0.015	抗草铵膦 glufosinate
Rieger et al. 2002	5 000	3 000	0～0.15	抗 ALS 除草剂 ALS herbicide resistant

2.3 转基因野生种和自播植物在农田里的生存和传播

转基因花粉源不一定总是来自油菜（*B. napus*）农田，有时候花粉来自于独立的或其他斑块区域的转基因油菜，这些并不是农民有意种植的转基因油菜，而是油菜农田里或附近区域的自播植物和转基因野生植物。这些自播植物在种植转基因油菜后数年里经常出现[33,37,48]，这些转基因植物可能来自于土壤种子库，因为果荚易开裂而使得种子落入土壤[37]。在油菜种植区外，如公路边、铁路旁等地方，转基因野生植物的出现频率与种植农田区相似，例如有研究发现在抽查的样本野生植株中大约有 66%是转基因的 [49]。这些调查研究结果表明转基因油菜能在种植农田区域内外长期存在并扩散传播（表 2），这也将使得这些植物种群成为物种之间基因流的一个重要花粉源[50,51]。

Jørgensen et al. 研究发现 6%～32%的自播植物属于在 4～17 年之前种植的植物种类。在加拿大，观测到抗除草剂的转基因油菜（*Brassica napus*）和芸薹（*B. rapa*）的杂交代和回交代能在农田里存在超过 6 年，即使没有除草剂（glyphosate）的选择压力[48]。在瑞典的一处油菜地里，1995 年种植了转基因油菜后一直再也没有种植油菜，但是 10 年之后在同一块田里依然发现了抗除草剂油菜自播植物幼苗，这说明转基因油菜的种子库能够在土壤中存活至少 10 年[52]。同样，在法国，以农田区域为研究尺度的多年研究结果发现，抗除草剂的转基因油菜种植 5～8 年后，依然在农田里发现了抗除草剂的油菜自播植物 [53]。

表 2 油菜自播植物和野生植株长期存在于油菜种植区域内或区域外

Long-term persistence of volunteers and feral plants in and/or outside of cultivated fields

参考文献	年份	研究点	类型	转基因	转基因率
Jorgensen 等，2007	4～17	油菜地	油菜变种	否	6%～32%
D'Hertefeldt 等，2008	10	油菜地	土壤种子库的抗除草剂幼苗	是	39%
Warwick 等，2008	6	油菜地	油菜 *B. napus* 与白菜 *B. rapa* 之间的抗除草剂杂交种	是	2.5%
Méssean 等，2007	5～8	普通油菜地	抗除草剂自播植物	是	0～18%

3 作物相对适合度

Halfhill et al.[54] 发现不管是否有 *Bt* 转基因的存在，杂交种比父母亲本产生更少的生物量和种子产量，并且表现出更弱的竞争力。Di et al. [55]发现在无虫害条件下，*Bt* 抗虫转基因油菜和野芥菜间杂交种的综合适合度介于亲本之间，高于油菜但低于野芥菜。

一般情况下，杂交种的形态特征表现介于亲本之间，例如油菜（*B. napus*）与稗草（*Hirschfeldia incan*）的杂交种叶片上的细毛多于油菜但比稗草少[20]；埃塞俄比亚芥（*B. carinata*）和芸薹（*B. rapa*）杂交种的形态特征，包括花的大小和形状、开花时间以及成熟期等都介于父母本之间[19]。作物或者野生种的细胞质经种间杂交等途径转入到杂交种后，将影响其后代的形态特征[18,20,56,57]。例如，埃塞俄比亚芥分别和油菜及野芥菜杂交后，进一步回交得到回交七代，发现含有油菜细胞质的植株比含有野芥菜细胞质的植株开花更晚、花丝更短、雌蕊更短、花粉数和种子数更少、花瓣更短更窄、而且花的颜色表现也不同；这意味着细胞质与花器官的发育有关；而且油菜的细胞质比野芥菜的细胞质对埃塞俄比亚芥植株性状的干扰更大[18]。

当油菜（*B. napus*）和野萝卜（*Raphanus raphanistrum*）杂交时，其杂交一代 F_1 主要是异源三倍体（allotriploids），而且繁殖率很低，有研究发现每棵植株平均仅产生 3.2 颗种子[25]，也有研究发现每棵植株平均产生 11 颗种子[58]。相对于其父母本，杂交一代 F_1 的出苗率显著更低、出苗更迟、生存率降低、干物质也更低[59]。然而，杂交一代 F_1 的后代（F_2，BC_1）生长发育很好并能产生可育的种子，而且与野萝卜不断回交后，繁殖率显著升高[25]。此外，含有野萝卜细胞质的后代，即以野萝卜为母本进行杂回交的后代，比含有油菜细胞质的后代表现出更高的适合度，高出 100 倍以上，这说明以野萝卜为母本的杂回交后代更容易在农田或自然环境下生存并繁殖[57]。这也说明从作物到野生种的转基因流发生之后，转基因也许能够在其后代稳定遗传，并在自然界中继续存在。

3.1 转基因的适合度影响

转基因从油菜转入到近缘野生种的过程也许是中性的，但是转基因可能升高或降低接受转基因植株及其后代的适合度（表 3）。转基因对适合度的影响依赖于许多因素，如转基因的特征、转基因的选择和代价、转基因在基因组内的转入方法和部位、接受转基因植株所在的种群组成等[60]。在抵抗虫害的一些研究中，人们发现抗虫的 *Bt* 转基因在虫害压力下能够使油菜表现出更高的适合度[61-63]。而且，*Bt* 转基因在无虫害下也没有表现出适合度代价，即因为表现出抗性特征而降低植株适合度，如在十字花科的油菜中[55,62,64]，和菊科的向日葵中[65]等都没有发现 *Bt* 转基因适合度代价。

表 3 转基因油菜和野生种间的杂交种与非转基因杂交种的相对适合度表现

The relative performance of transgenic crop-wild relative hybrids as compared to non-transgenic hybrids（modified according to Hails and Morley 2005）

油菜	野生种	杂交种	确定杂交种的方法	结论	参考文献
抗草铵膦（glufosinat）	*B. rapa*	F1	PCR 和除草剂筛选	杂交种的雄性适合度低，但磁性适合度高	Pertl et al.（2002）
抗虫（Bt）	*B. rapa*	F1	绿色荧光蛋白标记（GFP）	转基因杂交种的生物量、花朵数、繁殖力在虫害条件下都比野生种高，在没有虫害时，生物量、花朵数与繁殖力要低，但花朵数依然更高	Vacher et al.（2004）
抗虫（Bt）	*B. rapa*	BC2	绿色荧光蛋白标记（GFP）	在低虫害压力下，转基因和非转基因的回交二代之间的生物量没有差异	Mason et al.（2003）
高月桂酸酯（laurate）	*B. rapa*	F1	转基因插入位点	转基因杂交种的种子休眠率低于野生种，但是发芽率和生长与非转基因杂交种没有差异	Linder and Schmitt（1995）
高月桂酸酯（laurate）	*B. rapa*	F1	转基因插入位点	转基因杂交种的种子休眠率低于野生种，但比野生种生长更快更高	Linder et al.（1998）
抗草铵膦	*B. rapa*	BC3	PCR	花粉活力、种子产量和存活率都与亲本没有差异	Snow. et al.（1999）
抗草铵膦	*B. rapa*	BC1	PCR	花粉活力与野生种没有差异	Mikkelsen et al.（1996）

Snow et al. 发现抗除草剂转基因从油菜渐渗到芸薹后，后代没有因表达抗草丁膦除草剂（glyphosinate-resistance）特征而降低植株适合度。而且，Ammitzboll et al. 发现杂交种与转基因油菜亲本的抗草甘膦除草剂的转基因（glyphosate transgene）在信使核糖核酸（mRNA）上的表达没有差异，这意味着这种转基因的表达不受基因型背景差异的影响。当然，这样的结论也许不适合其他转基因。抗除草剂油菜（*Brassica napus*）和野萝卜（*Raphanus raphanistrum*）杂交后，按照遗传规律应该有一半为抗性植株，但是在接下来的后代中（G8-G11，第八到第十一代），仅仅只有大约 18% 的植株在除草剂的选择压力下表现出抗性[68]，这说明转基因在野萝卜种群中的表达并不稳定。Ford et al. [15]采用流式细胞仪和作物特异性引物标记，在一块种植油菜（*B. napus*）和甘蓝（*B. oleracea*）的土地中发现甘蓝野生种群中有一个三倍体的杂交一代 F1，9 个二倍体和两个近三倍体基因渐渗植株，这说明油菜基因有能力在自然条件下渐渗到甘蓝植株中。

这些结果说明，转基因能够通过杂交等途径从转基因作物渐渗到野生种群中。例如，在加拿大的一块六年前种植过抗草甘膦除草剂油菜的农田里，尽管在接下来的六年中没有喷施除草剂的选择压力，但是研究人员检测到了一株野生芸薹（*Brassica rapa*）能够抵抗草甘膦除草剂[48]，这说明抗除草剂转基因已经渐渗到野生芸薹中。在美国，向日葵栽培种的一些基因位点广泛存在于野生向日葵（*Helianthus annuus*）种群中，同区域的野生向日葵种群已经被杂交种群所取代[69,70]。Snow et al. 发现栽培萝卜的特异基因位点能够在野萝卜（*Raphanus raphanistrum*）和栽培萝卜（*R. sativus*）杂交代的四个种群中稳定遗传十年，这说明中性的或者有益的转基因很容易存在于野生种群中（表 4）。

表 4 作物等位基因在野生种群中的长期存在

Long-term introgression of crop alleles into weed populations

参考文献	存在年数	实验地点	位点类型	转基因	概率/%
Whitton et al.，1997	＞5	野生向日葵（*Helianthus annuus*）	栽培种的 RAPD 特异标记序列	否	42
Hansen et al.， 2001	11	油菜与野生种 *B. Rapa* 的混合地	物种 AFLP 特异标记序列	否	50
Snow et al.， 2001	3	人工种植的 4 个萝卜种群	白色花朵，萝卜特异位点	否	8～22
Snow et al.， 2010	10	萝卜的杂交种群	萝卜与野萝卜杂交种中的作物的特异位点	否	—

3.2 以抗虫转基因为例探讨抗性植株和敏感植株间的相对适合度

昆虫取食使得植株进行化学防御和形态防御导致一些表型的变化，因为植物对昆虫咬食的反应主要为产生化学物质如芥子油苷（glucosinolates）[72]，或者次生化学物质发生变化[73,74]，从而导致形态特征发生变化。例如，植物表皮毛状体的密度和数量[75,76]，叶片的粗糙度[77]、花朵的一些像萼片长度和宽度等性状[78]。野萝卜被菜粉蝶（*Pierisrapae*）咬食后产生的芥子油苷浓度比没有被咬食的植株高 55%[79]，而且被咬食后的植株新长成的叶片上有更多的表皮刚毛数量和更高的刚毛密度[76]。Cresswell et al. 发现剪叶后的油菜植株的花萼长度和宽度以及雄蕊的长度都比完整植株的短和窄。并且，Lehtilä and Strauss 显示野萝卜叶片的破坏降低了植株对授粉媒介的吸引力，因为野萝卜的花朵大小和数量都在最初的几周里有着显著的减少。由此可推断，在昆虫取食的压力下，对昆虫敏感的野生植株也许会有类似的反应，而抗虫转基因植株由于不会受到昆虫取食而将不会做出这样的反应。所以，在虫害压力下，这样的反应差异也许将导致转基因杂交后代的形态特征不同于其亲本，从而产生植物形态上的分化。

一般情况下，昆虫取食对植株适合度有负影响，降低其适合度[81,82]，但同时也有一些研究发现由于植株的补偿作用导致昆虫取食并没有显著降低适合度[76,83-86]。在植株生长早期阶段用人工剪叶诱导植株做出反应，降低昆虫在后期对诱导植株的取食，从而导致诱导植株（人工剪叶）比没有被诱导的植株产生更高的适合度[76,87]。所以，昆虫取食对植物生长发育的影响并不完全一样，还取决于许多其他因素，例如植物种类[82]、营养资源水平[82,85]、昆虫取食的阶段和时间[76,87]和昆虫种类[88]等。

抗虫 *Bt* 转基因是转入到大多数作物中最重要的转基因之一，因为 Bt 抗虫蛋白对世界上大部分作物的害虫是有毒的，如棉花、水稻、玉米、油菜和向日葵等作物上的主要害虫。

从 *Bt* 转基因作物到野生种的基因流和基因渐渗发生后，可能减少杂交种的昆虫取食，从而使得杂交种的适合度（如生存率、生长能力和繁殖力）高于那些没有 *Bt* 基因保护的野生种（Stewart et al.，1996；Ramachandran et al.，2000；Snow et al.，2003；Vacher et al.，2004）。而且，很多研究发现 *Bt* 转基因的表达是没有适合度代价的，例如在油菜及野生种等植物中的表达（Mason et al.，2003；Moon et al.，2007）、在向日葵里的表达（Snow et al.，2003），但是 Vacher et al.（2004）发现在没有虫害的条件下 *Bt* 杂交种比野生植株产生更少的种子数。

抗虫 *Bt* 转基因对野生植株适合度的影响取决于昆虫的取食与否，大多数研究结果显示当植株遭受对 *Bt* 敏感的昆虫时，抗虫转基因植物比非转基因表现出更高的适合度。Stewart et al. 发现在农田实验中当油菜遭受小菜蛾（*diamondback moth*）取食时，转基因油菜比非转基因油菜有更高的生物量并产生更多的种子量。Snow et al. 发现由于转基因向日葵比非转基因向日葵遭受更少的昆虫取食伤害，使得转基因向日葵产生更多的种子，比非转基因多 55%。Moon et al. 发现转入 *Bt* 基因后的芸薹（*B. rapa*）和由转基因油菜与芸薹形成的杂交种在温室的昆虫添加实验中都比非转基因油菜产生更高的生物量和更多的种子。然而，抗虫这个特性并不一定能提高植物的适合度，尤其是在没有虫害或者低虫害的条件下[62,90]。Sutherland et al. 发现模拟虫害（在子叶期机械破坏叶片）并没有影响油菜和芸苔及其之间的杂交种的适合度。Moon et al. [90]同时也发现在大田里由于实验时的害虫压力不大，所以 *Bt* 转基因植株并没有比敏感植株显示出显著更高的适合度。综合前人研究，这些结果说明 *Bt* 转基因能够在中度或高度的害虫压力下提高抗性植株的适合度，但是在低度或者没有害虫压力的情况下，抗性植株并没有适合度优势。

因为抗虫基因植株在虫害条件下具有适合度优势，所以在混合种群中，抗虫植株可能抑制敏感植株的生长发育。因此，抗性植株和敏感植株间的竞争关系将在一定程度上决定种群的动态变化。Ramachandran et al. [62] 发现在转基因和非转基因混合种植的种群中，相对于敏感植株，抗性植株是优势竞争者，具有竞争优势。然而，混合种群的动态变化还取决于许多其他因素，包括虫害程度、资源水平、抗性植株在种群中的相对比例等[92]。由于抗性植株（转基因作物或杂回交种）的入侵和殖民，随着抗性植株的增多，这两种植株类型（抗性和敏感）之间的竞争将减少，但是抗性植株之间的竞争（类型内部）将增加。所以，野生种在混合种群中的命运在一定程度上将取决于它和转基因植株之间以及它们各自内部的竞争关系。这样一来，混合种群或者达到一个包含转基因和野生种并以一个稳定比例存在的平衡状态，或者敏感野生种被抗性转基因植物所取代从而推动种群更替。关于这类种群动态变化的研究到目前为止还非常少，但是这样的研究对更好地了解转基因作物和近缘野生种间基因渐渗过程的进化生态学效应是非常关键和重要的。

4　基因渐渗

转基因作物和近缘野生种间杂回交数代后，转基因可能在后代稳定表达遗传，引起基因渐渗。基因渐渗也许将：①提高适应性进化的遗传变化[93,94]，②提高野生种的杂草性[95,96]，③因杂交优势或转基因新性状而提高植株竞争力或殖民能力[2,62]，④因适合度提高而增强入侵能力[97]，⑤提升适应新环境的能力[98,99]。当转基因从转基因作物渐渗到野生种去之后，这个外来基因将在几个不同的水平上对接收基因的野生植株产生影响：从可能发生的形态特征上的遗传同化到野生种群的动态变化。

4.1 基因渐渗对植株形态特征的影响

由于基因渐渗后改变了接收基因个体的遗传结构，所以它们的形态性状和行为表现也许会因此而改变，例如植株个体大小、叶片大小、花色、果实和种子特征和大小、种子休眠等。这些变化或者是与转基因直接相关，或通过接下来杂回交代的一系列进化过程而间接产生。例如，通过基因联合（hitch-hiking）或者通过抗虫和抗除草剂等新性状维持杂回交后代的丰富度。作物常常与其近缘野生种是性亲和的，但是它们之间表现出某些明显不同的形态表现和生活史特征。所以，由于作物和野生种的杂交，作物和野生种间的基因流和基因渐渗也许将成为一个模型系统来研究形态特征以及与适合度相关的生活史特征的快速进化。事实上，在一些研究中，形态特征已成为研究基因渐渗是否发生的一个重要检测指标（表 5）[100]。

表 5 形态特征作为检测作物和野生种之间发生自然杂交和基因渐渗的重要指标

Morphology was used to measure introgression in documented cases of natural hybridization and introgression of crops and wild relatives（One part of Table 1 in Jarvis and Hodgkin 1999）

作物种类	参考文献	用来检验基因渐渗的技术
白菜，芥菜，油菜 Cabbages，mustards，rapes （*Brassic* spp.）	Jorgensen & Andersen（1994）	形态学，细胞学，同工酶，RAPDs
	Perrino & Hammer（1985）	形态学
	Snogerup et al.（1990）	形态学
	Stace（1991）	形态学
	Worede（1986）	形态学
萝卜（*Raphanus sativus* L.）	Hammer & Perrino（1995）	形态学
	Klinger et al.（1992）	形态学，同工酶
	Klinger & Ellstrand（1994）	形态学

植物种群中形态特征的进化在一定程度上依赖于基因渐渗的发生和自然选择压力的方向和强度。例如，野萝卜的花色变化（植株中的一个单一位点）常常被用来作为栽培萝卜与野萝卜间基因流的一个间接证据[101-103]。在美国的加利福尼亚州，野萝卜（*Raphanus raphanistrum*）的花为黄色，而栽培萝卜（*R. sativus*）和其杂交种的花色大部分为白色和粉色[103,104]。花色能够影响昆虫的授粉：蝴蝶（*Pieris rapae*）和飞蛾更偏好给黄色花授粉[101,103]，而蜜蜂更愿意飞往白色和粉色花丛[103]。所以，当自然界中蜜蜂缺少时，授粉媒介也许能够导致白色花这个基因位点的丧失，尽管 Snow et al. 发现野萝卜的白色花这个基因位点能够在人工杂交种群中存在 10 年。另外，Campbell et al. 报道白色花与开花延迟有关，即白色花一般比其他花色开得更晚，而且这是一个可遗传的表型特征。所以，如果自然选择反对晚期开花，例如当作物收割时，晚期开花植株不能抽苔或者不能产生种子，这样的性状链接也能够导致白色花位点的丧失。同样，花的大小也能影响花色的出现频率。Lehtilä & Holmén Bränn 发现由大花而来的作物系比小花作物系植株更大，并且开花更晚，产生更多的花朵。因此，通过一些性状的相互关联，如果当授粉媒介选择黄色花而导致基因渐渗过程中白花基因位点丢失，那么种群中大花和大植株数目将减少。基于此，如果转基因与

白花位点相关联，那么当授粉媒介选择黄花的时候，转基因从作物到野生种的基因渐渗风险将降低，反之，基因渐渗风险将升高。

不仅基因渐渗能影响植株形态特征，反过来，形态特征也能影响作物与野生种之间的转基因渐渗。种子大小就是一个典型的例子。杂交种的种子大小也许阻碍转基因作物和野生种间的基因流和基因渐渗[16]，这是因为大多数作物和野生种间杂交种只产生小种子[22,23,26]。例如，油菜与大多数近缘野生种间的杂交只产生种子直径小于 1.6 mm 的小种子，像油菜与野芥菜（*B. juncea*）[11]、与芸薹（*B. rapa*）[107]、与稗草（*Hirschfeldia incana*）和与野萝卜（*R. raphanistrum*）[22,108]。Wei 和 Darmency 利用雄性不育油菜（*B. napus*）和五种野生种来获得杂交种，发现所有的杂交种种子都是小种子。由于种子内的储存物质影响幼苗的出苗率和幼苗大小，所以由小种子长成的幼苗表现出生长劣势[109,110]。这将降低由小种子长成的植株的适合度[111,112]，特别是在生境恶劣的条件下，如高密度、干旱、资源缺乏和虫害压力等。然而，这种种子大小影响可能还受遗传背景、生境条件等因素影响。

4.2 基因渐渗对种群动态的影响

基因流和基因渐渗在自然种群的进化过程中起着重要作用[94,113]，因为基因流能够抵消遗传漂变和近亲繁殖[114]、遗传分化[93]以及当地选择[113]在遗传变异上所引起的负作用，从而影响植物适应能力的进化。从短期来看，入侵的杂交种群也许会经历本地环境条件的强选择压力[115,116]。作物基因位点渗入到野生种群的长期基因渐渗也许影响野生杂草种群的进化路径，例如向日葵[117]、油菜[118]和野萝卜[119]。自然选择和种间杂交之间的交互作用也许将明显改变基因渐渗种群的动态过程，从而改变其生态入侵以及在新生境中建立的模式[113]。

一般情况下，作物和野生种间的杂交种群期望比亲本种群有更高的多样性表现。所以，在接下来的一系列进化过程中，在当地选择的压力下，杂交种群或者显著不同于亲本，或者与亲本越来越相似；但是，这个过程同样依赖于由遗传背景改变而引起的表型变化对植物适合度的影响：升高或降低适合度。如果表型变化有利于升高植物适合度，那么种间杂交能导致长期的基因渐渗。然而，杂交种杂草的表现型是向其野生杂草祖先种是分离还是聚集，以及是以一个怎样的速率进化等方面的研究到目前为止还十分少[99,120]。同时，基因流和基因渐渗在保持不同地理区域间作物与野生种的基因交流联合（cohesion）过程中起一个关键作用，例如野萝卜[104]、玉米[121]和水稻[122]等。当自然选择压力足够强大到能够战胜基因流或者当基因流受到限制时，由地理区域上的隔离而导致的种群分离就能发生[121,122]。另一方面，基因流和基因渐渗也许能联系有遗传差异和不同地理区域间的作物，阻碍各种群间的分化[104,123]；同时，同一地理区域内的基因流反而能进一步加剧不同区域种群间的分化。

Haygood et al. 通过利用野生种群不断接收基因渐渗作物花粉的模型，发现遗传同化涉及阈值界限和杂交优势，从而导致少量的外来基因入侵不仅导致基因（包括有害基因）在种群中的固定，而且导致野生种群的迅速减少甚至被取代，这是因为个体数量快速扩充（demographic swamping）能够导致“迁移萎缩”（migrational meltdown）。一些研究发现作物基因渐渗到野生种去后，有些杂交种也许已经取代了它们的亲本[98,125,126]，甚至导致了它们祖先的灭绝[120]。Campbell et al. 的研究结果显示作物基因渐渗后的野萝卜杂交种比亲本有更高的繁殖力和生存率。Hegde et al. [120]以形态特征和酶为证据发现杂交后代种取代了加利福尼亚州当地的两个萝卜种群，而且加利福尼亚野萝卜（杂交种）作为一个独立的进化单元并与它的亲生本分离开来，这说明杂交种不断的殖民行为不是来自于种群在本地的遗传变异而是来自于父母本特征的整合。

转基因的基因渐渗除了影响接收转基因植物的种群动态外，它还影响靶标和非靶标昆虫种群的影响。例如，在中国，有研究发现靶昆虫（*Helicoverpa armigera*）的密度和 *Bt* 转基因棉花的种植年份表现为一个负相关关系，无论是在高密度还是低密度区域，这说明 *Bt* 棉花的种植能减少靶昆虫（*Helicoverpa armigera*）种群的密度[127]。反过来，靶标昆虫的变化将影响转基因和非转基因植株的动态变化，因为虫害压力能影响抗虫植株和敏感植株的相对适合度和相对竞争能力[62,90]。所以，转基因作物和非转基因敏感植物之间的相互关系、植株与昆虫以及天敌的相互关系等都将是影响转基因作物和野生种间基因流和基因渐渗后果的重要因素。

参考文献

[1] C. James，Global Status of Commercialized Biotech/GM Crops：2012，ISAAA Brief，44 Ithaca，NY，2012.

[2] N.C. Ellstrand，H.C. Prentice，J.F. Hancock，Gene flow and introgression from domesticated plants into their wild relatives，Annu. Rev. Ecol.，Evol. Syst.，1999，30：539-563.

[3] T.T. Armstrong，R.G. Fitzjohn，L.E. Newstrom，A.D. Wilton，W.G. Lee，Transgene escape：what potential for crop-wild hybridization？，Mol. Ecol.，2005，14：2111-2132.

[4] M.L. Arnold，Natural hybridization and the evolution of domesticated，pest and disease organisms，Mol. Ecol.，2004，13：997-1007.

[5] L.H. Rieseberg，M.J. Kim，G.J. Seiler，Introgression between the cultivated sunflower and a sympatric wild relative，Helianthus petiolaris（Asteraceae），Int. J. Plant Sci.，1999，160：102-108.

[6] J.P. Lotsy，Evolution by means of hybridization，Nijhoff，Dordrecht，The Netherlands，1916.

[7] D.M. Anderson，Introgressive hybridization，John Wiley and Sons，N.Y.，1949.

[8] V. Grant，Plant Speciation，Columbus University Press，New York，1981.

[9] J.A. Scheffler，R. Parkinson，P.J. Dale，Frequency and distance of pollen dispersal from transgenic oilseed rape（*Brassica napus*），Transgenic Res.，1993，2：356-364.

[10] H.C. Becker，C. Damgaard，B. Karlsson，Environmental variation for outcrossing rate in rapeseed（*Brassica napus*），Theor. Appl. Genet.，1992，84：303-306.

[11] D.J. Bing，R.K. Downey，F.W. Rakow，Hybridization among *Brassica napus*，*B. rapa* and *B. juncea* and their two weedy relatives *B. nigra* and *Sinapis arvensis* under open pollination conditions in the field，Plant Breeding，1996，115：470-473.

[12] D.J. Bing，Potential of gene transfer among oilseed Brassica and their weedy relatives，University of Saskatchewan，Saskatoon，Saskatchewan，1991.

[13] D. Leckie，A. Smithson，I.R. Crute，Gene movement from oilseed rape to weedy populations - a component of risk assessment for transgenic cultivars，Aspects of Applied Biology，1993，35：61-66.

[14] S. Frello，K.R. Hansen，J. Jensen，R.B. Jørgensen，Inheritance of rapeseed（*Brassica napus*）- specific RAPD markers and a transgene in the cross *B.juncea* ×（*B.juncea* × *B. napus*），Theor. Appl. Genet.，1995，91：236-241.

[15] C.S. Ford，J. Allainguillaume，P. Grilli-Chantler，G. Cuccato，C.J. Allender，M.J. Wilkinson，Spontaneous gene flow from rapeseed（Brassica napus）to wild Brassica oleracea，Proc. R. Soc. B，2006，273：3111-3115.

[16] W. Wei，H. Darmency，Gene flow hampered by low seed size of hybrids between oilseed rape and five wild relatives，Seed Science Research，2008，18：115-123.

[17] D.J. Bing，R.K. Downey，F.W. Rakow，Assessment of transgene escape from *Brassica rapa*（*B. campestris*）into *B. nigra* or *Sinapis arvensis*，Plant Breeding，1996，115：1-4.

[18] C.T. Chang，R. Uesugi，K. Hondo，F. Kakihara，M. Kato，The effect of the cytoplasms of Brassica napus and B. juncea on some characteristics of B. carinata，including flower morphology，Euphytica，2007，158：261-270.

[19] B.R. Choudhary，P. Joshi，S. Ramarao，Interspecific hybridization between *Brassica carinata* and *Brassica rapa*，Plant Breeding，2000，119：417-420.

[20] E. Lefol，V. Danielou，H. Darmency，Predicting hybridization between transgenic oilseed rape and wild mustard，Field Crops Res.，1996，45：153-161.

[21] C.L. Moyes，J.M. Lilley，C.A. Casais，S.G. Cole，P.D. Haeger，P.J. Dale，Barriers to gene flow from oilseed rape（*Brassica napus*）into populations of *Sinapis arvensis*，Mol. Ecol.，2002，11：103-112.

[22] F. Eber，A.M. Chèvre，A. Baranger，P. Vallèe，X. Tanguy，M. Renard，Spontaneous hybridization between a male-sterile oilseed rape and two weeds，Theor. Appl. Genet.，1994，88：362-368.

[23] A. Baranger，A.M. Chèvre，F. Eber，M. Renard，Effect of oilseed rape genotype on the spontaneous hybridization rate with a weedy species an assessment of transgene dispersal，Theor. Appl. Genet.，1995，91：956-963.

[24] H. Ammitzbøll，R.B. Jørgensen，Hybridization between oilseed rape（*Brassica napus*）and different populations and species of Raphanus，Environ. Biosafety Res.，2006，5：3-13.

[25] H. Darmency，E. Lefol，A. Fleury，Spontaneous hybridizations between oilseed rape and wild radish，Mol. Ecol.，1998，7：1467-1473.

[26] A.M. Chèvre，F. Eber，H. Darmency，A. Fleury，H. Picault，J.C. Letanneur，M. Renard，Assessment of interspecific hybridization between transgenic oilseed rape and wild radish under normal agronomic conditions，Theor. Appl. Genet.，2000，100：1233-1239.

[27] M.A. Rieger，T.D. Potter，C. Preston，S.B. Powles，Hybridisation between *Brassica napus* L. and *Raphanus raphanistrum* L. under agronomic field conditions，Theor. Appl. Genet.，2001，103：555-560.

[28] A.M. Chèvre，F. Eber，A. Baranger，M.C. Kerlan，P. Barret，P. Vallée，M. Renard，Interspecific gene flow as a component of risk assessment for transgenic Brassicas，Acta Horticulturae，1996，407：169-179.

[29] G. Guéritaine，H. Darmency，Polymorphism for interspecific hybridisation within a population of wild radish（*Raphanus raphanistrum*）pollinated by oilseed rape（*Brassica napus*），Sex Plant Reprod，2001，14：169-172.

[30] M.A. Rieger，C. Preston，S.B. Powles，Risks of gene flow from transgenic herbicide-resistant canola（*Brassica napus*）to weedy relatives in southern Australian cropping systems，Aust. J. Agric. Res.，1999，50：115-128.

[31] R.G. FitzJohn，T.T. Armstrong，L.E. Newstrom-Lloyd，A.D. Wilton，M. Cochrane，Hybridisation within Brassica and allied genera：evaluation of potential for transgene escape，Euphytica，2007，158：209-230.

[32] M.S. Andersson，M.C. de Vicente，Gene flow between crops and their wild relatives，The Johns Hopkins University press，2010.

[33] L. Hall，K. Topinka，J. Huffman，L. Davis，A. Good，Pollen flow between herbicide resistant *Brassica napus* is the cause of multiple-resistant *B. napus* volunteers，Weed Sci.，2000，48：688-694.

[34] H.J. Beckie，S.I. Warwick，H. Nair，G. Seguin-Swartz，Gene flow in commercial fields of herbicide-resistant canola（*Brassica napus*），Ecol. Appl.，2003，13：1276-1294.

[35] A.L. Knispel，S.M. McLachlan，R.C. Van Acker，L.F. Friesen，Gene flow and multiple herbicide resistance in escaped canola populations，Weed Sci.，2008，56：72-80.

[36] G.S. Begg，S. Hockaday，J.W. McNicol，M. Askew，G.R. Squire，Modelling the persistence of volunteer oilseed rape（*Brassica napus*），Ecol. Model.，2006，198：195-207.

[37] M.J. Simard，A. Legere，D. Pageau，J. Lajeuness，S. WARWICk，The frequency and persistence of volunteer canola（*Brassica napus*）in Québec cropping system，Weed Technol.，2002，16：433-439.

[38] H. Saji，N. Nakajima，M. Aono，M. Tamaoki，A. Kubo，S. Wakiyama，Y. Hatase，M. Nagatsu，Monitoring the escape of transgenic oilseed rape around Japanese ports and roadsides，Environ. Biosafety Res.，2005，4：217-222.

[39] M. Aono，S. Wakiyama，M. Nagatsu，N. Nakajima，M. Tamaoki，A. Kubo，H. Saji，Detection of feral transgenic oilseed rape with multiple-herbicide resistance in Japan，Environ. Biosafety Res.，2006，5：77-87.

[40] C.G. Kim，H. Yi，S. Park，J.E. Yeon，D.Y. Kim，D.I. Kim，K.H. Lee，T.C. Lee，I.S. Paek，W.K. Yoon，S.C. Jeong，H.M. Kim，Monitoring the occurrence of genetically modified soybean and maize around cultivated fields and at a grain receiving port in Korea，Journal of Plant Biology，2006，49：218-223.

[41] I.H. Williams，A.P. Martin，R.P. White，The effect Of insect pollination on plant development and seed production in winter oil-seed rape（*Brassica napus* L.），The Journal of Agricultural Science，1987，109：135-139.

[42] A.M. Timmons，E.T. O'Brien，Y.M. Charters，S.J. Dubbels，M.J. Wilkinson，Assessing the risks of wind pollination from fields of genetically modified *Brassica napus*，Euphytica，1995，85：417-423.

[43] M.A. Rieger，Pollen-mediated movement of herbicide resistance between commercial canola fields，Science，2002，296：2386-2388.

[44] L. Cai，B. Zhou，X. Guo，C. Dong，X. Hu，M. Hou，S. Liu，Pollen-mediated gene flow in Chinese commercial fields of glufosinate-resistant canola（*Brassica napus*），Chin. Sci. Bull.，2008，53：2333-2341.

[45] C. Lavigne，E.K. Klein，P. Vallee，J. Pierre，B. Godelle，M. Renard，A pollen-dispersal experiment

with transgenic oilseed rape. Estimation of the average pollen dispersal of an individual plant within a field，Theor. Appl. Genet.，1998，96：886-896.

[46] A. Hüsken，A. Dietz-Pfeilstetter，Pollen-mediated intraspecific gene flow from herbicide resistant oilseed rape（*Brassica napus* L.），Transgenic Res.，2007，16：557-569.

[47] J.E. Cresswell，Spatial heterogeneity，pollinator behaviour and pollinator-mediated gene flow：bumblebee movements in variously aggregated rows of oil-seed rape，Oikos，1997，78：546-556.

[48] S.I. Warwick，A. Légère，M.J. Simard，T. James，Do escaped transgenes persist in nature? The case of an herbicide resistance transgene in a weedy *Brassica rapa* population，Mol. Ecol.，2008，17：1387-1395.

[49] Y. Yoshimura，H.J. Beckie，K. Matsuo，Transgenic oilseed rape along transportation routes and port of Vancouver in western Canada，Environ. Biosafety Res.，2006，5：67-75.

[50] T. Jørgensen，T.P. Hauser，R.B. Jørgensen，Adventitious presence of other varieties in oilseed rape（*Brassica napus*）from seed banks and certified seed，Seed Science Research，2007，17：115.

[51] N. Colbach，C. Durr，S. Gruber，C. Pekrun，Modelling the seed bank evolution and emergence of oilseed rape volunteers for managing co-existence of GM and non-GM varieties，European Journal of Agronomy，2008，28：19-32.

[52] T. D'Hertefeldt，R.B. Jorgensen，L.B. Pettersson，Long-term persistence of GM oilseed rape in the seedbank，Biol. Lett.，2008，4：314-317.

[53] A. Méssean，C. Sausse，J. Gasquez，H. Darmency，Occurrence of genetically modified oilseed rape seeds in the harvest of subsequent conventional oilseed rape over time，European Journal of Agronomy，2007，27：115-122.

[54] M.D. Halfhill，J.P. Sutherland，H.S. Moon，G.M. Poppy，S.I. Warwick，A.K. Weissinger，T.W. Rufty，P.L. Raymer，C.N. Stewart，Growth，productivity，and competitiveness of introgressed weedy Brassica rapa hybrids selected for the presence of Bt cry1Ac and gfp transgenes，Mol. Ecol.，2005，14：3177-3189.

[55] K. Di，C.N. Stewart，W. Wei，B.-c. Shen，Z.-X. Tang，K.-P. Ma，Fitness and maternal effects in hybrids formed between transgenic oilseed rape（*Brassica napus* L.）and wild brown mustard[*B. juncea*（L.）Czern et Coss.] in the field，Pest Manage. Sci.，2009，65：753-760.

[56] B. Zhang，C.M. Lu，F. Kakihara，M. Kato，Effect of genome composition and cytoplasm on petal colour in resynthesized amphidiploids and sesquidiploids derived from crosses between *Brassica rapa* and *Brassica oleracea*，Plant Breeding，2002，121：297-300.

[57] G. Guéritaine，M. Sester，F. Eber，A.M. Chèvre，H. Darmency，Fitness of backcross six of hybrids between transgenic oilseed rape（*Brassica napus*）and wild radish（*Raphanus raphanistrum*），Mol. Ecol.，2002，11：1419-1426.

[58] A.M. Chevre，F. Eber，A. Baranger，G. Hureau，P. Barret，H. Picault，M. Renard，Characterization of backcross generations obtained under field conditions from oilseed rape-wild radish F1 interspecific hybrids：an assessment of transgene dispersal，Theor. Appl. Genet.，1998，97：90-98.

[59] G. Guéritaine，S. Bazot，H. Darmency，Emergence and growth of hybrids between Brassica napus and Raphanus raphanistrum，New Phytol.，2003，158：561-567.

[60] F. Felber，G. Kozlowski，N. Arrigo，R. Guadagnuolo，Genetic and ecological consequences of transgene flow to the wild flora，Green Gene Technology：Research in an Area of Social Conflict，2007，107：173-205.

[61] C.N. Stewart，J.N. All，P.L. Raymer，S. Ramachandran，Increased fitness of transgenic insecticidal rapeseed under insect selection pressure，Mol. Ecol.，1997，6：773-779.

[62] S. Ramachandran，G.D. Buntin，J.N. All，P.L. Raymer，C.N. Stewart，Intraspecific competition of an insect-resistant transgenic canola in seed mixtures，Agron. J.，2000，92：368 - 374.

[63] D.K. Letourneau，J.A. Hagen，Plant fitness assessment for wild relatives of insect resistant crops，Environ. Biosafety Res.，2009，8：45-55.

[64] P. Mason，L. Braun，S.I. Warwick，B. Zhu，S. Neal，Transgenic Bt-producing *Brassica napus*：*Plutella xylostella* selection pressure and fitness of weedy relatives，Environ. Biosafety Res.，2003，2：263-276.

[65] A.A. Snow，D. Pilson，L.H. Rieseberg，M.J. Paulsen，N. Pleskac，M.R. Reagon，D.E. Wolf，S.M. Selbo，A Bt transgene reduces herbivory and enhances fecundity in wild sunflowers，Ecol. Appl.，2003，13：279-286.

[66] A.A. Snow，B. Andersen，R.B. Jorgensen，Costs of transgenic herbicide resistance introgressed from *Brassica napus* into weedy *B. rapa*，Mol. Ecol.，1999，8：605-615.

[67] H. Ammitzbøll，T.N. Mikkelsen，R.B. Jørgensen，Transgene expression and fitness of hybrids between GM oilseed rape and *Brassica rapa*，Environ. Biosafety Res.，2005，4：3-12.

[68] A. Al Mouemar，H. Darmency，Lack of stable inheritance of introgressed transgene from oilseed rape in wild radish，Environ. Biosafety Res.，2004，3：209-214.

[69] J. Whitton，D.E. Wolf，D.M. Arias，A.A. Snow，L.H. Rieseberg，The persistence of cultivar alleles in wild populations of sunflowers five generations after hybridization，Theor. Appl. Genet.，1997，95：33-40.

[70] C.R. Linder，I. Taha，G.J. Seiler，A.A. Snow，L.H. Rieseberg，Long-term introgression of crop genes into wild sunflower populations，Theor. Appl. Genet.，1998，96：339-347.

[71] A.A. Snow，T.M. Culley，L.G. Campbell，P.M. Sweeney，S.G. Hegde，N.C. Ellstrand，Long-term persistence of crop alleles in weedy populations of wild radish（*Raphanus raphanistrum*），New Phytol.，2010，186：537-548.

[72] A.A. Agrawal，C. Laforsch，R. Tollrian，Transgenerational induction of defences in animals and plants，Nature，401（1999）60-63.

[73] I.T. Baldwinn，Chemical changes rapidly induced by folivory，Chemical Rubber Company，Boca Raton，Florida，1994.

[74] R. Karban，I.T. Baldwin，Induced responses to herbivory，University of Chicago Press，Chicago，Illinois，USA，1997.

[75] R. Baur，S. Binder，G. Benz，Nonglandular leaf trichomes as short-term inducible defense of the gray alder，*Alnus incana*（L.），against the chrysomelid beetle，Agelastica alni L. Oecologia（Berlin），1991，87：219 - 226.

[76] A.A. Agrawal，Induced responses to herbivory in wild radish：effects on several herbivores and plant

fitness，Ecology，1999，80：1713－1723.

[77] G. Kudo，Herbivory pattern and induced responses to simulated herbivory in *Quercus mongolica* var. *grosseserrata*，Ecol. Res.，1996，11：283-289.

[78] J.E. Cresswell，C. Hagen，J.M. Woolnough，Attributes of individual flowers of *Brassica napus* L. are affected by defoliation but not by intraspecific competition，Ann. Bot.，2001，88：111-117.

[79] A.A. Agrawal，J.K. Conner，M.T.J. Johnson，R. Wallsgrove，Ecological genetics of an induced plant defense against herbivores：Additive genetic variance and costs of phenotypic plasticity，Evolution，2002，56：2206-2213.

[80] K. Lehtilä，S.Y. Strauss，Leaf damage by herbivores affects attractiveness to pollinators in wild radish，*Raphanus raphanistrum*，Oecologia，1997，111：396-403.

[81] J. Escarré，J. Lepart，X. Sans，J.J. Sentuc，V. Gorse，Effects of herbivory on the growth and reproduction of *Picris hieracioides* in the Mediterranean region，Journal of Vegetation Science，1999，10：101-110.

[82] W. Rogers，E. Siemann，Effects of simulated herbivory and resource availability on native and invasive exotic tree seedlings，Basic Appl. Ecol.，2002，3：297-307.

[83] S.Y. Strauss，A.A. Agrawal，The ecology and evolution of plant tolerance to herbivory，Trends Ecol. Evol.，1999，14：179-185.

[84] M.E. Gadd，T.P. Young，T.M. Palmer，Effects of simulated shoot and leaf herbivory on vegetative growth and plant defense in *Acacia drepanolobium*，Oikos，2001，92：515-521.

[85] C.V. Hawkes，J.J. Sullivan，The impact of herbivory on plants in different resource conditions a meta-analysis，Ecology，2001，82：2045-2058.

[86] E. Boalt，K. Lehtilä，Tolerance to apical and foliar damage：costs and mechanisms in *Raphanus raphanistrum*，Oikos，2007，116：2071-2081.

[87] A.A. Agrawal，Induced responses to herbivory and increased plant performance，Science，1998，279：1201-1202.

[88] S. Schooler，Z. Baron，M. Julien，Effect of simulated and actual herbivory on alligator weed，*Alternanthera philoxeroides*，growth and reproduction，Biol. Control，2006，36：74-79.

[89] C.N. Stewart，M.J. Adang，J.N. All，H.R. Boerma，C. Cardineau，D. Tucker，W.A. Parrott，Genetic transformation，recovery，and characterization of fertile soybean transgenic for a synthetic *Bacillus fhuringiensis* cry/Ac gene，Plant Physiol.，1996，112：121-129.

[90] H.S. Moon，M.D. Halfhill，L.L. Good，P.L. Raymer，C.N. Stewart，Characterization of directly transformed weedy Brassica rapa and introgressed *B. rapa* with Bt cry1Ac and gfp genes，Plant Cell Rep.，2007，26：1001-1010.

[91] J.P. Sutherland，L. Justinova，G.M. Poppy，The responses of crop-wild Brassica hybrids to simulated herbivory and interspecific competition：Implications for transgene introgression，Environ. Biosafety Res.，2006，5：15-25.

[92] H.J. Verkaar，Population dynamics- the influence of herbivory，New Phytol.，1987，106：49-60.

[93] J.A. Rattenbury，Cyclic hybridization as a survival mechanism in New Zealand forest flora，Evolution，1962，16：348-363.

[94] E. Postma, A.J. van Noordwijk, Gene flow maintains a large genetic difference in clutch size at a small spatial scale, 2005.

[95] H. Darmency, The impact of hybrids between genetically modified crop plants and their related species: introgression and weediness, Mol. Ecol., 1994, 3: 37-40.

[96] D. Pilson, H.R. Prendeville, Ecological effects of transgenic crops and the escape of transgenes into wild populations, Annual Review of Ecology, Evolution, and Systematics, 2004, 35: 149-174.

[97] C. Vacher, A.E. Weis, D. Hermann, T. Kossler, C. Young, M.E. Hochberg, Impact of ecological factors on the initial invasion of Bt transgenes into wild populations of birdseed rape (*Brassica rapa*), Theor. Appl. Genet., 2004, 109: 806-814.

[98] L.G. Campbell, A.A. Snow, C.E. Ridley, Weed evolution after crop gene introgression: greater survival and fecundity of hybrids in a new environment, Ecol. Lett., 2006, 9: 1198-1209.

[99] K.D. Whitney, R.A. Randell, L.H. Rieseberg, Adaptiveintrogression of herbivore resistance traits in the weedy sunflower *Helianthus annuus*, Am. Nat., 2006, 167: 794-807.

[100] D.I. Jarvis, T. Hodgkin, Wild relatives and crop cultivars detecting natural introgression and farmer selection of new genetic combinations in agroecosystems, Mol. Ecol., 1999, 8: S159-S173.

[101] M.L. Stanton, A.A. Snow, S.N. Handel, J. Bereczky, The impact of a flower-color polymorphism on mating patterns in experimental populations of wild radish (*Raphanus raphanistrum*), Evolution, 1989, 43: 335-346.

[102] A.A. Snow, K.L. Uthus, T.M. Culley, Long-term persistence of crop alleles in experimental populations of a common weed, Raphanus raphanistrum, Ecological Society of America Annual Meeting Abstracts, 2001, 86: 208.

[103] T.N. Lee, A.A. Snow, Pollinator preferences and the persistence of crop genes in wild radish populations (*Raphanus raphanistrum*, Brassicaceae), Am. J. Bot., 1998, 85: 333-339.

[104] S. Kercher, J.K. Conner, Patterns of genetic variability within and among populations of wild radish, Raphanus raphanistrum (Brassicaceae), Am. J. Bot., 1996, 83: 1416-1421.

[105] L.G. Campbell, A.A. Snow, P.M. Sweeney, When divergent life histories hybridize: insights into adaptive life-history traits in an annual weed, New Phytol., 2009, 184: 806-818.

[106] K. Lehtilä, K. Holmén Bränn, Correlated effects of selection for flower size in Raphanus raphanistrum, Canadian Journal of Botany, 2007, 85: 160-166.

[107] R.B. Jørgensen, B. Andersen, Spontaneous hybridization between oilseed rape (*Brassica napus*) and weedy *B. campestris* (Brassicaceae), Am. J. Bot., 1994, 81: 1620-1626.

[108] R. Chadoeuf, H. Darmency, J. Maillet, M. Renard, Survival of buried seeds of interspecific hybrids between oilseed rape hoary mustard and wild radish, Field Crops Res., 1998, 58: 197-204.

[109] N. Aparicio, D. Villegas, J.L. Araus, R. Blanco, C. Royo, Seedling development and biomass as affected by seed size and morphology in durum wheat, J. Agric. Sci., 2002, 139: 143-150.

[110] M. Westoby, D.S. Falster, A.T. Moles, P.A. Vesk, I.J. Wright, Plant ecological strategies: Some leading dimensions of variation between species, Annu. Rev. Ecol. Syst., 2002, 33: 125-159.

[111] J.C. Gardner, R.L. Vanderlip, Effect of seed size on developmental traits andability to tolerate drought in pearl millet, Trans. Kans. Acad. Sci., 1989, 92: 49- 59.

[112] M. Verdu，A. Traveset，Early emergence enhances plant fitness：a phylogenetically controlled metaanalysis，Ecology，2005，86：1385-1394.

[113] T. Lenormand，Gene flow and the limits to natural selection，Trends Ecol. Evol.，2002，17：183-189.

[114] D. Ebert，C. Haag，M. Kirkpatrick，M. Riek，J.W. Hottinger，V.I. Pajunen，A selective advantage to immigrant genes in a daphnia metapopulation，Science，2002，295：485-488.

[115] S. Sakai，Y. Harada，Sink-limitation and the size-number trade-off of organs：Production of organs using a fixed amount of reserves，Evolution，2001，55：467-476.

[116] F.W. Allendorf，L.L. Lundquist，Introduction：Population biology，evolution，and control of invasive species，Conserv. Biol.，2003，17：24-30.

[117] J. Whitton，D.E. Wolf，D.M. Arias，A.A. Snow，L.H. Rieseberg，The persistence of cultivar alleles in wild populations of sunflowers five generations after hybridization，Theor. Appl. Genet.，1997，95：33-40.

[118] L.B. Hansen，H.R. Siegismund，R.B. Jørgensen，Introgression between oilseed rape（*Brassica napus* L.）and its weedy relative *B. rapa* L. in a natural population，Genet. Resour. Crop Evol.，2001，48：621－627.

[119] A.A. Snow，K.L. Uthus，T.M. Culley，Fitness of hybrids between weedy and cultivated radish：Implications for weed evolution，Ecol. Appl.，2001，11：934-943.

[120] S.G. Hegde，J.D. Nason，J.M. Clegg，N.C. Ellstrand，The evolution of California's wild radish has resulted in the extinction of its progenitors，Evolution，2006，60：1187-1197.

[121] J. Ross-Ibarra，M. Tenaillon，B.S. Gaut，Historical divergence and gene flow in the genus Zea，Genetics，2009，181：1399-1413.

[122] X.-M. Zheng，S. Ge，Ecological divergence in the presence of gene flow in two closely related Oryza species（*Oryza rufipogon* and *O. nivara*），Mol. Ecol.，2010，19：2439-2454.

[123] T. Tokunaga，O. Ohnishi，Spatial autocorrelation analysis of allozyme variants within local sites of wild radish population，The Japanese Journal of Genetics，1992，67：209-216.

[124] R. Haygood，A.R. Ives，D.A. Andow，Consequences of recurrent gene flow from crops to wild relatives，Proc. R. Soc. B，2003，270：1879-1886.

[125] C.R. Linder，I. Taha，G.J. Seiler，A.A. Snow，L.H. Rieseberg，Long-term introgression of crop genes into wild sunflower populations，Theor. Appl. Genet.，1998，96：339-347.

[126] L.G. Campbell，A.A. Snow，Competition alters life history and increases the relative fecundity of crop-wild radish hybrids（*Raphanus* spp.），New Phytol.，2007，173：648-660.

[127] Y.-l. Gao，H.-q. Feng，K.-m. Wu，Regulation of the seasonal population patterns of *Helicoverpa armigera* moths by Bt cotton planting，Transgenic Res.，2010，19：557-562.

A 90-day Safety Assessment of Genetically Modified Glyphosate-Tolerant Soybean in Japanese Quails

日本鸟对抗草甘膦转基因大豆为期 90 天的安全评估

WANG Chang-Yong LIU Yan

（A Key Laboratory on Biosafety of National Environmental Protection，Nanjing Institute of Environmental Sciences，Ministry of Environmental Protection，Nanjing 210042，PR China）

王长永 刘燕

（国家环境保护生物安全重点实验室，南京环境科学研究所，南京 210042）

Abstract: Public concerns on the risk of genetically modified（GM）crops to the health of mammals including human calls for robust risk assessment. In the present study，we conducted a 90-day feeding trial with GM soybean suspension in Japanese quails. The animals were by gavage fed daily on either the GM soybean（XKD0101）expressing the CP4-EPSPS protein which confers high glyphosate tolerance or the parental non-GM soybean variety（XD9）while a third group of the animals was provided a commercial basal diet containing GM-free soybean. No adverse effect on clinical behavior or body weight was observed. The comparative toxicological examination of dissected tissues and blood samples showed，with few exceptions，no significant histological and biochemical effects during the 90-day exposure among the three groups. No sex or dose effects on the parameters examined were observed. Furthermore，another 14d recovery period did not influence those examined toxicological parameters. The results were consistent with most previous studies，indicating that a dose as high as 2 g kg^{-1} d^{-1} of glyphosate-tolerant soybean would not cause adverse effects on Japanese quails in prolonged exposure（90-d）. Based on a literature survey on the GM crops feeding trails targeted different foreign genes，we found that daily doses might define GM crops exposure and provide a link between exposure and potential risk. To conclude，our present study showed that GM soybean（XKD 0101）is as safe as its unmodified counterpart（XD 9）to Japanese quails.

Keywords: Genetically modified soybean; Feeding trial; Japanese quail; Toxicity; Adverse effects; Daily dose

A 90-day Safety Assessment of Genetically Modified Glyphosate-Tolerant Soybean in Japanese Quails

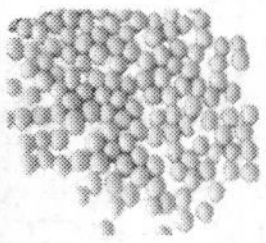

Dr. Wang Changyong and Liu Yan

Nanjing Institute of Environmental Sciences,
Ministry of Environmental Protection of China

1. Introduction

- There are an increasing number of studies that focused on potential health impacts of genetically modified crops and their products.
- Most studies did not find direct evidence of adverse effects related to the genetic modification, But some studies have reported the occurrence of unexpected adverse effects.
- Therefore it is necessary for more study cases to support the general scientific judgment about the safety of GM crops .
- At present study, we have examined the toxicological effects of GM soybean in a 90-day feeding trial of model animal Japanese quails .

2

2. Materials and Methods

2.1 Test materials and soybean suspension

The GM soybean (XKD0101) is a hybrid of Round-up Ready® Soybean (Event GST40-3-2) with local conventional soybean (XD9) and identified as containing the EPSPS gene.

Japanese quails were fed with the suspension of soybean flour by oral gavage. The suspension was prepared in different concentration :

high-dose: 400mg/ml (soybean flour/water)
medium- dose: 200mg/ml
low-dose: 100mg/ml

2.2 Experimental design

- The acclimated quails were randomly assigned to three groups with 10 male and 10 female each.

	High dose	Medium dose	Low dose
GM soybean: GM group	2g/kg/d	1g/kg/d	0.5g/kg/d
Local Conventional soybean: NGM group	2g/kg/d	1g/kg/d	0.5g/kg/d
MiliQ water: Rreference group	Equivalent Water	Equivalent Water	Equivalent Water

- Each group was fed with GM soybean, NGM one and Water, respectively, one time per day at a frequency of six days per week
- All animals had free access to water and the commercial basal diet over the feeding trial.
- Animals were weighed once every week and their exposure doses were adjusted accordingly.
- Their mortality, morbidity and other clinical signs of toxicity were observed twice daily.

- After 90-day exposure, 7 female/male quails were fasted overnight and then sacrificed by cervical dislocation for further analysis. The rest of animals were reared without oral administration of soybean for another 2 weeks and then harvested.

2.3 Haematology and blood biochemistry

Blood samples were collected for determination of the following parameters:

Blood biochemistry	Haematology
Total protein (TP) 总蛋白	Red blood cells (RBC) 红细胞
Spartate aminotransferase (AST) 天门冬氨酸氨基转换酶	White blood cells (WBC) 白细胞
Alanine aminotranferase (ALT) 丙氨酸氨基转换酶	Haemoglobin (HGB) 血红蛋白
Alkaline phosphatase (ALP) 碱性磷酸酶	Lymphocytes (LY) 淋巴细胞
Total bilirubin (TBIL) 总胆红素	Neutrophils (NE) 中性粒细胞
Albumin (ALB) 碱性磷酸酶	Platelets (PLT) 血小板计数
Urea 尿素氮	Clotting time (CT) 凝血时间
Creatinine (Cr) 肌酐	
Total cholesterol (T-CHO) 总胆固醇	
Triglyceride (TG) 甘油三酯	

2.4 Organ weights, gross necropsy and histopathology

All quails were sacrificed after blood collection and dissected for heart, brain, liver, kidney, lung, stomach, large and small intestine, epididymis, ovaries and uterus for a thorough necropsy and were processed for light microscopy.

The relative organ weights (RW) were calculated as

$$RW \leqslant \frac{weight_{(organ,\ g)}}{weight_{(body,\ g)}}$$

3. Results

3.1 Behavior and body weight

- Over the entire experiment (i.e., 13 weeks exposure and 2 weeks recovery), all surviving animals survived.
- There were no observable behavioral or clinical adverse effects in any groups.
- No significant differences in body weights of males or females were observed among the GM group, NGM group and reference group (Fig1).
- Animal growth was neither influenced by daily dose nor their sex. Female and male quail showed similar growth patterns (Fig1).

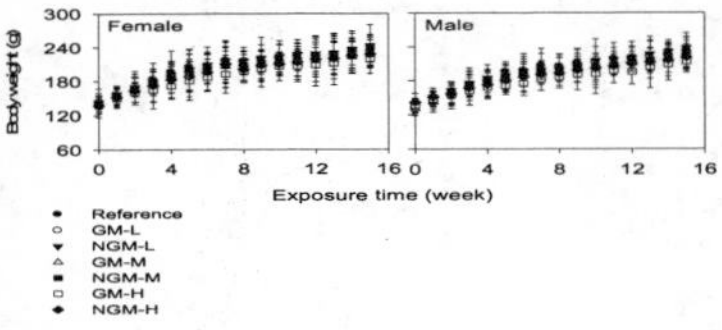

Fig. 1. Body weights of Japanese quails exposed for 13 weeks to genetically modified soybean (GM) and non-genetically modified soybean (NGM). Body weights for another 2 weeks recovery were also demonstrated. Values are means±SD. N=10 for 13 weeks exposure and N=3 for the recovery period.

3.2 Blood haematology

- No treatment-related significant differences were observed for 7 hematology parameters among the tested doses groups from either sex (Table 1).
- Furthermore, haematology was similar between 13 weeks exposure and the extra 2 weeks recovery (Table 1).

Values are means±SD

Parameters	Exposure (weeks)	N	Reference	NGM-Low	GM-Low	NGM-Middle	GM-Middle	NGM-High	GM-High
WBC (10^9/L)	13	14	25.35±3.89	25.53±6.31	23.36±2.74	23.34±5.26	24.60±6.92	24.32±3.57	27.22±6.09
	15	6	23.80±8.84	26.48±1.45	28.33±2.13	30.46±4.47	26.67±5.60	24.05±8.28	28.08±4.32
LY (10^9/L)	13	14	11.87±3.14	10.22±3.82	10.84±2.91	8.87±3.20	13.58±4.29	9.66±3.36	11.91±2.51
	15	6	12.83±5.99	8.58±2.91	13.78±3.85	12.00±1.74	12.32±3.38	9.18±4.23	15.95±4.85
NE (10^9/L)	13	14	12.09±2.81	16.12±4.91	12.61±3.50	14.12±3.54	11.02±3.54	14.67±4.16	15.31±4.84
	15	6	10.97±5.03	17.90±3.49	14.55±4.26	18.47±5.25	14.35±3.91	14.88±4.17	12.13±3.87
RBC (10^{12}/L)	13	14	3.28±0.69	3.34±0.72	3.44±0.47	3.16±0.85	3.57±0.46	3.42±0.45	3.45±0.58
	15	6	2.97±0.68	3.17±0.35	3.23±0.33	3.67±0.60	3.38±0.24	3.25±0.88	3.53±0.25
PLT (10^9/L)	13	14	73.00±28.9	53.92±21.9	57.18±19.9	76.91±29.9	56.33±19.5	65.23±17.2	76.47±20.7
	15	6	79.14±20.9	52.00±7.94	60.50±18.9	77.50±20.3	57.17±11.2	69.00±12.2	81.50±13.0
HGB (g/L)	13	14	123.41±24.3	125.67±24.8	129.62±17.3	118.59±31.7	131±18.05	127.59±15.4	128.78±21.6
	15	6	137.86±28.9	124.44±12.7	122.38±9.11	125.22±17.8	125.15±8.93	117.27±15.2	131.74±7.49
CT (sec)	13	14	33.00±7.94	32.71±7.82	32.29±7.05	32.43±7.96	31.83±4.36	35.00±6.10	34.67±6.22
	15	6	35.80±6.76	32.86±7.76	33.67±4.51	31.25±8.54	35.33±6.43	32.00±9.38	34.67±6.11

3.3 Blood biochemistry

- There were no treatment-associated differences in any of the 10 biochemistry response parameters for males and females consuming GM soybean at any of the dose levels (Table 2).

Values are means ± SD.

Parameters	Exposure (weeks)	N	Reference	NGM-Low	GM-Low	NGM-Middle	GM-Middle	NGM-High	GM-High
TBIL (μmol/L)	13	14	2.49±0.73	1.79±0.45	1.86±0.23	1.95±0.53	2.16±0.52	2.09±0.71	1.83±0.35
	15	6	2.54±0.81	1.98±0.28	2.30±0.79	1.92±0.52	2.13±0.55	2.50±0.68	1.96±0.56
TP (g/L)	13	14	29.42±7.46	29.77±7.45	33.23±6.66	32.67±6.86	31.53±3.49	32.77±4.99	31.70±5.71
	15	6	32.04±6.03	35.35±5.50	35.10±4.16	35.46±2.02	34.29±3.99	33.93±9.67	33.93±3.68
ALB (g/L)	13	14	13.09±3.15	13.26±3.35	14.33±2.59	14.36±2.75	14.08±1.37	15.03±2.06	14.25±2.31
	15	6	15.20±3.04	17.15±2.98	14.63±1.75	15.72±3.19	15.19±2.41	15.93±4.52	13.88±3.18
GLB (g/L)	13	14	16.35±4.35	16.51±4.23	18.93±4.46	18.32±4.34	17.46±2.54	17.75±3.18	17.46±3.50
	15	6	19.35±1.22	18.95±2.66	21.14±2.66	20.92±2.60	19.80±1.80	19.01±3.99	19.73±2.19
A/G	13	14	0.81±0.083	0.81±0.11	0.79±0.14	0.81±0.10	0.81±0.083	0.87±0.11	0.85±0.066
	15	6	0.73±0.12	0.70±0.08	0.70±0.11	0.70±0.07	0.70±0.08	0.69±0.08	0.73±0.14
AST (IU/L)	13	14	383.36±33.6	419.29±74.6	381.55±56.5	350±48.0	320.25±52.4	385.55±86.4	368.46±23.9
	15	6	374.80±18.73	389.00±30.34	356.89±43.45	349.00±37.10	356.71±49.15	381.33±111.35	367.11±60.79

- During 15 weeks, a significant difference of ALT and ALP were noted in the GM group and NGM group for low-dose, but their values were within normal range.

Values are means ± SD.

Parameters	Exposure (weeks)	N	Reference	NGM-Low	GM-Low	NGM-Middle	GM-Middle	NGM-High	GM-High
ALT (IU/L)	13	14	5.82±1.17	6.08±2.16	4.64±0.67	4.63±1.45	4.50±0.93	5.55±1.51	4.92±1.66
	15	6	5.80±1.30	5.80±1.64	4.44±1.51	4.40±0.55	4.43±1.40	5.89±1.96	5.22±1.64
ALP (IU/L)	13	14	251.36±31.7	214.07±69.9	312.45±101.2	253.69±67.8	243.50±73.7	220.27±50.1	253.23±69.1
	15	6	271.00±19.62	235.25±86.91	292.22±26.95	261.60±53.17	300.00±53.42	253.44±40.38	263. 44±43.78
Urea (mmol/L)	13	14	0.86±0.22	0.94±0.32	0.72±0.23	0.78±0.22	0.68±0.24	0.92±0.31	0.97±0.21
	15	6	0.93±0.24	0.75±0.13	0.69±0.24	0.91±0.24	0.66±0.22	0.80±0.23	0.64±0.19
Cr (μmol/L)	13	14	37.34±4.31	36.37±1.99	36.64±4.89	38.02±3.39	41.03±3.86	38.24±4.94	40.88±4.10
	15	6	35.33±4.12	34.70±3.56	33.27±4.94	35.14±3.66	36.13±2.87	34.39±3.37	37.17±3.43
TG (mmol/L)	13	14	0.73±0.17	0.74±0.23	0.81±0.27	0.83±0.27	1.02±0.37	0.80±0.24	0.93±0.31
	15	6	0.94±0.30	0.94±0.23	1.28±0.49	1.16±0.20	0.97±0.36	1.03±0.35	0.87±0.25
CHO (mmol/L)	13	14	5.89±1.36	5.73±1.45	6.32±1.63	5.33±1.22	5.59±0.80	5.75±1.29	6.49±1.81
	15	6	6.07±1.13	5.83±1.27	6.83±1.25	5.93±1.62	5.91±0.70	6.12±1.06	6.52±0.92

3.4 Relative weights of organs

- Relative organ weight (RW) was generally comparable among the three groups across all doses(Table 3) .
- However, sex-dependent RW in liver (p=0.004), lung (p=0.02) and stomach (p<0.0001) was observed at low dose for GM group, with 34-42% increase in female individuals.
- Such an increase in sex-dependent RW was not observed at other doses for GM group or NGM group, and was thus considered unrelated to the GM intake.

Values are means ± SD.

Organs	Exposure (weeks)	N	Reference	NGM-Low	GM-Low	NGM-Middle	GM-Middle	NGM-High	GM-High
heart	13	14	0.69±0.077	0.73±0.10	0.74±0.088	0.72±0.098	0.74±0.16	0.78±0.20	0.83±0.24
	15	6	0.73±0.073	0.81±0.062	0.87±0.13	0.78±0.071	0.86±0.11	0.81±0.066	0.87±0.21
liver	13	14	1.60±0.43	1.70±0.37	1.84±0.39	1.63±0.34	1.69±0.24	1.78±0.26	1.72±0.37
	15	6	1.73±0.53	1.85±0.43	2.25±0.60	1.83±0.21	2.12±0.31	1.89±0.33	1.97±0.52
brain	13	14	0.45±0.069	0.45±0.081	0.52±0.11	0.43±0.024	0.46±0.059	0.45±0.068	0.43±0.057
	15	6	0.42±0.049	0.37±0.062	0.45±0.11	0.38±0.050	0.39±0.073	0.48±0.12	0.41±0.10
lung	13	14	0.71±0.13	0.71±0.14	0.81±0.19	0.70±0.067	0.72±0.079	0.71±0.12	0.74±0.15
	15	6	0.28±0.086	0.73±0.11	0.76±0.080	0.75±0.17	0.81±0.16	0.78±0.13	0.75±0.23
kidney	13	14	0.27±0.091	0.22±0.060	0.25±0.044	0.23±0.060	0.27±0.069	0.31±0.078	0.26±0.048
	15	6	0.71±0.12	0.32±0.097	0.24±0.080	0.25±0.048	0.28±0.085	0.26±0.049	0.24±0.068
stomach	13	14	2.55±0.57	2.77±0.52	2.51±0.52	2.58±0.52	2.67±0.42	2.67±0.42	2.73±0.47
	15	6	2.71±0.61	2.52±0.28	2.86±0.66	2.46±0.61	2.33±0.53	2.91±0.33	2.41±0.53
uterus	13	7	0.52±0.17	063±0.084	0.59±0.058	0.46±0.054	0.55±0.091	0.58±0.059	0.53±0.050
	15	3	0.42±0.097	0.50±0.082	0.41±0.11	0.52±0.053	0.42±0.083	0.50±0.10	0.41±0.017
ovary	13	7	0.21±0.035	0.22±0.045	0.20±0.025	0.21±0.031	0.18±0.026	0.18±0.039	0.18±0.021
	15	3	0.22±0.068	0.19±0.057	0.23±0.050	0.19±0.032	0.25±0.058	0.19±0.027	0.26±0.071
testes	13	7	1.07±0.23	1.01±0.16	1.28±0.31	1.20±0.42	0.84±0.22	1.18±0.43	1.67±0.60
	15	3	1.16±0.28	0.85±0.034	0.98±0.30	1.40±0.15	1.43±0.20	1.35±0.63	1.37±0.29

3.5 Gross necropsy and histopathology

- Neither pathological lesions nor histopathological abnormalities were presented in dissected organs (e.g., heart, brain, liver, kidney, lung, intestine, ovaries and uterus) after 90-d exposure or by the end of the recovery period.
- Light microscopy observations of heart, brain, liver, kidney, lung, intestine, ovaries and uterus did not reveal any dose-related histological differences either after 90-d exposure or by the end of the recovery period.

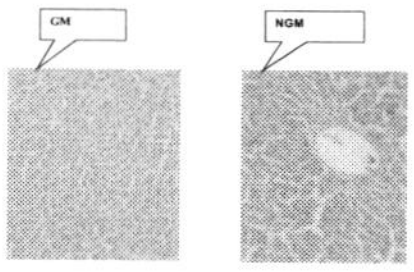

图1 灌胃末期高剂量组肝脏（100×）Liver

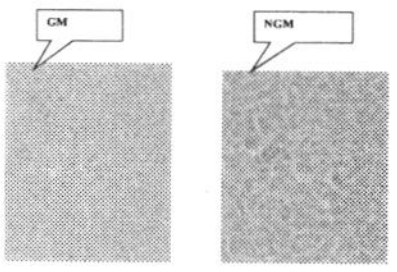

图2 恢复末期高剂量组肝脏（100×）Liver

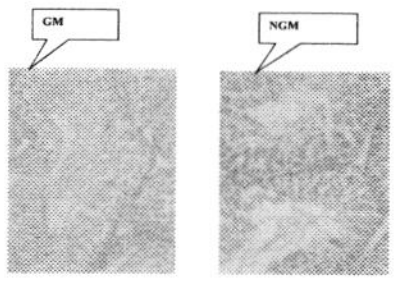

图5 灌胃末期高剂量组睾丸（100×）Testis

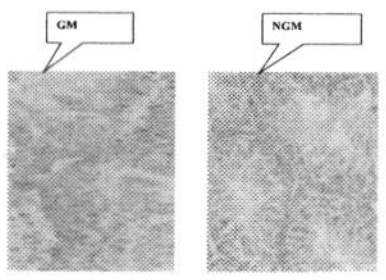

图6 恢复末期高剂量组睾丸（100×）Testis

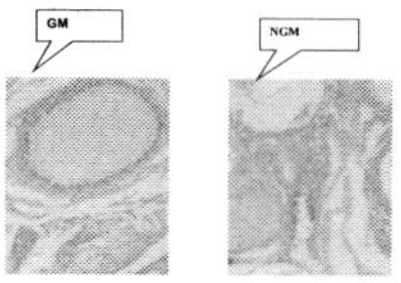

图3 灌胃末期高剂量组卵巢（100×）Ovary

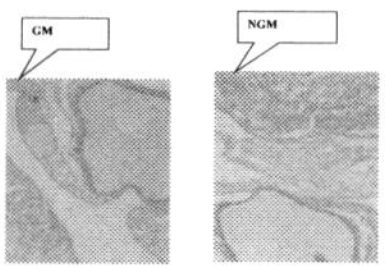

图4 恢复末期高剂量组卵巢（100×）Ovary

4. Conclusion

- **There were no toxicological significances effects in clinical and behavioral signs, organ weights, and gross and microscopic pathology between GM and conventional soybean fed quails after 90 days exposure.**
- **Another 14-d recovery period did not influence the toxicological parameters.**
- **The results are consistent with most previous studies, indicating that a dose as high as 2 g kg^{-1} d^{-1} of glyphosate-tolerant soybean XKD 0101 would not cause adverse effects on Japanese quails in prolonged exposure (e.g., 90-d).**
- **Further studies are warranted to examine the effects of higher doses of GM crops in a more chronic exposure.**

基于故障树模型的转基因作物实验室风险分析研究

Laboratory Risk Analysis and Research of GM Crop Based on FTA Model

杨君[1] 王国豫[2]

（1 大连理工大学生命科学与技术学院；2 大连理工大学哲学系，大连 116024）

YANG Jun[1] WANG Gou-Yu[2]

（1. School of Life Science and Biotechnology Da Lian University of Technology；
2.School of Marxism Studies Dalian University of Technology）

摘 要：转基因作物的产生经历了科学研究与环境释放两个基本阶段，在转基因作物获准进入商业化环境释放之前的科学研究阶段，需要经历实验（基因操作）、温室栽培、生产性小试和中试等过程。由于生物技术本身的不确定性，加之在科学研究阶段，伴随实验作物的生长，不可避免地涉及环境释放问题，因此在转基因作物的科学研究阶段就已经可能产生潜在的生物风险，并有可能带来对环境和健康的危害后果。本文尝试利用故障树模型对我国转基因作物在研究阶段的风险因素予以分析，构造了一个含有 17 个中间事件和 30 个底事件的转基因作物实验室生物风险故障树，通过布尔代数化简法求其最小割集，共得到 106 组最小割集，根据不同的基本事件在 106 组割集中出现的频率大小，初步定性地获得各底事件在转基因实验室生物风险事故发生中的结构重要度 $I_\varphi(n)$，并分析了影响转基因作物实验室生物风险故障的主要原因。

关键词：转基因作物；实验室；风险因素；故障树

1 引言

转基因作物的产生经历了科学研究与环境释放两个基本阶段，在转基因作物获准进入商业化环境释放之前的科学研究阶段，需要经历实验（基因操作）、温室栽培、生产性小试和中试等过程。由于生物技术本身的不确定性，加之在科学研究阶段，伴随实验作物的生长，不可避免地涉及环境释放问题，因此在转基因作物的科学研究阶段就已经可能产生潜在的生物风险，并有可能带来对环境和健康的危害后果。近年来，以病毒感染为代表的生物实验室获得性感染的例子时有发生，使得病原微生物实验室的生物安全问题受到重视[1,2]，同时也反映出生物实验室风险是生物安全研究中不容忽视的内容。

早在 2004 年 4 月，我国就发布了《实验室生物安全通用要求》（GB 19489—2004）国家标准，这一标准是在 2003 年 5 月非典流行期间提出并制定的，对保障实验室生物安全和依法管理病原微生物实验室发挥了重大作用。2009 年 7 月 1 日，在此基础上修订的《实验室生物安全通用要求》（GB 19489—2008）正式实施，进一步规范了生物学实验室（尤其是病原微生物学实验室）的基本生物安全管理要求，但其中对于转基因生物实验室的生物安全问题并未涉及。

由于我国农业转基因安全管理采取的是前审后批的原则，根据《农业转基因生物安全管理条例》（第六条）和《农业转基因生物安全评价管理办法》（第十一条）的规定，审批前的安全管理应由从事农业转基因生物研究与试验的单位负责。目前在我国开展转基因作物相关研究工作的单位涉及面广，分布广泛，研究的转基因作物对象多样，研究水平和条件能力参差不齐[3]。随着国家转基因专项的逐步深入，这种局面将进一步复杂化。一方面，转基因作物研究方兴未艾，而另一方面，关于转基因作物实验研究阶段中的风险问题却并没有得到足够的重视，相关研究还很缺乏。由于国家在对于转基因作物的风险调控中，研究阶段的风险控制是处于“自检自查”的水平，相关研究机构的监管缺位，将使转基因作物研究因缺乏有效地监管和约束在研究之初就可能产生一系列的生物风险。

故障树模型是一种灵活和系统化的分析工具，在复杂系统的故障诊断中具有重要价值[4]。作为一种“自上而下（Top-down）”的风险分析工具，故障树利用图形化模型路径，通过定义一个系统可能导致的一个可预知的，或不可预知的顶事件（不希望发生的事件，故障或失效），然后确定可能导致该项事件的全部假设，如系统内可能发生的部件失效、环境变化、人为失误等因素（各种底事件），以及系统失效之间的逻辑联系（发生路径），并利用标准逻辑符号“或（OR）”和“与（AND）表示各部分之间的依赖性和可靠性参数。因此，一个故障树模型，就是一个表现为图形模式的所有可能导致顶层事件发生的平行和连续事件的组合。

故障树模型广泛用于工业生产系统、矿山、交通、航天等复杂控制系统中的危险性分析、评价与控制[5-7]；也被用于生物入侵[8]和转基因食品的健康风险分析[9]。在本文中，我们尝试利用故障树模型对我国转基因作物在研究阶段的风险因素予以分析，构造了一个含有 17 个中间事件和 30 个底事件的转基因作物实验室生物风险故障树，通过布尔代数化简法求其最小割集，共得到 106 组最小割集，根据不同的基本事件在 106 组割集中出现的频率大小，初步定性地获得各底事件在转基因实验室生物风险事故发生中的结构重要度 $I_{\varphi}(n)$，并分析了影响转基因作物实验室生物风险故障的主要原因。

2 转基因作物实验室的风险因素分析

转基因生物实验室的生物风险既包含生物学实验室的一般风险，又因其研究对象和研究方法的特殊性，而存在其特有的风险。总体来说，转基因生物实验室的潜在风险源包括以下四个方面：生物学实验材料与对象，生物学实验操作，生物学实验室废弃物，生产性实验。

2.1 生物材料、试剂风险

在转基因作物研究中需要涉及使用和配置的实验材料与试剂包括：

分子生物学和生物化学操作所必需的生物活性材料，包括组织、细胞和微生物菌种、质粒、载体以及病毒等；

各种化学试剂，其中包括有毒、有腐蚀性的生化试剂，如氯仿等有机溶剂、溴化乙锭（EB）、丙烯酰胺及其结合物、各种酸碱溶液、染料、抗生素、细胞培养基（液）、洗脱液、抗体、放射性同位素等。

转基因作物的试验对象主要包括如烟草、拟南芥等模式植物，以及玉米、大豆、棉花等经济作物。

2.2 实验操作风险

转基因生物实验中涉及的实验操作风险来自研究人员的操作失误和器材设备风险两个方面。

实验器械与耗材主要包括塑料制品，如各种吸头、吸管、离心管、注射器、手套、培养皿等一次性耗材；玻璃制品，如各种培养皿、试管、吸管、玻片、盖片、常用容器、过滤器皿等易损易碎材料；金属物品，如注射针头、刀片等器具；以及常用设备，如离心机、水浴锅、烘箱、灭菌锅、超净台、电转仪、电泳仪、扩增仪等，及其所涉及用电、用水及设备使用风险。

2.3 废弃物风险

由转基因生物实验产生的废弃物主要包括：

生物活性材料类，如组织、细胞和微生物（细菌、真菌和病毒等）及其培养物（如含有筛选药物、抗生素、有毒代谢物、外源基因残留物等）；

实验对象，如转基因植物植株、花粉、果实、种子等；

生化试剂类，如有毒物品及其他实验废弃物，如重金属、氰化物、溴化乙锭（EB）、丙烯酰胺、甲酰胺及其结合物、酸碱溶液、有机溶剂、染料、抗生素、同位素、凝胶电泳、培养基（液）、洗脱液等，以及转基因操作中的残液、缓冲液等。

实验耗材类，如各种吸头、吸管、离心管、注射器、手套、培养皿等塑料用品；各种培养皿、试管、吸管、载玻片、盖玻片、常用容器、过滤器皿玻璃制品；注射针头及刀片等金属物品等。

2.4 生产性实验意外释放风险

在转基因植物研究过程中，处于目的基因克隆与功能研究、表达载体建立与功能鉴定、突变体构建等目标所进行小规模种植，及由此而产生的潜在释放风险，如实验室盆栽、温室种植，或田间小试过程中可能发生的基因逃逸、水平基因转移、种子散落或遗失等。

3 故障树模型（Fault Tree Analysis，FTA）的建立与描述

利用故障树分析原理，对转基因作物实验室生物风险的影响因素进行分析，分别以农业转基因实验室风险为顶事件（T），推导发生失效事件的基本影响事件（x），再根据事件间的逻辑关系，构造出故障树，从而对其进行安全因素重要度分析。转基因作物实验室生物风险故障树的构造如图 1-4 所示，共包含 30 个底事件。故障树的符号说明见表 1。

表 1　故障树的事件与逻辑门符号

Table1 Event and gate symbols used in the fault tree

逻辑门或事件名称	符号	说明
与门	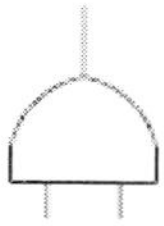	AND
或门		OR
顶事件	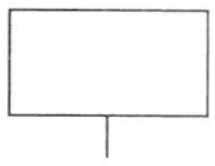	T
中间事件	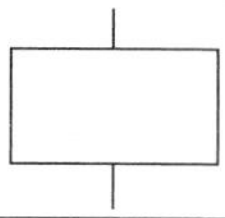	$T_n, n = 1, 2, 3, \cdots, n$
底事件		$x_n, n = 1, 2, 3, \cdots, n$

如图 1 所示，转基因作物实验室可能的生物安全故障发生的必要原因在于存在一系列的转基因生物风险，同时监管层对可能出现的实验室风险未监管或监管不力。在国家标准《实验室生物安全通用要求》（GB 19489—2008）中明确指出，对现有实验室生物安全事故分析表明，超过 90%的生物安全事故是管理问题导致的。因此，建立系统的实验室生物安全管理体系是保证实验室安全不可缺少的手段。从转基因作物实验室风险的角度，主要考虑三个方面，包括：①转基因生物实验室的废弃物处理不当，及由此可能造成污染。②转基因操作过程中的生物材料、试剂或废弃物可能对人畜造成感染，如转基因操作过程中，操作人员可能接触危险试剂、仪器、材料等，因防护不足而带来的感染；转基因生物实验废弃物的随意排放，可能使一些人员，包括清洁人员、拾荒人员，以及以人类生活垃圾为食的野生动等，在未知危害的情况下与未经处理的活性生物材料被动接触。③在基因操作或者田间试验过程中可能发生的基因逃逸。

图 2 中分析了导致转基因实验室废弃物污染的可能底事件。这些可能的因素主要包括，转基因操作中涉及的生物活性材料（如微生物菌体、病毒、植物材料），以及可能残留有生物活性材料的一次性耗材（如塑料吸头、针头、离心管、培养瓶、手套等），由于未经灭活，或者随意倾倒而进入水体，或者混入土壤，或者混入生活垃圾中。比如，目前很多高校实验室的下水道是与居民的下水道或者城市的排污管道相通的，此类污染物通过下水道后极易形成交叉污染，最后流入河中或者渗入地下，长期积累可能造成不可估量的危害。再如，我国目前研究机构的实验垃圾并没有实行分类处理，一般直接混入生活垃圾中，同样可能因为活性生物材料的残留而带来交叉污染。并且，由于部分实验废弃物是塑料、玻璃或者金属材质，也是部分拾荒者收集的对象，并因此可能带来被动的接触和感染。

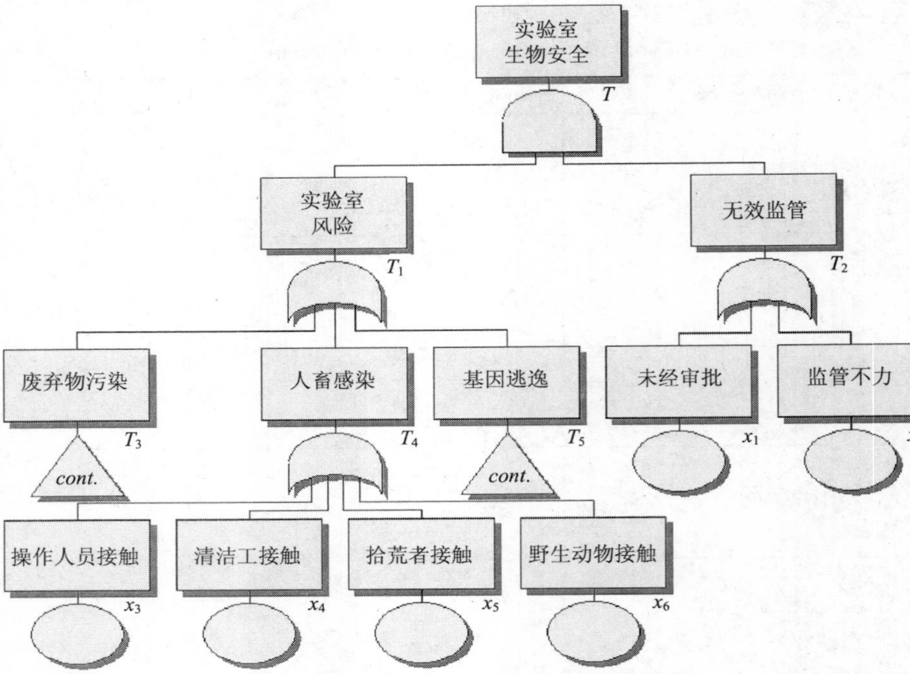

图 1　转基因作物实验室生物风险故障树：总图

Fig.1　FAT for bio-risk of GM crop laboratory：general layout

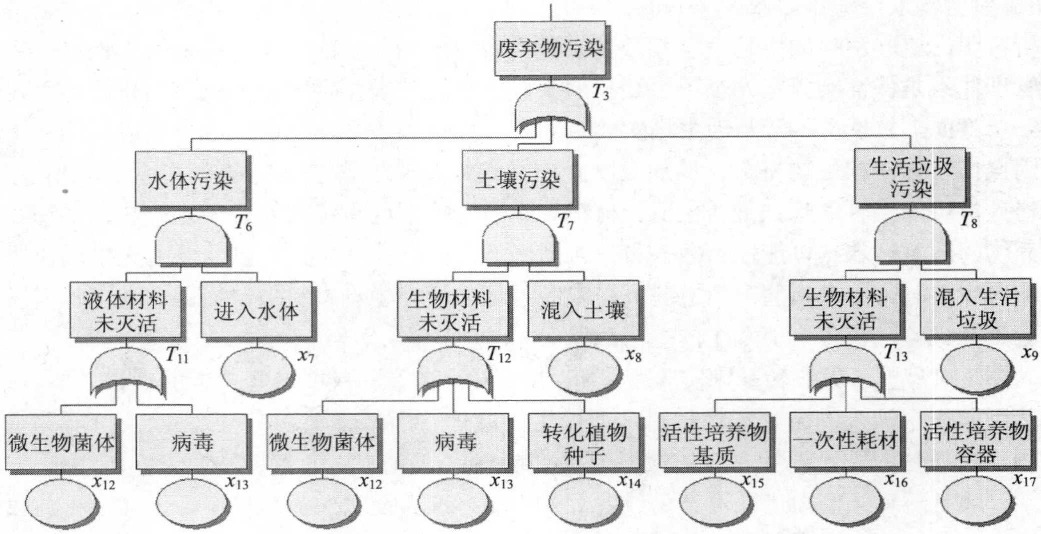

图 2　转基因作物实验室生物风险故障树：续图——废弃物风险

Fig.2　FAT for bio-risk of GM crop laboratory：waste risks

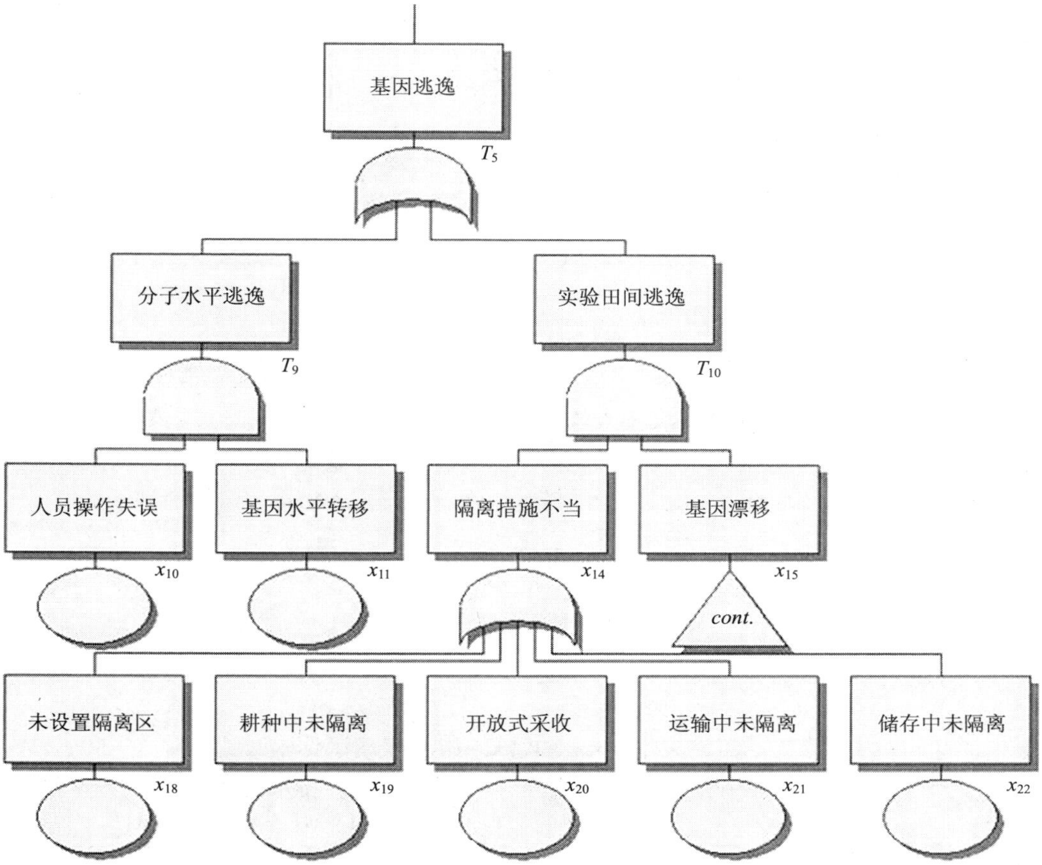

图 3　转基因作物实验室生物风险故障树：续图——基因逃逸

Fig.3　FAT for bio-risk of GM crop laboratory：gene escape

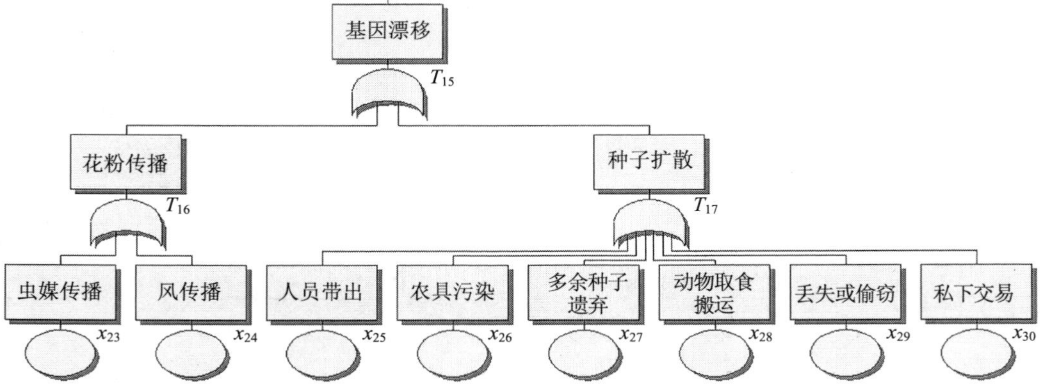

图 4　转基因作物实验室生物风险故障树：续图——基因漂移

Fig.4　FAT for bio-risk of GM crop laboratory：gene flow

图 3 中分析了引起转基因实验室基因逃逸的可能底事件。这些可能的因素主要包括，在实验室分子操作过程中，因操作人员失误而造成的菌种或基因混杂等，以及由此可能造成的水平基因转移等；或者在温室或试验田间实验过程中，由于未采取隔离措施，或隔离措施不当，而造成的基因漂移。隔离措施是农业转基因生物安全管理中的重要因素，隔离设置应在不同水平考虑，如在温室或试验田间设置隔离缓冲区、庇护所、隔离带以及在耕种期内实施技术隔离，如花期去雄、去花、套袋、花期不遇；或在采收、转运和加工过程中实施的隔离，如种子净化、分级分离、专用车辆、独立生产线等。应避免常规农业生产中常见的，如种植密度大、品种混种、操作人员不着工作服、种子的露天晾晒、农机具混用、植株采收后随意堆放等隔离不当的行为。

图 4 中分析了转基因实验室因隔离措施不当而引起基因漂移的可能底事件。这些可能的因素主要包括因虫媒或风媒造成的转基因作物花粉传播；或者由于人员通过衣物、鞋等夹带、或农机具携带，或多余种子不当遗弃，或由于种植区内的动物取食搬运，或由于丢失或偷窃，或私下交易而使处于研究阶段的未经审批的转基因作物种子流入市场等造成的种子扩散。

4 结果与讨论

4.1 故障树最小割集求解

图 1～图 4 所示转基因实验室生物安全故障树中，表明了影响顶事件 T 的 30 个基本事件（表 2）的相互逻辑关系。根据故障树的分析方法，通过布尔代数化简法求其最小割集，定性地获得基本事件对顶上事件的影响程度，为转基因实验室的安全管理提供参考。

表 2 可能影响顶事件的 30 个基本事件

Table 2 The 30 basic initiating events that may influence the top event

底事件	说明
x_1	转基因作物研究工作未备案，无监管
x_2	监管不力
x_3	操作人员接触
x_4	保洁人员接触
x_5	拾荒人员接触
x_6	野生生物接触
x_7	转基因操作废弃物进入水体
x_8	转基因操作废弃物混入土壤
x_9	转基因操作废弃物混入生活垃圾
x_{10}	分子操作中失误
x_{11}	水平基因转移
x_{12}	废弃物中含有微生物菌体活体
x_{13}	废弃物中含有病毒载体活体
x_{14}	废弃物中含有转化植物种子
x_{15}	废弃物中含有活性生物材料基质（如培养基、培养液、凝胶等）

底事件	说明
x_{16}	废弃物中含有被活性生物材料污染的一次性耗材（如针头、枪头、离心管、手套等）
x_{17}	废弃物中含有培养活性材料的容器（如培养皿、三角瓶、试管等）
x_{18}	试验田未设置隔离区
x_{19}	耕种中未采取隔离措施
x_{20}	采收中未采取隔离措施
x_{21}	运输中未采取隔离措施
x_{22}	保存中未采取隔离措施
x_{23}	转基因作物花粉虫媒传播
x_{24}	转基因作物花粉风传播
x_{25}	转基因作物种子通过工作人员（如衣物或鞋子）意外带出试验田
x_{26}	转基因作物种子通过农机用具被意外带出试验田
x_{27}	多余种子随意遗弃
x_{28}	转基因作物种子被田间动物（昆虫、蚂蚁、鼠等）取食搬运
x_{29}	转基因作物种子丢失或者被偷窃
x_{30}	转基因作物种子私下交易

利用布尔代数化简法求得事故树最小割集，过程如下：

$$
\begin{aligned}
T &= T_1T_2 \\
&= (T_3+T_4+T_5)(x_1+x_2) \\
&= (T_6+T_7+T_8+x_3+x_4+x_5+x_6+T_9+T_{10})(x_1+x_2) \\
&= (x_7T_{11}+x_8T_{12}+x_9T_{13}+x_3+x_4+x_5+x_6+x_{10}x_{11}+T_{14}T_{15})(x_1+x_2) \\
&= [x_7(x_{12}+x_{13})+x_8(x_{12}+x_{13}+x_{14})+x_9(x_{15}+x_{16}+x_{17})+x_3+x_4+ \\
&\quad x_5+x_6+x_{10}x_{11}+(x_{18}+x_{19}+x_{20}+x_{21}+x_{22})(T_{16}+T_{17})](x_1+x_2) \\
&= [x_7x_{12}+x_7x_{13}+x_8x_{12}+x_8x_{13}+x_8x_{14}+x_9x_{15}+x_9x_{16}+x_9x_{17}+ \\
&\quad x_3+x_4+x_5+x_6+x_{10}x_{11}+(x_{18}+x_{19}+x_{20}+x_{21}+x_{22})(x_{23}+x_{24}+ \\
&\quad x_{25}+x_{26}+x_{27}+x_{28}+x_{29}+x_{30})](x_1+x_2) \\
&= (x_7x_{12}+x_7x_{13}+x_8x_{12}+x_8x_{13}+x_8x_{14}+x_9x_{15}+x_9x_{16}+x_9x_{17})+ \\
&\quad x_3+x_4+x_5+x_6+x_{10}x_{11}+x_{18}x_{23}+x_{18}x_{24}+x_{18}x_{25}+x_{18}x_{26}+x_{18}x_{27}+ \\
&\quad x_{18}x_{28}+x_{18}x_{29}+x_{18}x_{30}+x_{19}x_{23}+x_{19}x_{24}+x_{19}x_{25}+x_{19}x_{26}+x_{19}x_{27}+ \\
&\quad x_{19}x_{28}+x_{19}x_{29}+x_{19}x_{30}+x_{20}x_{23}+x_{20}x_{24}+x_{20}x_{25}+x_{20}x_{26}+x_{20}x_{27}+ \\
&\quad x_{20}x_{28}+x_{20}x_{29}+x_{20}x_{30}+x_{21}x_{23}+x_{21}x_{24}+x_{21}x_{25}+x_{21}x_{26}+x_{21}x_{27}+ \\
&\quad x_{21}x_{28}+x_{21}x_{29}+x_{21}x_{30}+x_{22}x_{23}+x_{22}x_{24}+x_{22}x_{25}+x_{22}x_{26}+x_{22}x_{27}+ \\
&\quad x_{22}x_{28}+x_{22}x_{29}+x_{22}x_{30})(x_1+x_2)
\end{aligned}
$$

将该式完全展开后，可以得到 106 组最小割集。最小割集代表了顶事件发生的路径数量，每一组割集由不同的基本事件（底事件）组成。不同的基本事件在 106 组割集中出现的频率大小反映了各底事件在转基因实验室生物风险事故发生中的影响程度，即结构重要度，定义为 $I_{\varphi}(n), n=1,2,3,\cdots,30$。

根据经典的故障树分析原理，我们确定利用最小割集排列结构重要度的基本原则如下：

（1）当最小割集中的底事件数目不等时，少事件割集的基本事件重要度大；

（2）当最小割集中的底事件数目相同时，出现频次高的基本事件重要度大；

（3）少事件割集内出现频次少的基本事件，与多事件割集内出现频次高的基本事件相比，少事件割集的低频次基本事件的重要度一般大于或等于多事件割集内的高频次基本事件。

根据这一原则，分析本文构造的故障树的结构重要度为：

$$
\begin{aligned}
&I_{\varphi}(1)=I_{\varphi}(2)>I_{\varphi}(3)=I_{\varphi}(4)=I_{\varphi}(5)=I_{\varphi}(6)\geqslant I_{\varphi}(18)=I_{\varphi}(19)=\\
&I_{\varphi}(20)=I_{\varphi}(21)=I_{\varphi}(22)>I_{\varphi}(23)=I_{\varphi}(24)=I_{\varphi}(25)=I_{\varphi}(26)=\\
&I_{\varphi}(27)=I_{\varphi}(28)=I_{\varphi}(29)=I_{\varphi}(30)>I_{\varphi}(8)=I_{\varphi}(9)=I_{\varphi}(7)>I_{\varphi}(12)=\\
&I_{\varphi}(13)>I_{\varphi}(10)=I_{\varphi}(11)=I_{\varphi}(14)=I_{\varphi}(15)=I_{\varphi}(16)=I_{\varphi}(17)
\end{aligned}
$$

4.2 转基因实验室生物风险事故原因分析

综合上述 106 组事件组合，转基因作物实验室生物安全故障发生的原因有如下方面：

（1）对于转基因作物实验室安全问题无监管，或监管不足，是转基因作物实验室生物安全故障发生的最主要原因；

（2）由于转基因作物实验操作过程中，或由此产生的产品或废弃物等对研究人员、非研究人员和野生生物所造成的直接感染、污染或危害，是转基因作物实验室生物安全故障发生的重要原因；

（3）由于隔离措施不当，或者未设置隔离区，并可能经由转基因作物的花粉传播或种子扩散而造成的基因逃逸，是转基因作物实验室生物安全故障发生的主要原因；

（4）由于转基因生物实验室废弃物处置不当，而可能造成的对地下水、土壤等的污染，是转基因作物实验室生物安全故障发生的较大原因；

（5）由于基因操作过程中的人员失误、由于菌种或基因混杂造成的水平基因转移、残留活性生物材料的实验废弃物（如塑料吸头、针头、离心管、培养瓶、手套等）混入生活垃圾并由此带来交叉污染等，也会引起转基因作物实验室生物安全故障。

4.3 故障树最小径集求解

将图 1～图 4 中构建的故障树中的与门变成或门，或门变成与门，则故障树即变成转基因作物实验室生物风险控制系统的成功树。通过对成功树求解获得最小径集，可以获得保证转基因作物实验室生物安全的管理预防措施。

成功树最小径集的求解过程如下：

$$
\begin{aligned}
T' &= T_1' + T_2' \\
&= T_3'T_4'T_5' + x_1'x_2' \\
&= T_6'T_7'T_8'x_3'x_4'x_5'x_6'T_9'T_{10}' + x_1'x_2' \\
&= (T_{11}' + x_7')(T_{12}' + x_8')(T_{13}' + x_9')x_3'x_4'x_5'x_6' \\
&\quad (x_{10}' + x_{11}')(T_{14}' + T_{15}') + x_1'x_2' \\
&= (x_{12}'x_{13}' + x_7')(x_{12}'x_{13}'x_{14}' + x_8')(x_{15}'x_{16}'x_{17}' + x_9') \\
&\quad x_3'x_4'x_5'x_6'(x_{10}' + x_{11}')(x_{18}'x_{19}'x_{20}'x_{21}'x_{22}' + T_{16}'T_{17}') + x_1'x_2' \\
&= (x_{12}'x_{13}' + x_7')(x_{12}'x_{13}'x_{14}' + x_8')(x_{15}'x_{16}'x_{17}' + x_9')x_3'x_4'x_5'x_6' \\
&\quad (x_{10}' + x_{11}')(x_{18}'x_{19}'x_{20}'x_{21}'x_{22}' + x_{23}'x_{24}'x_{25}'x_{26}'x_{27}'x_{28}'x_{29}'x_{30}') + x_1'x'
\end{aligned}
$$

将上式完全展开后，可以获得系统成功树的最小径集有 33 组，表示为：$P_n, n = 1, 2, 3, \cdots, 33$。

$P_1 = \{x_1, x_2\}$

$P_2 = \{x_3, x_4, x_5, x_6, x_7, x_8, x_9, x_{10}, x_{18}, x_{19}, x_{20}, x_{21}, x_{22}\}$

$P_3 = \{x_3, x_4, x_5, x_6, x_7, x_8, x_9, x_{11}, x_{18}, x_{19}, x_{20}, x_{21}, x_{22}\}$

$P_4 = \{x_3, x_4, x_5, x_6, x_9, x_{10}, x_{12}, x_{13}, x_{14}, x_{18}, x_{19}, x_{20}, x_{21}, x_{22}\}$

$P_5 = \{x_3, x_4, x_5, x_6, x_9, x_{11}, x_{12}, x_{13}, x_{14}, x_{18}, x_{19}, x_{20}, x_{21}, x_{22}\}$

$P_6 = \{x_3, x_4, x_5, x_6, x_8, x_9, x_{10}, x_{12}, x_{13}, x_{18}, x_{19}, x_{20}, x_{21}, x_{22}\}$

$P_7 = \{x_3, x_4, x_5, x_6, x_8, x_9, x_{11}, x_{12}, x_{13}, x_{18}, x_{19}, x_{20}, x_{21}, x_{22}\}$

$P_8 = \{x_3, x_4, x_5, x_6, x_7, x_9, x_{10}, x_{12}, x_{13}, x_{14}, x_{18}, x_{19}, x_{20}, x_{21}, x_{22}\}$

$P_9 = \{x_3, x_4, x_5, x_6, x_7, x_9, x_{11}, x_{12}, x_{13}, x_{14}, x_{18}, x_{19}, x_{20}, x_{21}, x_{22}\}$

$P_{10} = \{x_3, x_4, x_5, x_6, x_7, x_8, x_{10}, x_{15}, x_{16}, x_{17}, x_{18}, x_{19}, x_{20}, x_{21}, x_{22}\}$

$P_{11} = \{x_3, x_4, x_5, x_6, x_7, x_8, x_{11}, x_{15}, x_{16}, x_{17}, x_{18}, x_{19}, x_{20}, x_{21}, x_{22}\}$

$P_{12} = \{x_3, x_4, x_5, x_6, x_{10}, x_{12}, x_{13}, x_{14}, x_{15}, x_{16}, x_{17}, x_{18}, x_{19}, x_{20}, x_{21}, x_{22}\}$

$P_{13} = \{x_3, x_4, x_5, x_6, x_8, x_{10}, x_{12}, x_{13}, x_{15}, x_{16}, x_{17}, x_{18}, x_{19}, x_{20}, x_{21}, x_{22}\}$

$P_{14} = \{x_3, x_4, x_5, x_6, x_8, x_{11}, x_{12}, x_{13}, x_{15}, x_{16}, x_{17}, x_{18}, x_{19}, x_{20}, x_{21}, x_{22}\}$

$P_{15} = \{x_3, x_4, x_5, x_6, x_{11}, x_{12}, x_{13}, x_{14}, x_{15}, x_{16}, x_{17}, x_{18}, x_{19}, x_{20}, x_{21}, x_{22}\}$

$P_{16} = \{x_3, x_4, x_5, x_6, x_7, x_8, x_9, x_{10}, x_{23}, x_{24}, x_{25}, x_{26}, x_{27}, x_{28}, x_{29}, x_{30}\}$

$P_{17} = \{x_3, x_4, x_5, x_6, x_7, x_8, x_9, x_{11}, x_{23}, x_{24}, x_{25}, x_{26}, x_{27}, x_{28}, x_{29}, x_{30}\}$

$P_{18} = \{x_3, x_4, x_5, x_6, x_7, x_{10}, x_{12}, x_{13}, x_{14}, x_{15}, x_{16}, x_{17}, x_{18}, x_{19}, x_{20}, x_{21}, x_{22}\}$

$$P_{19}=\{x_3,x_4,x_5,x_6,x_7,x_{11},x_{12},x_{13},x_{14},x_{15},x_{16},x_{17},x_{18},x_{19},x_{20},x_{21},x_{22}\}$$
$$P_{20}=\{x_3,x_4,x_5,x_6,x_9,x_{10},x_{12},x_{13},x_{14},x_{23},x_{24},x_{25},x_{26},x_{27},x_{28},x_{29},x_{30}\}$$
$$P_{21}=\{x_3,x_4,x_5,x_6,x_9,x_{11},x_{12},x_{13},x_{14},x_{23},x_{24},x_{25},x_{26},x_{27},x_{28},x_{29},x_{30}\}$$
$$P_{22}=\{x_3,x_4,x_5,x_6,x_8,x_9,x_{10},x_{12},x_{13},x_{23},x_{24},x_{25},x_{26},x_{27},x_{28},x_{29},x_{30}\}$$
$$P_{23}=\{x_3,x_4,x_5,x_6,x_8,x_9,x_{11},x_{12},x_{13},x_{23},x_{24},x_{25},x_{26},x_{27},x_{28},x_{29},x_{30}\}$$
$$P_{24}=\{x_3,x_4,x_5,x_6,x_7,x_9,x_{10},x_{12},x_{13},x_{14},x_{23},x_{24},x_{25},x_{26},x_{27},x_{28},x_{29},x_{30}\}$$
$$P_{25}=\{x_3,x_4,x_5,x_6,x_7,x_8,x_{10},x_{15},x_{16},x_{17},x_{23},x_{24},x_{25},x_{26},x_{27},x_{28},x_{29},x_{30}\}$$
$$P_{26}=\{x_3,x_4,x_5,x_6,x_7,x_9,x_{11},x_{12},x_{13},x_{14},x_{23},x_{24},x_{25},x_{26},x_{27},x_{28},x_{29},x_{30}\}$$
$$P_{27}=\{x_3,x_4,x_5,x_6,x_7,x_8,x_{11},x_{15},x_{16},x_{17},x_{23},x_{24},x_{25},x_{26},x_{27},x_{28},x_{29},x_{30}\}$$
$$P_{28}=\{x_3,x_4,x_5,x_6,x_{10},x_{12},x_{13},x_{14},x_{15},x_{16},x_{17},x_{23},x_{24},x_{25},x_{26},x_{27},x_{28},x_{29},x_{30}\}$$
$$P_{29}=\{x_3,x_4,x_5,x_6,x_8,x_{10},x_{12},x_{13},x_{15},x_{16},x_{17},x_{23},x_{24},x_{25},x_{26},x_{27},x_{28},x_{29},x_{30}\}$$
$$P_{30}=\{x_3,x_4,x_5,x_6,x_8,x_{11},x_{12},x_{13},x_{15},x_{16},x_{17},x_{23},x_{24},x_{25},x_{26},x_{27},x_{28},x_{29},x_{30}\}$$
$$P_{31}=\{x_3,x_4,x_5,x_6,x_{11},x_{12},x_{13},x_{14},x_{15},x_{16},x_{17},x_{23},x_{24},x_{25},x_{26},x_{27},x_{28},x_{29},x_{30}\}$$
$$P_{32}=\{x_3,x_4,x_5,x_6,x_7,x_{10},x_{12},x_{13},x_{14},x_{15},x_{16},x_{17},x_{23},x_{24},x_{25},x_{26},x_{27},x_{28},x_{29},x_{30}\}$$
$$P_{33}=\{x_3,x_4,x_5,x_6,x_7,x_{11},x_{12},x_{13},x_{14},x_{15},x_{16},x_{17},x_{23},x_{24},x_{25},x_{26},x_{27},x_{28},x_{29},x_{30}\}$$

4.4 转基因实验室生物安全管理措施分析

从上述 33 个最小径集，可以分析得出以下转基因实验室生物安全管理的思路和措施：

（1）P_1 成功路径中基本事件 x_1 和 x_2 说明，在转基因作物实验室研究之初即介入管理和风险控制，国家、地方和研究机构的生物安全管理部门也应加大对转基因作物实验室生物安全问题的重视和研究，增加经费投入，加强法规法制建设和人员队伍的培训与储备，强化全社会范围内的风险交流和教育，在全过程实施有效地监管，对转基因作物实验室研究过程实时监控，及时反馈和调控，可以做到对转基因生物实验室安全事故的及时发现、及时处理。高校、农科院、研究所以及农业生物技术公司的研发中心等转基因作物研究机构必须担负生物安全管理的责任，成为转基因作物实验室生物安全调控的首道防线，是从源头上控制转基因作物生物风险因素。

（2）从 P_2～P_{10} 路径的基本事件 x_3～x_{17} 表明，在转基因作物实验操作过程中，应加强安全教育和风险交流，对实验废弃物应专门处理，如专门地点存放、专人管理、严格分区、分类、对危害性高的实验废弃物的存放和处理应记录存档；废弃物集中处理站从设计上就要考虑防止集中存放后的二次污染，应适当考虑增设安全监控和排风系统。含有活性生物材料（如菌种、病毒、试验中的转基因植物种子等）的废弃物应经过严格的灭活处理，且废弃物和高风险试剂、材料等应张贴安全警示标志，避免转基因作物实验过程中及其产品和废弃物对直接接触的操作人员、可能被动接触的非操作人员和野生生物的感染以及由于直接倾倒或者随意丢弃等造成对环境水体、土壤等的污染、或由于不当操作而造成的菌种和基因的交叉污染。

（3）从成功路径 P_{11}～P_{15}，以及 P_{18}，P_{19} 中的基本事件 x_{18}，x_{19}，x_{20}，x_{21}，x_{22} 可以得出，在转基因作物实验田间或温室进行生产性实验的全过程必须设置严格合理的隔离措施，如隔离区标志、隔离带、缓冲带、庇护所、清洗区、灭活处置区、隔离存储区等；了解转基因作物的生长发育性质；了解其花期、花粉传播和种子萌发特性；了解实验田区域内的生物多样性特征等，并据此制定适宜的隔离措施。

（4）从成功路径 P_{16}，P_{17}，P_{20}～P_{33} 中的基本事件 x_{23}～x_{28} 可以得出，在转基因作物实验室操作以及田间或温室生产性实验过程中，应严格遵守操作规范，实行隔离措施，防止因转基因作物花粉虫媒传播、风媒传播而造成的基因逃逸；或因工作人员（如衣物或鞋子）、农机用具意外将转基因作物种子带出试验田；应加强转基因作物种子的登记和保管，试验田生产获得的多余种子严禁随意晾晒或遗弃，田间散落的种子应及时回收，以免经田间动物（昆虫、蚂蚁、鼠等）取食搬运而移出实验区；应加强实验区的安保力度、加强种植区工作人员的安全责任和安全意识教育，以防发生转基因材料或种子的丢失或者被偷窃或出现转基因作物种子的私下交易，使未经审批的转基因作物种子流入市场或食物链。

5　结论

综上，转基因作物从研究到商业化释放一般需要经历较长的时间。获准进入商业化释放的转基因作物品种需要经过较为严格的申报和风险评估过程。但大多数转基因作物的研究，及相应开展的小规模生产性实验（如温室或田间小试），往往取决于研究人员或者研究团队的兴趣和条件，并不必须通过申报和相应的风险评估程序，从而出现生物风险管理的空白。现阶段国家在对于转基因作物的风险调控中，需要加强相关研究、加强对研究机构和实验室生物安全监管，对相关生物风险因素早介入、早控制。

致谢：2012 年度国家社科基金重大项目（12&ZD117）。

参考文献

[1] 徐静，阎斌伦，李晓英. 高校生物实验室安全问题初探[J]. 实验室科学，2009，12（6）：168-169.

[2] 徐静年，苏建茹，郭奋. 高校实验室污染不容忽视[J]. 实验技术与管理，2005，22（11）：130-132.

[3] 杨君，杨德礼. 农业转基因实验室生物风险研究进展[J]. 安徽农业科学，2011，39（1）：23-25.

[4] 金龙哲，宋存义. 安全科学原理[M]. 北京：化学工业出版社，2004.

[5] 邵延峰，薛红军. 故障树分析法在系统故障诊断中的应用[J]. 中国制造业信息化，2007，36（1）：72-74.

[6] 吴进煌. 舰空导弹发动机意外点火故障树分析[J]. 海军航空工程学院学报，2006，21（6）：653-656.

[7] 王莎莎，孙继银，李琳琳. 故障决策树模型在诊断专家系统中的应用[J]. 计算机应用，2005，25（s1）：293-294.

[8] HAYES K R. Identifying hazards in complex ecological systems—Part 1：Fault tree analysis for ballast water introductions[J]. Biological Invasions，2002（4）：235-249.

[9] National Research Council. Environmental Effects of Transgenic Plants：The Scope and Adequacy of Regulation[M]. Washington：National Academy Press，2002.

转 *Bt* 基因作物对天敌影响研究进展*

Influence of Transgenic *Bt* Crops on Natural Enemy

赵彩云，李俊生，吕凤春

（中国环境科学研究院，北京，100012）

ZHAO Cai-Yun，LI Jun-Sheng，LV Feng-Chun

（Chinese Research of Academy Environmental Sciences，Beijing，100012）

摘　要：随着生物技术的广泛运用，利用生物技术将抗虫性状转入植物体内，培育抗虫作物品系成为抵御虫害、降低农药环境污染的主要方式。转基因抗虫作物大规模种植虽然在一定程度上控制了靶标害虫，但同时改变了害虫演化地位，导致次生害虫大规模爆发；并且作物体内表达的 Bt 毒素有可能沿着食物链传递，对天敌类群构成潜在威胁。本文基于捕食性天敌和寄生性天敌与作物之间的关系，综述了目前国内外大田和室内的研究结果，分析了转基因作物对天敌的种群动态、生物学参数的影响，尤其着重于 *Bt* 基因沿着食物链在各营养级的传递，探讨 Bt 毒素沿食物链传递的效率，以及 Bt 毒素对天敌影响机制。

关键词：转 *Bt* 基因作物；捕食性天敌；寄生性天敌

Abstract: Many insect-resistant varieties recombined by DNA technology provided an effective tool for pest control and reducing environmental harm caused by pesticides. Although transgenic *Bt* crops can effectively control bollworms，many researches indicated that the cultivation of *Bt* crops could change the evolution status of pests, and then affect the natural enemies. Based on the effects of transgenic *Bt* crops to every trophic levels in the food chain from *Bt* crops to pests to natural enemies，we analyzed the effects of Bt toxin on population and biological development of predator and parasitoid，and the transfer efficiency of Bt toxin in the food chain was also discussed.

Keywords: Transgenic *Bt* crops; predator; parasitoid

转基因 DNA 重组技术消除了物种遗传隔离的限制，运用基因技术获得的转基因抗虫品系作物与传统作物品系相比具有快速、高效抗虫、且可以获取不同物种抗虫基因的优势[1]。1987 年美国将苏云金杆菌（*Bacillus thuringiensis*，简称 *Bt*）体内杀虫晶体蛋白基因成功导

* 基金项目：转基因重大专项（2013ZX08011-002）.

作者简介：赵彩云，女，1977 年生，博士，副研究员，研究方向为生物安全，E-mail：zhaocy@craes.org.cn.

入植物细胞，标志着第一例转基因植物的诞生[2]。自从 1996 年转基因作物商业化种植以来，转基因抗虫作物的大面积推广，尽管抗虫作物可以降低化学杀虫剂的使用量，对环境与人类健康有益，并增加农民收入[3,4]，但是与任何技术一样转基因抗虫作物也存在潜在的环境影响[1]。转基因抗虫作物主要的生态环境影响之一就是其对非靶标生物，尤其是对天敌类群潜在生态影响[5]。本文综述了国内外转基因作物对天敌昆虫大田种群动态、室内生理特征影响研究的结果，探讨了 *Bt* 毒素通过食物链对天敌昆虫的影响以及 Bt 毒素沿着食物链的传递。

1 转 *Bt* 基因作物对捕食性天敌的影响

我国大规模种植的主要是转 *Bt* 基因棉花，研究表明棉田内的主要捕食性天敌有瓢虫、蝽、隐翅虫、草蛉、蜘蛛和捕食螨等[6]。不同方法田间系统调查发现转 *Bt* 棉花能增加大多数捕食性天敌的种群数量，也有部分捕食性天敌种群数量略有下降，但均未达到显著水平。崔金杰等[7]用目测法统计百株棉的主要捕食性天敌龟纹瓢虫（*Propylaea japonica*），小花蝽虫（*Orins minutus*）和草间小黑蛛（*Erigonidium gramicicolum*）的种群数量变化特征研究发现，*Bt* 棉田内天敌龟纹瓢虫和草间小黑蛛的数量比常规棉田有所增加，小花蝽数量比常规不施药棉田略有下降，但均未达到显著水平；Bt 棉田天敌数量显著高于常规施药棉田。郦卫弟等[8]利用背负式吸虫器取样法研究华北地区转基因棉田和常规棉田的天敌种群动态得出相似的结论。杯诱法研究华北地区转 *Bt* 基因棉田与常规棉田内的步甲天敌种群动态，转 *Bt* 基因棉田内的步甲物种丰富度显著高于常规棉田，但物种多样性、香农威纳指数、均匀度指数没有明显差异；对优势物种的分析结果表明转 *Bt* 抗虫棉田内的捕食性甲虫后斑青步甲（*Chlaenius posticalis*）种群数量低于常规棉，而转 *Bt* 抗虫棉田内的蠋步甲（*Dolichus halensis*）、黄斑青步甲（*Chlaenius micans*）、单齿蝼步甲（*Scarites terricola*）的种群数量高于常规棉，但差异不显著（赵彩云等，未发表数据）。

尽管大田研究未发现转 *Bt* 基因作物对捕食性天敌种群动态有明显影响，但 Losey 等[9]在科学杂志上发表的 *Bt* 玉米花粉能够影响大斑蝶（*Danaus plexippus*）生长和发育的文章引发科学家运用不同的方法深入研究转 *Bt* 基因作物对非靶标昆虫，包括天敌的影响。大量研究表明 *Bt* 基因并通过食物链转 *Bt* 基因作物—害虫—天敌对捕食性天敌产生明显的副作用[11-18]：以转 *Bt* 棉花上采集的蚜虫饲养丽草蛉两代，结果表明转 *Bt* 基因棉花对幼虫期和茧期的死亡率、幼虫和茧的发育历期、茧重及成虫性比均无显著差异[18]。刘杰等[17]对蜘蛛的生长发育及捕食行为研究表明，转 *Bt* 棉花对草间钻头蛛（*Hylyphantes graminicola*）、八斑鞘腹蛛（*Coleosoma octomaculatum*）的发育历期和体重没有显著影响，对捕食行为也没有显著差异。García 等[14]研究了转 *Bt* 基因玉米通过食物链对隐翅虫（*Atheta coriaria* Kraatz）的成虫和幼虫消化生理学影响，结果表明无明显影响；针对棉花—粉纹夜蛾—瓢虫的三级营养关系研究结果表明，转 *Bt* 基因棉花与常规棉花相比，通过食物链对瓢虫的存活率、发育时间、体重和成虫寿命均没有显著影响[16]。也有部分研究结果表明转 *Bt* 基因作物对某些捕食性天敌会造成不利影响[19-23]：Ponsard 等[19]对短小暗色花蝽（*Orius tristicolor*）和大眼蝉长蝽（*Geocoris pallidipennis*）的室内饲养观察结果却与田间调查结果大相径庭，认为短小暗色花蝽和大眼蝉长蝽成虫取食以 *Bt* 棉为食的甜菜夜蛾后，能显著降低两种天敌的寿命。对捕食性天敌拟水狼蛛（*Pirata subpiraticus*）的研究结果与此相似，室内饲养试验表明取食转 *Bt* 基因水稻害虫的（*Cnaphalocrocis medinalis*）的天敌拟水狼蛛发育历期延长，进一步研究表明可能与 Bt 蛋白与害虫中肠细胞结合，导致天敌食物质量下降有关[21]；郭建英等[22]也认为食物质量下降是导致 Bt 棉区内龟纹瓢虫成虫体长、体宽、体厚和体积均显著降低的原因，室内卵块饲养试验也表明 Bt 棉区龟纹瓢虫幼虫取食采自转 *Bt* 棉株上的棉蚜后，幼虫发育历期延长，雌成虫比率下降且羽化比率明显下降，羽化成虫体重显著降低。Yu et al.[1]综述近年的研究成果表明造成捕食性天敌生物学特征变化的主要因素来源于其食物是否对 Bt 蛋白敏感。

为进一步探明 Bt 毒素与捕食性天敌之间的关系，有的学者直接用转 *Bt* 基因作物花粉饲喂捕食性天敌，比如 Smis[24]在饲料中掺入纯晶体蛋白模拟转 *Bt* 基因棉花的毒素表达，结果表明对集栖瓢虫（*Hippodamia covergens*）和普通草蛉（*Chrysoperla carnea*）没有明显影响；Romeis et al.直接以 1 mg/mL 高浓度的 Bt 蛋白（此浓度为一般初级消费者害虫体内的 10 000）直接饲喂捕食性天敌，结果表明对草蛉 2、3 龄幼虫没有明显影响，且对其捕食能力也没有显著影响[26]；Gonzaler-zamora et al.直接用不同浓度的 Cry1Ac、Cry1Ab、Cry2Ab 三种 Bt 蛋白饲喂棉铃虫幼虫，以棉铃虫幼虫饲养小花蝽，结果表明 Bt 蛋白对小花蝽的发育时间、繁殖率、捕食量、若虫存活率等特征均没有显著影响[27]。用转 *Bt* 基因玉米花粉直接饲喂胡瓜钝绥螨（*Neoseiulus cucumeris*），结果表明雌螨的发育时间延长了 9%，繁殖力减少了 17%，生长发育略微延缓，但是由于导致捕食性天敌发育迟缓的因素有很多种，很难归因于 Bt 毒素本身[25]。总之，目前没有证据表明 Bt 蛋白或者转 *Bt* 基因作物直接能对捕食性天敌造成显著影响。

2 转 *Bt* 基因作物对寄生性天敌的影响

与捕食性天敌相比，寄生性天敌与转 *Bt* 基因作物的联系更为紧密，大部分寄生性天敌成虫直接取食寄主，有些则取食植物花粉或花蜜[28]。棉铃虫的寄生性天敌中侧沟绿茧蜂（*Microplitis* spp.），拟澳洲赤眼蜂（*Trichogramma confusum*）和齿唇姬蜂（*Campoletis chlorideae*）小菜蛾蛹寄生性天敌蝶蛹金小蜂（*Pteromalus puparum*）的寄生率、生长、发育和种群数量明显下降，推断为棉铃虫等靶标昆虫因取食 *Bt* 棉常常不能完成生长发育而中毒死亡导致种群密度下降间接影响寄生性天敌的种群数量[29-31]。国外对寄生天敌长距茧蜂（*Microplitis cingulum*）、小茧蜂（*Cotesia marginiventris*）、黑唇姬蜂（*Campoletis sonorensis*）等寄生性天敌的研究得出相似的结论[32-33]。对以 *Bt* 棉花上的鳞翅目害虫为寄主的 6 种寄生蜂寄生行为研究表明，寄生蜂对取食了 Bt 毒素且发育迟缓的寄主表现非常敏感，这进一步说明抑制寄生蜂生长发育的主要原因是寄主营养质量差而非对 Bt 杀虫蛋白产生中毒反应[34]；也有研究表明转 *Bt* 基因作物对寄生性天敌没有显著影响，比如，Maria 等[35]用室内种植转 *Bt* 基因棉花饲喂棉铃虫，研究赤眼蜂（*Trichogramma pretisoum*）寄生棉铃虫的寄生率以及产卵行为，发现没有显著影响；有的学者直接用 Bt 蛋白+水混合液饲喂澳洲赤眼蜂（*Trichogramma confusum*），研究发现 Bt 蛋白对澳洲赤眼蜂的寿命、寄生卵的粒数、子代羽化率、性比等都没有显著影响。Liu et al.[36]饲养试验表明，同样用转 *Bt* 基因花椰菜饲喂抗性品系和敏感品系的小菜蛾（*Plutella xylostella*），岛弯尾姬蜂（*Diadegma insulare*）

对敏感品系小菜蛾的寄生率明显下降且无法羽化，而对抗性品系小菜蛾的寄生率、发育历期以及成虫体重和蛹重均未有显著变化，可见导致研究结果差异的主要原因在于害虫对 Bt 毒素的敏感性，寄生性天敌种群下降主要是由于取食了对 Bt 毒素比较敏感的寄主，寄主种群以及 Bt 毒素导致棉铃虫等寄主的营养状况变化而造成的，与 Bt 毒素本身没有直接的关系。

因此，转 *Bt* 基因作物对寄生性天敌的影响与捕食性天敌相似，种群数量以及生物学特征的变化可能主要来源于对敏感性害虫取食。

3 Bt 毒素沿着食物链的传递以及影响机制

部分研究表明 Bt 毒素会沿着食物链在不同营养级残留：对取食转基因玉米的禾蓟马（*Frankliniella tenuicornis* Uzel）不同时期体内的 *Cry1Ab* 毒素进行定量分析，研究表明 Bt 毒素在禾蓟马的幼虫和成虫体内含量最高，但是成中体内的 Bt 毒素在 24 小时内降解 97%，在不进食的预蛹期和蛹期没有检测到 Bt 毒素的存在[37]。Zhang et al.针对棉花—棉蚜—瓢虫的三级营养关系结果表明在蚜虫、瓢虫体内检测到 Bt 毒素的残留，且瓢虫体内 Bt 残留量随着取食时间的增加而增加，且在新孵化的幼虫体内也检测到了 Bt 毒素的残留[38]。Liu et al.在岛弯尾姬蜂和小菜蛾体内同样检测到 Cry1Ac[36]。Meissle et al. 定量检测了 Bt 玉米，海灰翅夜蛾 2 龄幼虫，步甲 3 龄幼虫体内 3 个不同营养级的 Cry1Ab 含量，结果表明 Bt 毒素会沿着食物链传递，且呈现出逐级递减的趋势[39]；Alvarez-Alfageme et al. 研究表明在初级营养级中的 Bt 含量越高，在高级营养级中残留量也越高[40]。也有部分研究表明 Bt 毒素不会在天敌体内残留，Howald et al.研究表明转基因油菜表达的 Cry1Ac 在黄翅菜叶蜂（*Athalia rosae* Linnaeus）幼虫、蛹和成虫没有残留，但在该种粪便内有残留[41]；Jorge et al. 用取食转基因棉花的甜菜夜蛾幼虫饲喂大眼长蝽幼虫和成虫，发现成虫体内没有 Cry1Ac 蛋白[42]。由于一方面 Bt 毒素具有水溶性特征，能溶解在初级或次级消费者体液中并随体液排出体外[43]，另一方面部分天敌的中肠内没有与 Bt 毒素结合位点，比如拟水狼蛛体内没有此结合位点[21]，因此由于研究对象不同，在不同天敌体内的残留也不同。

4 展望

大量的研究探索转 *Bt* 基因作物对天敌类群的影响，但目前的研究存在如下问题：①室内饲养时间较短，有的仅 1～2 d，且大多数仅 1～2 个世代，很难代表田间天敌不间断的长期接触 Bt 毒素的实际情况，也难以正确评估其慢性生态毒理效应；②大田内的食物网实际更为复杂，用单一的 Bt 作物—植食者—天敌来研究也不够全面；③不同类群天敌对 Bt 毒素的结合位点也存在差异。因此今后应选取不同类群有代表性的天敌物种，大田实验与室内试验相结合，从种群、群落、生态系统和食物网等方面长期、全面和系统研究，以更客观、全面的评价转 *Bt* 基因作物对天敌的影响。

参考文献

[1] Yu HL，Li YH and Wu KM. Risk assessment and ecological effects of transgenic Bt crops on non-target organisms[J]. Journal of Integrative Plant Biology，2011：1-47.

[2] Vaeck M，Reynaerts A，Hō fte H，et al. Transgenic plants protected from insect attack[J]. Nature，1987，328：33-37.

[3] Hutchison WD，Burkness EC，Mitchell PD，Moon RD，Leslie TW，Fleischer SJ，Abrahamson M，Hamilton KL，Steffey KL，Gray ME，Hellmich RL，Kaster LV，Hunt TE，Wright RJ，Pecinovsky K，Rabaey TL，Flood BR，Raun ES. Areawide suppression of European corn borer with Bt maize reaps savings to non-Bt maize growers[J]. Science，2010，330：222-225.

[4] Tabashnik BE. Communal benefits of transgenic corn[J]. Science，2010，330，189-190.

[5] Romeis J，Lawo NC，Raybould A. Making effective use of existing data for case-by-case risk assessments of genetically engineered crops[J]. J. Appl. Entomol.，2009，133：571-583.

[6] 李号宾，吴孔明，姚举，等. 新疆莎车县棉田自然天敌数量动态[J]. 昆虫知识，2007，44：（2）：219-222.

[7] 崔金杰，雒郡瑜，王春义，等. 转双价基因棉田主要害虫及其天敌的种群动态[J]. 棉花学报，2004，16（2）：94-101.

[8] 郦卫弟，吴孔明，陈学新，等. 华北地区转 Cry1Ac+CpTI 和 Cry1A 基因棉棉田害虫和天敌昆虫的群落结构[J]. 农业生物技术学报，2003，11（6）：494-499.

[9] Losy JE，Rayor LS，Cater ME. Transgenic pollen harms monarch larvae[J]. Nat. Biotechnol.，1999，399：214.

[10] Lewandowski A，Górecka J. Effect of transgenic maize MON 810 on selected non-target organisms：the bird cherry-ort aphid（*Rhopalosiphum padi* L.）and its predator-green lacewing（*Chrysoperla carnea* Steph.）[J]. Vegetable Crops Res. Butt.，2008，69：21-30.

[11] Meissle M，Romeis J. The web-building spider *Theridion impressum*（Araneae：Theridiidae）is not adversely affected by Bt maize resistant to corn rootworms[J]. Plant Biotechnol. J. 2009，7：645-656.

[12] Ålvarez-Alfageme F，Ortego F，Castañera P. *Bt* maize fed-prey mediated effect on fitness and digestive physiology of the ground predator *Poecilus cupreus* L.（Coleoptera：Carabidae）[J]. J. Insect Physiol.2009，55：144-150.

[13] Álvarez-Alfageme F，Bigler F，Romeis J. Laboratory toxicity studies demonstrate no adverse effects of Cry1Ab and Cry3Bb1 to larvae of *Adalia bipunctata*（Coleoptera：Coccinellidae）：the importance of study design[J]. Transgenic Res. 2010，DOI 10.1007/s11248-010-9430-5.

[14] García M，Ortego F，Castañera P，Farinós GP. Effects of exposure to the toxin Cry1Ab through *Bt* maize fed-prey on the performance and digestive physiology of the predatory rove beetle *Atheta coriaria*[J]. Biol. Control.，2010，55：225-233.

[15] Li YH，Romeis J. *Bt* maize expressing Cry3Bb1 does not harm the spider mite，*Tetranychus urticae*，or its ladybird beetle predator，*Stethorus punctillum*[J]. Biol. Control，2010，53：337-344.

[16] L YH，Romeis J，Wang P，et al. A comprehensive assessment of the effects of Btcotton on Coleomegilla maculata demonstrates no detrimental effects by Cry1Ac and Cry2Ab[J]. PlosOne，2011，6（7）：22-85.

[17] 刘杰，陈建，李明. 转 *Bt* 棉花对蜘蛛生长发育和捕食行为的影响[J]. 生态学报，2006，26（3）：945-949.

[18] 郭建英，万方浩，董亮，等. 取食转 *Bt* 基因棉花上的棉蚜对丽草蛉发育和繁殖的影响[J]. 昆虫知识，2005，42（4）：149-154.

[19] Ponsard S，Gutierrez AP，Mills NJ. Effect of Bt-toxin in transgenic cotton on the adult longevity of four heteropteran predators[J]. Environ. Entomol.，2002，31（6）：1197 - 1205.

[20] Zhang GF，Wan FH，Liu WX，Guo JY. Early instar response to plant-delivered *Bt*-toxin in a herbivore（*Spodoptera litura*）and a predator（*Propylaea japonica*）[J]. Crop Prot.，2006，25：527-533.

[21] Chen M，Ye GY，Liu ZC，Fang Q，Hu C，Peng YF，Shelton AM. Analysis of Cry1Ab toxin bioaccumulation in a food chain of *Bt* rice，an herbivore and a predator[J]. Exotoxicology，2009，18：230-238.

[22] 郭建英，周洪旭，万方浩，等. *Bt* 棉种植时期和作物布局方式对龟纹瓢虫生长发育和繁殖的影响[J]. 植物保护学报，2008，35（2）：137-142.

[23] Lawo NC，Wäckers FL，Romeis J. Characterizing indirect prey-quality mediated effects of a *Bt* crop on predatory larvae of the green lacewing，*Chrysoperla carnea*[J]. J. Insect Physiol. 2010，56：1702-1710.

[24] Sims SR. Bacillus thuringiensis var. knrstaki（Cry1Ac）protein expressed in transgenic cotton：effects on beneficial and other non-target insects[J]. Southwestern Entomologist，1995，20（4）：493- 500.

[25] Obrist LB，Klein H，Dutton A，Bigler F. Assessing the effects of *Bt* maize on the predatory mite *Neoseiulus cucumeris*[J]. Exp. Appl. Acarol.，2006，38：125-139.

[26] Romeis J，Dutton A，Bigler F. Bacillus thuringiensis toxin（Cry1Ab）has no direct effect on larvae of the green lacewing Chrysoperla carnea（Stephens）（Neuroptera：Chrysopidae）[J]. Journal of insect Physiology，2004，50：175-183.

[27] Gonzalez-zamora JE，Camunez S，Avilla C. Effects of Bacillus thuringiensis Cry toxins on developmental and reproductive characteristics of the predator Orius albidipennnis（Hemiptera：Anthocoridae）under laboratory conditions[J]. Environmental Entomology，2007，36（5）：1246-1253.

[28] Jervis MA，Kiddneds M. Insect natural enemies[M]. London：Chapman & Hall，1996：375-394.

[29] 崔金杰，夏敬源. 转 *Bt* 基因棉对天敌种群动态的影响[J]. 棉花学报，1999，11（2）：84 － 91.

[30] 杨益众，余月书，任璐，等. 转基因棉花对棉铃虫天敌寄生率的影响[J]. 昆虫知识，2001，38（6）：435-437.

[31] Chen M，Zhao JZ，Shelton AM，Cao J，Earle ED. Impact of single-gene and dual-gene *Bt* broccoli on the herbivore *Pieris rapae*（Lepidoptera：Pieridae）and its pupal endoparasitoid *Pteromalus puparum*（Hymenoptera：Pteromalidae）[J]. Transgenic Res.，2008，17：545-555.

[32] Ramirez-Romero R，Bernal JS，Chaufaux J，Kaiser L. Impact assessment of *Bt*-maize on a moth parasitoid，*Cotesia marginiventris*（Hymenoptera：Braconidae），via host exposure to purified Cry1Ab protein or *Bt*- plants[J]. Corp Prot.，2007，26：953-962.

[33] Sanders CJ，Pell JK，Poppy GM，Raybould A，Garcia-Alonso M，Schuler TH. Host-plant mediated effects of transgenic maize on the insect parasitoid *Campoletis sonorensis*（Hymenoptera：Ichneumonidae）[J]. Biol. Control.，2007，40：362-369.

[34] Romeis J，Meissle M，Bigler F. Transgenic crops expressing *Bacillus thuringiensis* toxins and biological control[J]. Nat Biotechnol.，2006，24：63-71.

[35] Moraes MCB，Laumann RA，Aquino MFS，Paula DP，Borges M. Effect of Bt genetic engineering on

indirect defense in cotton via a tritrophic interaction[J]. Transgenic Research，2010，DOL：10. 1007/s11248-010-9399-0.

[36] Liu XX，Chen M，D Onstad，Roush R，Shelton AM. Effect of *Bt* broccoli and resistant genotype of *Plutella xylostella*（Lepidoptera：Plutellidae）on development and host acceptance of the parasitoid *Diadegma insulare*（Hymenoptera：Ichneumonidae）[J]. Transgenic Res.，2010，DOI 10.1007/s11248-010-9471-9.

[37] Obrist LB，Klein H，Dutton A，*et al*. Effect s of Bt maize on Frankliniella tenuicornis and exposure of thrips predators to prey mediated Bt toxin[J]. Netherlands Entomological Society，2005，115：409-416.

[38] Zhang GF，Wan FH，LÊvei GL，Liu WX，Guo JY. Transmission of Bt toxin to the predator Propylaea japonica（Coleoptera：Coccinellidae）through its aphid prey feeding on transgenic Bt cotton[J]. Environ. Entomol.，2006，35（1）：143 - 150.

[39] Meissle M，Vojtech E，Poppy GM. Effect of Bt maize-fed prey on the generalist predator *Poecilus cupreus* L.（Coleoptera：Carabidae）[J]. Transgenic Research，2005，14：123-132.

[40] Ålvarez-Alfageme F，Ferry N，Castañera P，Ortego F，Gatehouse AMR. Prey mediated effects of *Bt* maize on fitness and digestive physiology of the red spider mite predator *Stethorus punctillum* Weise（Coleoptera：Coccinellidae）[J]. Transgenic Res.，2008，17：943-954.

[41] Howald R，Zwahlen C，Nentwig W. Evaluation of Bt oilseed rape on the non-target herbivore *Athalia rosae*[J]. Entomologia Experimentalis et Applicata，2003，106：87-93.

[42] Jorge B，Torres，Ruberson J R. Interact ions of Bt cotton and the omnivorous big eyed bug *Geocoris punctipes*（Say），a key predator in cotton fields[J] . Biological Control，2006，39：47- 57.

[43] McShaffrey D. Leaders in environmental activism-trophic levels[EB/OL].（2001-09-18）http://www. marietta. Edu/mcshaffd/lead/trophic.pdf. pp 1-6. Accessed January 20，2008（1995）.

从转基因大讨论看社会风险管理

Social Risk Management and Control for GMOs

尹帅军
（海南海洋安全与合作发展研究院特约研究员）
YIN Shuai-Jun
Visiting Researcher Representative of the Public

摘　要：2009 年底中国农业部颁发转基因水稻和玉米的安全证书。2010 年初两位重量级院士宣称，转基因主粮将在 3 ~ 5 年内推上中国人的餐桌。此后中国国内针对转基因问题开展了广泛的讨论。2011 年 9 月消息，转基因主粮商业化推迟 5 ~ 10 年。2013 年 4 月，中国科学院学部主席团发布《关于负责任的转基因技术研发行为的倡议》，劝说转基因科研人员“以对人类社会发展高度负责任的态度，加强职业操守，规范科研行为，履行社会责任，积极与社会沟通，促进转基因技术良性发展。”2013 年 6 月 13 日，农业部又批准进口三种转基因大豆。2013 年 7 月 13 日，公众代表首次举办转基因与食品安全国际研讨会。转基因问题大讨论的影响力可见一斑。从转基因的大讨论过程，我们可以对中国的社会风险管理进行一番管中窥豹。

Abstract: In the late 2009，China Ministry of Agriculture issued security certificates of Genetically-Modified （GM） rice and corn. Early in the year of 2010，two great famous academy members claimed that genetically-modified food would be put onto our dinner tables of Chinese in the next 3 to 5 years. From then on，China has been witnessing the extensive and in-depth debates over the issue of GMOs. In September 2011，the commercialization of Genetically-Modified crops would be postponed to the next 5 to 10 years. In April 2013，the Presidium of academicians in Chinese Academy of Sciences released one initiative entitled” to do responsible research on GM technologies，which persuades the related researchers to take highly responsible attitudes towards the development of human society and enhance their professional ethics. The initiative also urges them to standardize their research behaviors，fulfill their social responsibility as well as publicize the results in time. Only in such way，GM technologies will develop thoroughly well. On June 13 of 2013，China Ministry of Agriculture approved the import of three kinds of genetically-modified soybean. On July 13 of 2013，the first International Symposium on GMO and Food Safety was held by representatives of the public. Above all，the effects of the great debates on GM safety have produced overwhelmingly influence，which will also remind us its necessary to discuss the social risk management of GM in china.

2009年11月27日，国家农业部颁发了两种转基因水稻、一种转基因玉米的安全证书。此事成为一个标志性事件。在此之前，转基因问题已经是公众和媒体关注的热点。在此之后，中国掀起了一场大范围关于转基因问题的讨论热潮。海内外人士上书全国人大和国家领导人，主推派和反对派激烈辩论，网络、报刊、电视积极参与。直到今天，辩论还在持续激烈进行。

转基因问题大讨论的热潮，一方面表明国家部门的透明度正在增加，社会主义民主建设正在成长，另一方面也表现出公众积极参政议政的高度热情。在大讨论过程中暴露出的我国种业安全、粮食安全、农业安全、环境安全、生物安全、经济安全、科研体制安全、国家安全等诸多领域的风险，使我国可以及时应对。另一方面，在大讨论过程中暴露出政府部门与公众沟通交流的诸多问题，促使我们思考应该如何建立更有效的群众参政议政平台，如何引导社会舆论健康发展。转基因大讨论的整个过程，总的来说健康积极，热烈而理性，有关部门应该创造更多条件让公众参与讨论。

1 转基因大讨论的过程和主要事件

1.1 转基因大讨论的参与方

回顾最近三年多的大讨论，我们可以发现，转基因主粮商业化主推派主要是各级农业部门、农业院校、农业科研机构及跨国公司。他们在国家重大科技专项的扶持下，从事转基因技术的研究和推广。

“十二五”期间，国家对转基因品种的研发支持超过100亿元，常规育种只有1.8亿元，相差近百倍。搞常规育种没有活路。为了生计，大多数农业机构和科研人员只能支持转基因。当然，在这类机构中也有少数科学家对转基因主粮推广持谨慎态度。

一些从事有机农业、生态农业、常规育种的专家则对转基因农业、化工农业持深刻质疑态度。

转基因主粮商业化反对派包括了体制内外很多人。既包括科学家，如农业专家、医学家、环境学家、生物学家，也包括经济学家、社会学家、律师、政府官员、军人，以及各行各业的社会公众。反对派的观点基本可以概括为，支持国家进行转基因科学技术研究，但是反对草率商业化推广转基因主粮。未经充分安全评价的转基因技术，有可能对全体国民造成危害，甚至灭绝种族。

质疑的声音首先发端于报刊，但是生长壮大却是在网络。关注社会问题的网站迅速成为其核心。国内外关于转基因的研究资料、新闻报道、书籍迅速集中起来，既包括老资料，也包括最新资料。资料不仅包含转基因研发技术、转基因食品安全、转基因生态安全、转基因审批问题、转基因非法种植和流通、转基因历史、转基因跨国公司，还包括医学、农业、生态、国家科研机制、法律、经济安全、国家安全等。资料汇总筛选的速度不仅迅速，而且极为广泛。每当主推派发出一个声音之后，迅速就会有许多网友回应。一些社会公众成为转基因问题专家。他们虽然不能在实验室中造出转基因，但是其知识面的宽广却超过转基因技术专家。在网络的持续发酵下，报刊、杂志、电视也开始做起关于转基因的话题。

从舆论力量对比来看，在社会公众占据主导地位的网络上几乎是一片倒的反对声音，在报刊、杂志上反对声音占上风，在官方主导的电视上则是平分秋色。

1.2　从白热化到转折点

尽管转基因问题的争论曾一度达到白热化程度，但是事实上，其自始至终其实都处于和平状态。虽有多次质问抗议主管部门、跨国公司、转基因主推派学者的行为，但是并未出现过激行为。公众的情绪之所以会白热化，主要原因还是因为主管部门缺少回应。每当主推派发出一个声音，公众就会有针锋相对的质疑，但是这些疑问却总是无法得到回应和解答。主管部门沉默的时候居多，致使恐慌情绪一度蔓延。

因为得不到主管部门的回应，公众便将目光转向上级单位，全国人大和国家领导人。2010 年两会期间，多份海内外人士联名信上递全国人大和国家领导人。在公众的持续推动下，重庆和福建地方政府做出回应，相继发布文件，根据国家法规对转基因进行严格监管，打击转基因的非法种植、流通泛滥行为。还有一些地方政府发布文件，在幼儿园和大中小学中不再使用转基因油。杭州市禁止中小学、幼儿园食用转基因油，乌鲁木齐禁止学校食用转基因油、西安市禁止学校食用转基因大豆油。一些北京市民积极活动，要求北京各大高校给学生换非转基因油食用。

2010 年 7 月 16 日，公众代表前往农业部进行抗议。2011 年 9 月 9 日、2013 年 1 月 7 日公众代表又两次去农业部进行抗议。

2010 年 9 月，国际先驱导报记者金微报道，山西、吉林等地出现动物异常现象。此事引起国家相关部门重视。

在公众压力下，农业部从 2010 年 5 月开始进行转基因检查风暴，2010 年底结束，27 个玉米品种、6 个水稻品种、7 个大豆品种被勒令退出市场，其中多个品种为违规转基因。科学家瞒天过海，将转基因品种以常规品种名义申报并通过。事后并未受到什么处罚。我国常规作物品种在审定时，并不对其进行转基因检测，所以无法确知其是否是转基因。这给一些不法科学家留了漏洞。从 2000 年到 2008 年，通过国家合法机构审定的玉米品种就达到 3 150 个，仅仅对已审定通过的玉米品种做一次全面检测，很难检测出品种的真实性和是否含有转基因成分，而农业部的转基因检查风暴只是对种子市场做“抽检”而已。因此很难说清楚中国到底有多少转基因品种在违法种植。

2010 年 11 月 26 日，转基因水稻安全证书的获得者张启发院士，在中国农业大学做讲座。提问阶段，一位年轻人向张启发提了三个尖锐问题，张启发院士无法回答。抗议者写了篇文章《三个问题吓走张启发，农大讲座仓皇收场》。张启发院士原定于 2010 年 12 月召开一次与公众的沟通会进行科普，后来因故取消。

农业部新闻发言人曾指出，“农业部将本着对广大消费者高度负责的态度，积极稳妥地推进转基因生物技术的研发与应用”。

2010 年 12 月爆出新闻，农业部机关幼儿园不吃转基因。幼儿园网站的“营养配餐及食品安全”一栏明确指出，幼儿园的“鱼肉类食品统一由为农业部机关食堂供货的优质水产品公司供货，食用油采用非转基因油”。

2011 年 2 月，新华社高级记者从亚平上书，温家宝总理做出重要批示。

2011 年两会期间，一些公众又将转基因安全、种族安全、食品安全三份提案交给任远征等老革命后代，由他们交给人大、政协和国家领导人。九大常委都在第一时间做出批示。几百张转基因视频光碟资料《转基因的前世今生》被带入人大和政协会场，引发震动。光碟综合了国内外几十部转基因纪录片、专题片的视频资料。该视频在网络播放时，毫无推荐和置顶，三个月时间点击量却突破 1 000 万。可谓是公众参与转基因生物安全管理的代表作。

在此之后，转基因讨论开始出现转折点。2011 年 4 月，由农业部、环保部、科技部和卫生部组成的四部委联合调查组，调查发现东北地区非法种植转基因。这是官方首度承认国内存在非法种植转基因。

2011 年 5 月 3 日上午，国务院召开一次关于转基因水稻商业化问题征求意见座谈会。著名玉米专家、育种界元老佟屏亚院士在会议上做了长篇发言，反对转基因主粮的草率商业化。他在会上提问之后，转基因科学家无以回答。

1.3 政府部门与公众的互动

2011 年 4 月 28 日至 29 日，环保部和国家民委批准召开“生物安全国际论坛第四次研讨会”。会议邀请了国内外专家、跨国公司代表、多个政府部门官员和公众代表。主推派和反对派在会场上激烈辩论。这次半官方性质的会议取得了良好效果。遗憾的是，农业部派出的代表职位太低、知识面太窄，无法回答很多公众和科学家的疑问。牧川、金微、顾秀林、陈一文文章《11 次激烈交锋——生物安全国际论坛上的转基因》记录了会议经过。

2011 年 8 月，环保部主管的《环境教育》杂志刊登文章《转基因在中国》。文章指出目前转基因的食品安全、环境安全、社会经济安全、粮食安全、公众利益等都未有定论，且我国的转基因审批和管理不完善，公众的知情权和参与权未得到充分保障。

2011 年 8 月 23 日，农业部举办“全国科技记者转基因培训班”，对记者进行科普宣传，宣传推广转基因知识。

2011 年 9 月 9 日，公众代表前往农业部抗议。农业部首度承认先玉 335 玉米含启动子。2011 年初，环保部曾检测出先玉 335 玉米含有启动子。但是农业部认为，是不是含有启动子，就一定说明其是转基因品种，这还需要进一步研究。

2011 年 9 月底传出消息，未来 5～10 年中国将不会推行以水稻、小麦为主的转基因主粮商业化。一是因为各界对转基因安全性的质疑，二是中国在转基因主粮的研究、推广、监管乃至后期企业运作等方面都还不成熟。不过转基因玉米商业化将有可能适时推进。

2011 年 10 月 25 日《中国国防报》刊登文章《让生物战争走进国防视野》。文章从转基因的大讨论引出基因战争和生物国防话题：“ 在全球金融危机尚未结束、国际形势复杂多变的现阶段，中国国家安全和中华民族的生存发展是第一位的头等大事。因此，采取以下措施或是未雨绸缪之策：树立新时期生物国防的战略地位，制订生物国防计划，提高全民生物国防意识；坚持统一集中管理，把水和粮食等关系到人民健康和民族安全的战略产业和命脉领域，牢牢掌握在国家公共部门手中；重视基因技术研究，集中国家力量力争走在世界基因研究前列，切实增强生物国防的科技储备和生物战争的应对能力。

2012 年 1 月 14 日，由国家重大专项办公室组织在海南三亚开会，农业部、科技部有关领导，戴景瑞、武维华、赖锦盛、吴孔明、林敏、万建民、黄大昉、贾士荣、彭于发、陈茹梅在海南三亚开会，研讨如何加快我国转基因主粮商业化进程。

2012 年 2 月 21 日，国务院法制办公布《粮食法（征求意见稿）》，面向社会各界征求意见。意见稿规定，“转基因粮食种子的科研、试验、生产、销售、进出口应当符合国家有关规定。任何单位和个人不得擅自在主要粮食品种上应用转基因技术”。这项法规（草案）的颁布，是各方博弈的成果，是一个很大的进步。

不过著名玉米专家、育种界元老佟屏亚院士却指出，“中国现有的转基因生物安全法律法规尚不完善，监管权责不很明确，急需加快转基因生物安全法的立法进程，加大对转基因作物及相关产品的监管力度”。

2013 年 4 月，中国科学院学部主席团发布《关于负责任的转基因技术研发行为的倡议》，劝说转基因科研人员“以对人类社会发展高度负责任的态度，加强职业操守，规范科研行为，履行社会责任，积极与社会沟通，促进转基因技术良性发展”。

1.4 转基因作物的监管漏洞

转基因主粮的监管漏洞很多。2012 年 1～5 月欧盟共 19 次通报，从中国进口的大米制品涉非法转基因。欧盟发布《决定》，拒绝从中国进口转基因稻米。媒体报道，转基因水稻种子和稻米已经遍布湖北、江西、安徽、江苏、四川、湖南、河南、浙江等地种子市场和食品市场。转基因水稻已形成规模种植，农民将收获的转基因大米混在普通大米中进入流通市场。农业部多次重申“从未批准任何转基因粮食作物种植”。据中国农业大学戴景瑞院士称我国转基因玉米已经种植 6 000 万亩。

非主粮的转基因作物监管漏洞更多。2010 年，转基因小米张杂谷已种植 400 多万亩。当时新闻报道，计划在未来 10～15 年推广到 1 亿亩。张杂谷的审批，有没有做过食品安全试验、环境安全试验不得而知。

在蔬菜瓜果中，转基因作物监管漏洞更多。国内已经无法找到非转基因的木瓜种子，原因是国内某些科研机构将转基因木瓜种子外流，转基因木瓜通过花粉传播，污染了非转基因木瓜，导致目前国内再也找不到非转基因的木瓜种子。在既成事实面前，不得不将转基因木瓜合法化。中国已经批准了转基因番茄、甜椒、木瓜等蔬菜瓜果种植，这些作物的审批有没有做过充分的食品安全试验、环境安全试验仍然不得而知。

2012 年 8 月，媒体爆出“黄金大米”事件。美国研究机构勾结国内机构，拿中国小学生做实验，试吃先正达公司研发的转基因“黄金大米”。2012 年 12 月 6 日，中国疾控中心等单位联合发布转基因“黄金大米”事件的调查结果，对一些人进行了处理。该事件再一次暴露中国转基因监管的漏洞。有网友指出《“黄金大米”事件毫不吃惊，中国转基因监管状如马蜂窝》，他说在 2010 年已经看到该新闻。他认为不仅转基因的种植和流通领域存在很多非法行为，转基因的法律法规本身也有很多漏洞。

2012 年两会期间，政协委员、中国军事科学学会常务理事兼副秘书长罗援少将发出警告：“要高度重视国家生物安全问题，警惕转基因物种的无序迅速扩散和外国插手中国疫苗生产过程，警惕敌国以转基因物种和特种疫苗等为武器，针对中国人口发动新型战略打击。”

1.5 转基因作物的立法漏洞

事实上，不仅转基因的种植和流通领域存在很多非法行为，有关转基因作物的法律法规存在很多漏洞不足。

2013 年 5 月 27 日召开的“生物安全国际论坛第五次研讨会”上，一位公众代表指出我国的转基因立法存在很多漏洞，而且还是非常明显、非常大的漏洞。

首先，关于常规种子审批的有关规定，并不要求对其进行转基因检测，所以根本不可能知道其是否是转基因种子。一些知名科学家借此蒙混过关，将转基因作物以常规种子的名义申报通过。（补充说明，这个漏洞早在 2010 年就有新闻报道。）

其次，最近几年，我国每年进口 5 000 多万 t 转基因大豆，却不对转基因大豆的草甘膦农药残留进行检测，因为根本没有有关的检测标准。国标规定国内生产的绝大部分粮食类农作物产品“农药最大残留限量”低于 1 mg/kg。美国环保部的转基因大豆草甘膦残留量标准却是 20 mg/kg（转基因大豆在美国主要用于动物饲料和燃料）。阿根廷转基因大豆的草甘膦残留量高达 17 mg/kg。从巴西、阿根廷、美国进口的转基因大豆，其草甘膦农药残留是中国国内检测标准的 20 倍左右。（补充说明，这个漏洞早在 2010 年底被陈一文发现。）

最后，由转基因作物与非转基因作物杂交之后产生的新的作物品种，相关规定中，没有要求对其进行转基因作物的食品安全、环境安全试验和审批，而是把它看做常规品种来审批。此漏洞将转基因作物合法化。许多公司，就是借此漏洞将国内国外的转基因作物通过杂交手段，合法化生产。（补充说明，杜邦先锋公司的转基因玉米先玉 335，其公司负责人曾经说过，先玉 335 在美国是转基因品种。但是中国国内相关部门却否认先玉 335 玉米是转基因品种。其基础或许就是中国的转基因法规。中国还有多少以常规种子名义蒙混过关的转基因种子，不得而知。）

农业部官员在听取了公众代表的观点之后指出，这位公众代表对转基因问题的研究非常透彻。他说，在过去的工作中他也接触过这些问题，农业部正在对这些问题进行研究。如果公众代表有什么可行性的报告和措施，可以通过相关渠道递交给农业部和相关部门。

公众代表指出，这些都是很明显的立法漏洞，只需要把法规的漏洞改变即可。

1.6 公众要求政府信息公开、质询跨国公司

2002 年，农业部批准进口孟山都抗草甘膦转基因大豆，至今该大豆已经占据中国 80% 市场，近几年每年都进口 5 000 万 t 以上。

根据《转基因食品卫生管理办法》（2002 年 7 月 1 日起施行，2007 年 12 月 1 日废止）、《新资源食品管理办法》（2007 年 12 月 1 日起施行），新上市的转基因食品需要卫生部的审批。

2011 年 11 月 21 日，公众代表杨晓陆向卫生部递交政府信息公开申请表，要求卫生部出具采用美国孟山都抗草甘膦转基因大豆品系加工的三种产品的审批文件。这三种产品分别是，转基因大豆油、采用“化学浸出”工艺加工生产的转基因大豆油、其他含有转基因大豆成分的食品。2011 年 11 月 23 日，卫生部回复《政府信息不存在告知书》，卫生部未受理和审批过这三种产品。

2011 年 9 月 30 日，顾秀林等四人向农业部递交政府信息公开申请表，要求农业部出具文件：美国孟山都公司转基因大豆出口中国的全套申请文件、农业部批准其进口申请的全套审批文件，还有中美双方的科学实验报告。

根据国务院《农业转基因生物安全管理条例》（2001 年 5 月 23 日颁布实施）、《农业部

转基因进口安全管理办法》（2001 年 7 月 11 日农业部通过，2002 年 3 月 20 日施行），对于进口的转基因产品需要三方面的科学实验文件和审批文件：

①输出国家或者地区经过科学试验证明对人类、动植物、微生物和生态环境无害；

②经[中国]农业转基因生物技术检测机构检测，确认对人类、动植物、微生物和生态环境不存在危险；

③农业部委托的技术检测机构出具的对人类、动植物、微生物和生态环境安全性的检测报告。

2011 年 11 月 4 日顾秀林获准进行查询，在农业部提供的文件夹中，没有看到任何有关“三方”科学实验报告的内容。

农业部提供的孟山都公司的申请文件包括四份，“美国食品和药品管理局关于食用安全批文”（1996 年）、“美国农业部动植物检疫局关于环境安全批文”（1994 年）、一页纸的申请表、“毒理学研究动物试验报告”。这些文件内容并不符合中国法律规定。

“美国食品和药品管理局关于食用安全批文”的主要内容是，将孟山都公司的申请文件和结论重复一遍：

“……你公司（孟山都）的结论是：此一新大豆品种和现有市售大豆品种，在产品成分、安全性以及其他相关指标上没有物质上的不同，因此上市前无须进行复审及批准。本次咨询的所有材料均已归档于 BNF001，由市场准入批准办公室保管。”

“根据这次咨询的有关数据的描述及提供的信息，新的大豆品种依照 21 CFR 170.30（f）（2）的含义似乎没有显著不同。因此关于此产品我们没有其他要问的问题。但是，你们也知道，保证孟山都公司营销的食品安全、有益健康，并遵循其他相关的法律与法规，是贵公司的责任。”（生物技术公司咨询回函，BNF 000001. 美国 FDA Biotechnology Consultation Agency Response Letter BNF No. 000001）

美国食品和药品管理局并没有对孟山都公司的产品进行科学试验和检测，只是把孟山都公司的结论复述一遍，而后说明孟山都公司对它的产品负有全部责任，潜台词是若出了问题，美国食品和药品管理局无须负责。

孟山都的转基因大豆在美国种植仅 4 年后，2000 年，超级杂草就被发现。2000 年开始已经有学术论文专门讨论超级杂草的抗性。因此孟山都公司向中国农业部出具的“美国农业部动植物检疫局关于环境安全批文（1994 年）”，应被视为失效文件，不符合中国的法律要求。

在“毒理学研究动物试验报告”的文件中，并不包括独立的第三方单位做的动物实验，也没有孟山都公司的“毒理学研究动物实验报告”。没有任何包含了记录数据、进行分析的原始报告。

顾秀林查询后认为美国孟山都公司向中国农业部申请获批的抗除草剂转基因大豆 40-3-2 安全证书存在较大争议，不符合有关规定。

2011 年 10 月 14 日、11 月 15 日，公众代表两次去益海嘉里公司递交质询函，要求其出具金龙鱼“化学浸出”转基因大豆油是合法转基因食品的相关部门的审批文件复印件。益海嘉里公司未能提供文件。

1.7 科普之争

农业部有高级官员认为，“在转基因食品问题上，要加强科普推广”。力推转基因作物的黄大昉院士也曾表达同样的观点：“公众的恐慌是因为科学素养不够。”还有多位农业部高级官员称民众质疑转基因食品源于无知。公众的恐慌仅仅是因为不懂科学吗？事实上，不管从理论角度，还是从科学实验和实践角度，都证明转基因食品存在安全隐患。

2012 年 9 月 19 日，法国科学家塞拉利尼在《食品与化学毒理学》杂志发表研究报告，用抗除草剂的 NK603 转基因玉米喂养的大鼠，致癌率大幅度上升。文中附有照片，大鼠身上长满肿瘤，肿瘤和大鼠的脖子差不多大。试验结果一公布，引起全球关注，将转基因争论推向高潮。俄罗斯、英国、奥地利、澳大利亚、美国、意大利等国科学家都曾做过转基因作物的喂养实验，得出结论认为存在安全问题，但这些实验都是 2008 年之前，并未引起公众关注。

从国内已知的信息来看，积极推广转基因的专家学者至今只做过 90 d 小白鼠喂养转基因大米试验、8 d 抗虫 Bt 毒蛋白灌胃试验，既没有做过大型哺乳动物实验，也没有做过灵长类动物试验，更没有做过类似医学药品推广前的临床试验。国内的诸多实验室很少去做转基因的长期喂养试验、环境试验。倒是有三个青年硕士生初生牛犊不怕虎，做了这类实验。

中科院植物所首席研究员蒋高明教授介绍了这三个学生的实验论文。湖南师范大学硕士论文《转基因水稻致小鼠肠腺细胞病变增生》。上海师范大学硕士黄同学等人的研究，《转 *Bt* 基因水稻对土壤微生物的影响》，该论文指出种植转 *Bt* 基因水稻的时间越长，种植的强度越大，对土壤氨氧化细菌的影响也就越大。中国科学院海洋研究所陶同学硕士论文：《饲料中转基因成分检测及对水产动物安全性评价的初步研究》及相关论文《进口转基因大豆（Roundup Ready）对吉富罗非鱼生长和生理的影响》，该论文指出将含有 30%转基因豆粕的罗非鱼饲料，7 周后导致罗非鱼肌体内部产生炎症。研究显示，在罗非鱼心脏、肝脏、肠、胃、卵巢、精巢、脑、鳃丝、脾脏、胆囊、肌肉等不同部位的 DNA 中都能检测到外源基因的存在。该研究还指出，采用巢式 PCR 检测方法对市场上的豆制品进行检测，结果发现，市场上有 46.5%的豆制品被检测出含有转基因成分。

1.8 转基因基本理论遭遇质疑之一

2011 年 7 月 6 日，文汇报发表著名分子生物学家曹明华的文章《“转基因”之争在美国》。曹明华指出，目前的转基因食物产业所基于的“科学”，是已经过时了的生物学理论。

转基因技术刚出现时，生物学家以为，真核生物（如植物、动物和人）的基因编码规律与原核生物（如细菌）是一样的，即一个基因只编码一个特定的蛋白质。按这一传统的遗传学模型，生物学家曾估算：人体中的蛋白质约有 10 万个或更多，因此他们预测在人类 DNA 中的基因约有 10 万个。

而在 2000 年 6 月 26 日，整个科学界在震惊中发现：人类基因总共只不到 3 万个。更令人困惑的是，比人低等得多的杂草却可以有 2.6 万个基因。那么，大多数基因都不只编码一个蛋白质，有些基因可以产生许多不同的蛋白质，比如果蝇，它的一个基因可以产生 38 016 个不同的蛋白质分子。

人与杂草间的特性、功能等有那么巨大的差异，而在基因数量上却并没有呈现出数量级的差异——那么一定是有些什么东西错了。“一个基因只编码一个特定的蛋白质”的理论是错误的。

在这一更新了的分子生物学模型面前，转基因食物产业的主要根基动摇了！

而中国的有些转基因专家，在 2011 年的今天，还在用 20 世纪 80 年代对于原核生物所适用的基因学理论来认识真核生物（包括农作物、动物、人）。他们的知识更新跟不上时代发展，由此可能引发严重后果。

现实实验中我们已经看到了很多类似的例子。科学家曾将人的生长激素基因转到猪的身上，目的是要制造出极快速生长的猪来，可没料到，所产生的小母猪居然没有肛门。科学家又制作了转基因酵母，目的是增加酿酒产量——但很惊讶地发现：这种酵母中原有的一种自然毒素——可能致癌的因子被意外地提升了 40～200 倍。这个实验的研究者不由得感叹：“看来公众对转基因食物的恐惧是有道理的……”他们指出，在这一实验中，还并没有转入任何异类基因，而只是——将酵母自身的基因多转了几个拷贝进去……

在曹明华发表该文之后，2011 年 7 月 11 日，人民日报发表署名饶毅的文章《转基因在美国的遭际》，支持推广转基因。7 月 14 日，《文汇报》又发表曹明华文章《有一种误解亟须辨明》，继续进行学术探讨。7 月 18 日，《人民日报》再发署名饶毅的文章《转基因是现代科技的必然》。7 月 21 日，《南方周末》刊登署名“柯贝”的文章《对转基因的无知与偏见》，大力支持转基因。柯贝文章点名批评曹明华和蒋高明。为此，曹明华写了一篇反驳文章，希望《南方周末》能够发表，未能如愿。为此曹明华发表了《致〈南方周末〉社领导的一封公开信》，蒋高明发表了《我对转基因的真实态度》。美籍华裔生命科学家和环保专家刘实也发了文章《我对主粮转基因之争的看法和建议》，反对柯贝的文章。2011 年 12 月，反转派专家顾秀林又发表文章《道理讲的清、利益讲不清》。

挺转派代表人物、反转派代表人物之间的直接交锋，十分有助于政府和公众认识真理。

1.9 转基因基本理论遭遇质疑之二

2011 年 9 月，南京大学生命科学学院张辰宇教授的团队研究成果发表在国际知名刊物《细胞研究》上，颠覆了转基因安全的神话。“植物的微小核糖核酸（microRNA）可以通过日常食物摄取的方式进入人体血液和组织器官。并且，一旦进入体内，它们将通过调控人体内靶基因表达的方式影响人体的生理功能，进而发挥生物学作用”。

美国《环境科学》杂志为此刊文指出：“植物微小核糖核酸（microRNA）能调控人体生理过程，这一发现也说明人体是高度综合的生态系统。张辰宇说，这一发现还能帮助我们理解共同进化机制，即某一物种的基因改变可以触发另一物种的改变。比如，人类开始驯养奶牛之后，机体才开始进化出消化牛奶中的乳糖的功能。人类培育的植物，有没有可能也在改变人类呢？张辰宇的研究又一次提醒我们，自然界中没有孤立的存在。”

美国著名杂志《大西洋月刊》为此刊文《转基因食品有非常真实的危险》，该文指出：“如果该研究通得过科学的评审——这是关键性门槛——它将会颠覆许多领域的研究。它意味着，我们吃下去的不只是维他命、蛋白质和能量，还有基因调节因子。”

张辰宇教授的研究颠覆了世界最大的转基因公司孟山都公司的理论基础，孟山都公司宣称：“测试转基因食品对人类的安全性是完全没有必要、也没有意义的。”这是个有利于商业的说法，其理论依据是 1960 年前后的遗传学，即“中心法则”遗传学，中心法则假说是："DNA 主导细胞，从 DNA 到细胞之间存在一种单方向运行的指令控制链。"但是真实的世界却是，除了 DNA，微小 RNA 也参与到生物的进化当中。"

该论文指出，孟山都和其他公司用"敲除"某个 RNA 的方法，可以开发出杀虫转基因植物，既然 RNA（核糖核酸）干扰原理可以用于杀虫，那么这个能致死昆虫的生物学机理，在人类身上应该也是同样存在的。

随着时间的推移，越来越多的公众要求更为权威、客观、公正的实验，要求科学家具备真正的科学态度。

中国妇联执委曾发布报告，《中国不孕不育患者数已超 5 000 万》。类似的报道层出不穷，《南京捐精者合格率仅三成》《广西在校大学男生精液抽检过半不合格》《我国每年新增胃癌 40 万，占全世界 42%》《黑龙江大豆协会：转基因与肿瘤高度相关》。我国公民的健康状态可以说是每况愈下，尤其是青年人的体质状况，更是呈现逐年下降的趋势。这和我国非法种植和流通的转基因有多大关系不得而知。国家机构、科学家如能对此做一番全面、系统、客观的研究，不仅于国于民有利，还能大幅度提高中国的科技水平。

1.10 农业部批准进口 3 种转基因大豆

在国内转基因讨论愈来愈热烈的关口，2013 年 6 月 13 日，农业部批准发放了巴斯夫农化有限公司申请的抗除草剂大豆 CV127、孟山都远东有限公司申请的抗虫大豆 MON87701 和抗虫耐除草剂大豆 MON87701×MON89788 三个可进口用作加工原料的农业转基因生物安全证书。

此事再次成为公众和媒体关注的热点。一些公众激烈批评农业部，而农业部多位高级官员则再次回应，民众质疑转基因食品是源于无知。

农业部官员的说法显然无法说服公众，反而带来了更激烈的反应。许多人指出农业部的做法拯救了孟山都公司。

1.11 转基因科学家反水

恰在此时，长期从事转基因研究的"挺转派"科学家曾庆平指出，他依然支持转基因研究，但是转基因主粮的商业化却建议暂缓。原因有四点：

（1）转基因技术本身的局限性（无法分离并导入高产基因）使之难以解决粮食安全问题；

（2）转基因抗病虫作物一旦推广，可能像抗生素导致耐药性一样，因病虫的适应性进化而导致生态灾难；

（3）转基因果蔬及其他农产品、副产品的直接食用或饮用存在过敏风险；

（4）转基因目标蛋白的潜在催化功能可能让有害小分子化合物经过家畜、家禽进入人体，导致难以预计的后果。

因此，转基因食品的推广应用必须慎之又慎。

这位科学家的说法，其实与之前许多公众代表、科学家的说法相一致。

1.12 公众举办转基因与食品安全国际研讨会

2013 年 7 月 13 日，来自美国、英国、法国、印度、新西兰等国家的国际顶尖转基因危害研究科学家齐聚北京，召开转基因与食品安全国际研讨会。这是第一次由民间主办的转基因国际会议。国内反转派专家、国际专家、普通民众、政府官员、老革命后代进行了很好的交流。

来自美国的胡伯博士确认了一条重要消息，2011 年，中国国家质量监督检验检疫总局的科学家给胡伯写信，称在中国进口的转基因大豆中发现了胡伯博士在美国转基因大豆、玉米与流产动物死胎中发现的新病原体。这种病原体可能就是导致不孕不育的重要因素。

2　转基因大讨论存在的问题

公众参与转基因生物风险管理主要分为三个渠道。首先是转基因问题的主管单位。在实践中，因为缺乏主管单位和公众的沟通交流平台，公众无法参与管理，便导致了第二个渠道的扩大，公众不得不找主管单位的上级单位。上级单位主要表现为人大和国家领导人。人大在转基因辩论的进程中发挥了重大作用，不过客观地讲，人大的功能并没有充分发挥出来。公众参与管理的第三个渠道是媒体，借助媒体发出自己的声音，并评价、批评其他媒体的观点。我们主要从这三个方面来分析，如何才能让公众更好地参与管理国家事务。

2.1　公众与转基因生物风险管理主管单位

（1）转基因法规缺少公众参与条款

中国是联合国《生物多样性公约》和《卡塔赫纳生物安全议定书》缔约国，《生物多样性公约》和《卡塔赫纳生物安全议定书》都有关于公众参与的条款。

中国国内的环境和资源保护法规，基本都有关于公众参与管理和监督的条款。1999 年，由国家环保总局、农业部、科技部、教育部、国家林业局、中科院、国家药品监督管理局发布的《中国国家生物安全框架》，特别提出建立生物安全管理的“公众参与制度”。

《中华人民共和国行政许可法》第 46 条规定：“法律、法规、规章规定实施行政许可应当听证的事项，或者行政机关认为需要听证的其他涉及公共利益的重大行政许可事项，行政机关应当向社会公告，并举行听证。”

转基因安全证书的发放、直接关系社会公众的健康安全，公众有权参与到证书颁布的过程中。

而 2001 年国务院发布的《农业转基因生物安全管理条例》中，没有涉及公众参与的条款。农业部于 2002 年发布的三个管理办法《农业转基因生物安全评价管理办法》、《农业转基因生物进口安全管理办法》和《农业转基因生物标识管理办法》，也没有涉及公众参与的条款。

我国《立法法》第 34 条第 1 款规定：“列入常务委员会会议议程的法律案，法律委员会、有关的专门委员会和常务委员会工作机构应当听取各方面的意见。听取意见可以采取座谈会、论证会、听证会等多种形式。”第 58 条规定：“行政法规在起草过程中，应当广泛听取有关机关、组织和公民的意见。听取意见可以采取座谈会、论证会、听证会等多种形式。”

我国的生物安全立法工作，应该切实保障公众对立法决策和立法结果的直接影响力。

（2）决策过程为少数转基因技术专家垄断

农业部下属的国家农业转基因生物安全委员会是颁布转基因安全证书的机构。委员会组成58人，参与实际讨论过程的常常只有10多人。组成人员主要是从事转基因研究的人士，其中多人正在申请转基因粮食专利和安全证书。也就是说自己申请、自己审批，自身的利益牵涉其中，难以保证其科学性、客观性。委员会中生物技术专家远远多于生态专家、医学专家和其他学科专家，也没有来自民间的社会团体和公众代表。委员会形成的结论也未征求相关部门意见。委员会的讨论过程和结果以及行政主管部门的决策不公开，透明度差，缺少公众参与和监督。

（3）问题发生后，缺少主管部门和公众的沟通交流平台

转基因大讨论的过程，充分暴露了我国种业、粮食、农业、生态环境、经济、科研体制、国家安全等诸方面的问题。通过此次讨论我们发现，中国在转基因主粮的研究、推广、监管乃至后期企业运作等方面还都不成熟，我们对于转基因安全性还有待进一步认识。这对于我们及早发现问题、解决问题意义重大，这是利国利民的好事。如果主管单位能够及时与公众建立起沟通交流的平台，更有效地发现问题、改正问题，一定会变被动为主动，提高政府职能部门的权威。遗憾的是，主管部门未能建立起来沟通交流平台。另外，主管部门的信息发布渠道比较滞后。

（4）公众法律诉诸渠道不畅通

国家要求对转基因生物及产品进行标识，但尚有许多不标识的情况，当公众购买了未经标识的产品后，却不能与生产者和销售者打官司，不能得到赔偿或补救。若是出现环境损害和健康损害后，难以诉诸法律。

2.2 公众与人大

人大是国家最高权力机关，对人民负责，受人民监督。转基因大讨论过程中，人大发挥了重大作用，人民的声音因此有了表达的渠道。公众看到国家最高权力机关内部也在激烈争论转基因问题，这在很大程度上稳定了公众的恐慌情绪，避免了非理性事件的发生。

这种社会关注度极高的大众安全议题，人大不应急于表态，人大可以考虑邀请双方的专家、公众代表共同讲座，这有助于稳定公众的恐慌情绪，也为正确决策提供依据。

2.3 公众与媒体

转基因讨论的热烈进行，表明中国的媒体环境正在变得愈来愈宽松。不过在讨论过程中，也还是多次发生了主管部门干涉新闻自由的事情。比如，山东电视台的转基因专题片《餐桌上的世纪悬念》在播出一期后，被要求停播；长江文艺出版社的书籍《转基因粮食凶猛》已经出版，等待发行，却被要求收回仓库；一些网站的转基因专题，被要求删除。新闻自由是有其前提的，那就是国家法律、社会伦常。转基因的大讨论，总的来说是健康的，应该允许其健康发展。

还需要指出一点，在激烈辩论过程中，代表国家喉舌的新闻媒体不应该急于表态。人民日报、新闻联播等媒体就是这样的国家喉舌。在媒体上应该允许两方面的声音同时存在，让两方面的声音在同一家媒体上针锋相对的辩论。新闻联播不适宜做辩论节目，但是人民日报可以做。如果国家喉舌拒绝辩论，不仅容易导致公众的恐慌情绪，也会使国家部门陷入被动，难以更改自己的决策主张。

综合以上内容可以看出，目前公众参与转基因生物风险管理存在诸多障碍。与此相对，

在欧洲一些国家，公众参与管理的形式非常多样。从简单的公告、说明会、记者发布会到复杂的咨询委员会、专家审查委员会、公众审查委员会、听证会、座谈会，还有非正式小型聚会、社区组织说明会、发行手册简讯、邮递名单、小组研究、民意调查、全民公决、设立公共通信站、发函征求意见等。另外，也有一些常规渠道之外的抗议活动。

我们可以根据中国国情，借鉴西方国家的一些好的做法，取其精华弃其糟粕，完善我们的公众参与制度，让人民真正参与到国家事务的民主管理当中。

3 公众如何更好参与转基因生物安全风险管理

目前，转基因问题在国内还在激烈争论当中。转基因大讨论的过程，充分暴露了我国种业、粮食、农业、生态环境、经济、科研体制、国家安全等诸方面的问题。通过此次讨论我们发现，中国在转基因主粮的研究、推广、监管乃至后期商业运作等方面还都不成熟，我们对于转基因安全性还有待进一步认识。这对于我们及早发现问题、解决问题意义重大，这是利国利民的好事。主管单位应该建立与公众有效沟通交流的渠道，才能有效地发现问题、改正问题。变被动为主动，使得部门利益和国家利益、人民利益一致。

通过对转基因大讨论这一公众事件的总结发现，我国亟待建立公众参与转基因生物风险管理的长效机制，从立法、决策、执行、监督、善后的一系列机制。

首先国家应确立和完善公众参与生物安全立法的参与权，使得法律法规能够真正代表人民利益、国家利益。

其次国家应确立和完善公众的知情权。知情权是公众参与的前提和基础，应通过立法直接保证公众的转基因生物安全知情权，并应具体规定知情权的行使方式、程序以及权力受到侵害后的法律后果与救济程序，从而提高生物安全管理的透明度。主管部门应保证公众能够获得相关材料和信息，并执行严格的标识制度，保证消费者购买食品的知情权和选择权。

再次国家应确立和完善公众的决策参与权。重大决策应征询公众意见，包括主管部门之外的其他部门、安委会之外的其他学科专家、非政府组织和广大人民的意见。

然后国家应该确立和完善公众对行政执法的监督权。实践表明，很多非法种植、非法流通的事实、很多存在的问题都是公众首先发现并提出的。公众直接监督生物安全行政执法活动，可以帮助行政机关更好的执法。

最后应建立和完善公众的法律诉讼权，当公众购买了未经标识的转基因产品，当环境安全、身体健康受到侵害时，公众可以拿起法律武器维护自身权益。

参考文献

[1] 薛达元. 转基因生物安全与管理. 北京：科学出版社，2009：318-337.

[2] 薛堃，尹帅军，汪文蓉，薛达元. 转基因生物风险管理中的公众参与. //中国科学院. 2012 高技术发展报告. 北京：科学出版社，2012：326.

双重风险下的中国转基因水稻研究

Double Risk Research of GM Rice in China

俞江丽
（绿色和平东亚办公室部门副经理）
YU Jiang-Li
（Assistant Manager of Greenpeace East Asia）

前言

粮食安全一直都是中国政府发展农业的重中之重。据农业部副部长李家洋介绍，2030年前后，我国人口将增加 2 亿以上，粮食总需求量将达到 7 亿 t 以上，单产需增加 40%以上[1]。由于资源环境的刚性约束，有一种观点认为只能依靠转基因技术保障未来 16 亿人口的粮食安全。2008 年，中国批准了总金额高达 200 多亿元的转基因生物新品种培育重大专项资金，旨在发展本国的转基因技术，突破当前农业发展中的重大瓶颈及发达国家的“科技要挟”，保障国家粮食主权和粮食安全。然而，跨国生物技术巨头无限期的专利控制和科研监管体系的漏洞，却令中国的转基因水稻科研面临双重风险。

绿色和平于 2012 年对中国转基因水稻科研面临的风险从两个角度进行了调查，包括对中国转基因水稻（包括抗鳞翅目害虫的克螟稻）研发涉及的国外专利调查，以及对八个省市市场上出售食品的抽样调查。调查结果表明：由于转基因技术专利被国外生物技术公司控制且无限期延长，中国的转基因水稻无法实现知识产权的独立，也就是说，一旦中国开始商业化种植受国外专利控制的转基因作物，国家的粮食主权将面临潜在风险[2]。另外由于国家对涉及转基因技术的科研单位监管不力，操作规程控制松懈，转基因水稻在一些省份的非法种植和流通的情况仍然存在。转基因水稻突破监管释放入环境，不仅对非转基因水稻等生物造成不可逆的基因污染，还让消费者在不知情的情况下失去选择权，暴露于食品安全的潜在风险中。此外，转基因水稻进入市场流通环节，已经给中国的粮食出口造成困境；而这种情况一旦扩大，就有可能涉嫌侵犯国外专利，甚至引起法律诉讼[3]。

1 “完美”的专利之网

在转基因作物的研发过程中，中国政府一直强调自主知识产权的重要性。由国家发展和改革委员会起草的《生物产业发展“十一五”规划（2006—2010）》将生物产业的发展目标定位为“具良好竞争前景的战略产业”，对其加以高度重视，并特别强调要发展具有自主知识产权的成果（专利权归属国内）[4]。

2008 年，中国批准实施了总金额高达 200 多亿元的转基因生物新品种培育重大专项，从财政上支持转基因技术的研发。2010 年中央 1 号文件提出：“继续实施转基因生物新品种培育科技重大专项，抓紧开发具有重要应用价值和自主知识产权的功能基因和生物新品种，在科学评估、依法管理基础上，推进转基因新品种产业化。”[5] 2012 年，国家发展和改革委员会在《全国农村经济发展“十二五”规划》中强调，要重视农业知识产权创造和运用，加大农业知识产权保护力度，维护好农业技术市场秩序。

然而，绿色和平于 2012 年对中国转基因水稻研发涉及的专利再次进行调查并发现，现状与政府的要求大相径庭。首先，一旦转基因作物涉及国外专利，将逃不开“专利网络”的控制，其影响将远远超出 20 年的专利保护期。此外，《材料转移协议》也对国内转基因科研具有无限期的专利制约。

1.1 *Bt* 水稻与绵延不绝的 *Cry* 基因专利

从 1996 年转基因作物大规模商业化至今，*Bt* 抗虫性状一直是全球范围内转基因作物的主要性状。在中国，2009 年获得转基因生物安全证书的转基因水稻也不例外。因此，绿色和平调查人员通过搜索美国和欧洲专利局的数据库，对目前全球范围内抗虫转基因作物常用的 *Cry* 基因系列涉及的 20 项专利进行了研究，并特别关注了 Cry1Ab 抗虫基因涉及的专利延长情况。

调查发现：孟山都每两、三年就会对其核心专利 US6，017，534 进行专利保护延长，而延长的手段主要是对基因序列进行小范围改动和修饰，变为“新”专利。以核心专利 US6，017，534 为例，权利要求书要求保护核心专利 US6，017，534 所涉及的 5 个 *Bt* 蛋白，而要实现对该专利所有权的延期，孟山都只需在“新”专利中使用略微更改甚至完全一样的蛋白即可。[附件二对专利涉及基因 DNA 和蛋白序列进行了详细比较。通过对比可见，“新”专利的基因序列仅有小范围的改动和修饰，蛋白质（氨基酸）的序列也仅有小改动，甚至完全一样。]

围绕核心专利，孟山都对涉及核心 *Bt* 蛋白的基因序列、相应转基因植物的获得方法等一系列方法和技术进行以“专利网络”形式进行保护，同样实现专利保护期的延长[6]。其中，专利 A（US 7，618，942）用于保护使用核心 *Bt* 蛋白的方法，专利 B（US7，304，206）用于保护相应转基因植物的转化，专利 C（US7，455，981）用于保护检测该核心 *Bt* 蛋白的方法（详细信息见附件一）。在实际应用中，一旦要使用该基因，必须涉及核心蛋白的方法和技术等，因此就涉及专利 A，B 和 C，也就是说，即使只用到技术中的一个步骤也会触及专利。所以这种“专利网络”的形式，也是专利持有人惯用的保护专利的方式，同样起到延长专利的作用。

由此可见，专利保护的有效期远远不止 20 年，每一粒转基因水稻种子，也远远不只涉及一两项专利，期待以专利过期来保障自主知识产权只能是一厢情愿。对于中国来说，一旦受到了国外专利的控制，就会逃不出专利的陷阱，将长期受制于人。一旦中国开始商业化种植受国外专利控制的转基因作物，由于没有独立自主的知识产权，对最终产品和品系没有控制权，势必会影响到长期的粮食安全，并威胁粮食主权。

1.2 “克螟稻”背后看不见的手

除了 *Bt* 转基因水稻，中外合制的“克螟稻”也是一个饱受专利控制的转基因品种。“克螟稻”研究由浙江农业大学（现浙江大学）原子核农业科学研究所植物保护系，以及加拿大渥太华大学生化系农业生物技术实验室共同完成。（研究论文见《转基因水稻“克螟稻”选育》[7]和《农杆菌介导的苏云金杆菌抗虫基因 cry1A（b）和 Cry1A（c）在水稻中的遗传转化及蛋白表达》[8]。）“克螟稻”研究得到浙江省科委“九五”重大项目和美国洛克菲勒基金水稻生物技术项目资助。该转基因稻种系采用农杆菌介导法将 Cry1A（b）导入外源抗虫基因粳稻品种秀水 11，并经过多代选育而成，对二化螟、三化螟等鳞翅目害虫具有抗性。

绿色和平调查发现克螟稻的研发涉及多项国外专利。

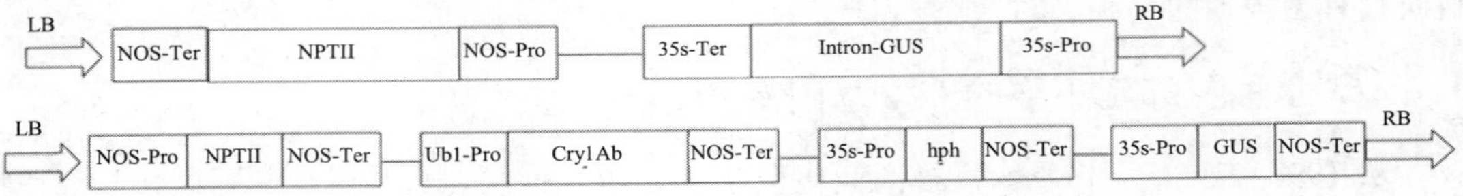

图 1 转基因水稻“克螟稻”所用载体结构图

根据公开发表的论文信息，克螟稻的基因研发可能涉及的国外专利有：该研究采用的标准遗传转化方法为农杆菌转化法（2 项专利），遗传转化载体中所涉及的标准元件包括启动子 CaMV35 s，Ubiquitin，终止子 NOS，标记基因 GUS、NptII、Hph 等。

本研究所用到的目的基因为 Cry1A（b），同样可能涉及多项国外专利，有 US5625136、US6051760、US5689052 等 5 项专利。

基于上述公开的信息，绿色和平检索专利数据库发现，有 10 项国外专利与“克螟稻”相关，此外还有文献中没有提到的研发过程中所用方法等。因此，如将所有相关的基因、基因调控元件、技术和方法等都列出，那么该转基因水稻涉及的专利可能达 12 项以上，均被国外生物技术公司包括孟山都，先正达，先锋等持有。（具体涉及的国外专利列表和详细信息参见附件一。）

根据“克螟稻”的研发论文《农杆菌介导的苏云金杆菌抗虫基因 cry1A（b）和 cry1A（c）在水稻中的遗传转化及蛋白表达》和《Transcriptional silencing and developmental reactivation of cry1ab gene in transgenic rice》[9]，该转基因水稻在遗传转化过程中所用的载体 pKUB 来自加拿大渥太华大学生化系，那么这种遗传元件的迁移即受到《材料转移协议》（合约）的制约。主要影响为，一旦接受方将材料应用于商业用途，就会受到提供方的制约，如拒绝商业授权，出售材料所有权等，而这些制约在材料原产国和接受国都有效，且不受时间限制。（具体协议的规定和产生的影响参见 2009 年绿色和平报告《谁是中国转基因水稻的真正主人》）。

据调查发现[10]，该研究合作方加拿大渥太华大学生化系的 Altosaar 教授的主要工作内容之一就是和各国签订材料转移协议，目前至少已与 25 个国家就 28 种农作物签订了材料转移协议，其中有 40 个实验室的协议是关于水稻的。“克螟稻”所使用的基因 Cry1Ab 不仅在 Altosaar 教授所签订协议中，而且为孟山都所控制，不难看出，转基因水稻“克螟稻”将面临《材料转移协议》和孟山都专利权的双重制约。

2 问题频现的转基因科研监管

目前我国没有批准任何转基因水稻的商业化种植，但大量的非法转基因水稻和转基因大米已在种子市场、田间和产品中出现。仅绿色和平就曾在湖北、湖南、江西、福建和广东发现转基因水稻种子及含有转基因水稻（*Bt*63）成分的米制品等。在转基因水稻未商业化的前提下，任何市场上流通的转基因水稻只可能来自于学术科研单位。显然，是转基因科研监管的不力导致了转基因试验材料流出实验田，进入消费市场。

2.1 转基因——中国出口水稻的“污点”

绿色和平在 2012 年着重关注了“克螟稻”的非法流通状况。作为转基因水稻品种，“克螟稻”尚未获得转基因生物安全证书及商业化生产资格，任何商业化种植和流通都属违法行为。《南方周末》调查显示[11]，前几年，克螟稻等转基因水稻就在浙江上虞出售，同时不少当地农民也自己留种进行种植。可见，一旦转基因水稻突破实验室的边界，就会由于种子公司的介入，私自育种及农民留种而加速流通和扩散。

2009 年初，浙江省检验检疫科学技术研究院动植检实验室就曾在浙江省出口欧盟米制品中多次检出转基因成分。当时的数据显示，2009 年 10 月份到 2010 年 4 月份的转基因阳性检出率高达 17.2%，检出的转基因成分为克螟稻的可能性较高。

此外，欧盟早在 2006 年就发现部分从中国进口的米制品含有未经批准的转基因稻米成分。根据欧盟委员会每周对外公布的欧盟食品饲料快速警报（RASFF）[12]，从 2006 年 9 月起至 2013 年 1 月底，欧盟有包括法国、德国、奥地利、希腊、塞浦路斯、意大利和英国在内的 17 个国家发出了总共 159 个涉及中国米制品含有转基因成分的通报，产品主要涉及米粉、大米蛋白等。其中，通报的含有转基因成分的中国米制品中有 12 个产品含有未经批准的转基因水稻克螟稻（KMD1）成分。由此可见，中国出口的米制品已经被未经国家批准商业化种植的克螟稻等转基因水稻所污染。

2011 年 12 月 23 日，欧盟委员会发布了《对中国出口大米制品中含有转基因成分采取紧急措施的决定》。根据该决定欧盟 27 国将对中国 25 种米制品采取强制性转基因成分检测，并依据检测结果采取退货和销毁处理措施。据了解，欧盟最新的管控系统能够检测出大米产品中的 26 种转基因物质，中国官方必须在向欧盟出口前对每一批米制品提交检验报告，表明是否含有转基因成分。而且，欧盟成员国还要加强抽样和检测的频率，使抽样和检测覆盖所有中国进口的米制品。这是对中国米制品实施的史上最为严苛的入境检查[13]。

2.2 转基因水稻地下流通实例——绿色和平 2012 年食品转基因检测结果

近年来，绿色和平一直在全国范围内进行食品中转基因成分的检测，屡次检出含有转基因成分的食品。2012 年年初至今，绿色和平针对北京、湖北、安徽、广东、四川、江苏、福建和浙江市面上可能含有转基因成分的大米、米粉、婴儿食品、大豆、豆制品及速冻食品进行随机取样，共购买76份样品，并送至第三方实验室进行转基因成分检测。

检测结果显示，9份样本呈阳性，其中6份样本含转基因水稻成分。也就是说，非法转基因成分污染的检出率达到7.9%。问题样本分别为：取自湖北武汉和安徽六安的4份大米样本检出转基因成分，并确认为*Bt*63转基因水稻；取自广东的2份米粉样本检出转基因成分，并确认为KMD1（克螟稻）转基因水稻；来自北京的3份豆奶粉检出转基因大豆成分，并确认为Roundup ready转基因大豆。检出转基因成分的详细结果见表1（全部76个样本信息见附件三）。

表1　绿色和平2012年转基因检测阳性结果信息

取样时间	取样地点	样品名称	样品信息	检测结果
2012/2/25	卜蜂莲花超市武汉市洪山区武珞路店	梁湖大米（香米）	产地：湖北 生产厂家：武汉市江夏区联发米厂汤氏米业	NOS、*Bt*63阳性
2012/4/1	百佳超市广州市黄埔分店	陈村猪骨浓汤过桥米线批号：G20120312A1、20130111前食用	制造商：佛山市顺德区春晓食品有限公司； 地址：佛山市顺德区陈村镇仙涌工业区大道东北6号； 委托生产商：高要市春晓食品有限公司（代码：G）； 地址：高要市蛟塘镇沙田工业园； 产地：广东省肇庆市	NOS、KMD1阳性
2012/4/17	乐购超市北京大成东店	乐购维他豆奶粉批号：2011/12/19	制造商：常州新区怡泰食品有限公司； 地址：常州市新北区天山路8号； 产地：江苏省常州市	35s、NOS、RR阳性
2012/4/17	乐购超市北京大成东店	维真维他型加钙豆奶粉批号：2012.10.12	制造商：江苏统业保健食品有限公司； 地址：江苏省姜堰经济开发区经二路； 产地：江苏泰州	35s、NOS、RR阳性
2012/4/17	乐购超市北京大成东店	南方纯豆粉批号：20110813 8326N	制造商：广西南方黑芝麻食品股份有限公司； 地址：广西容县荣州镇侨乡大道8号； 产地：广西玉林市	35s、NOS、RR阳性
2012/3/10	六安满天星超市南门店安徽省六安市皖西大道144号	散称丰良优大米	产地：六安寿县	NOS、*Bt*63阳性
2012/3/10	西商便民菜店兴美店安徽省六安市明珠步行街	周寨精制大米	产地：寿县周寨 厂商：安徽省寿县周寨米面有限公司	NOS、*Bt*63阳性
2012/3/10	刘记粮油安徽省六安市云路街菜市场	丝苗米	产地：寿县周寨 厂商：安徽省寿县周寨米面有限公司	NOS、*Bt*63阳性
2012/7/19	广东省肇庆市高要市蛟塘镇华顺达百货蛟塘店	陈村牌过桥米线（香菇炖鸡味）	制造商：佛山市顺德区春晓食品有限公司； 生产商：高要市春晓食品有限公司（代码：G）	NOS、KMD1阳性

3 结论和建议

转基因生物的研发需要长期大量的资金投入，也正因为如此，生物技术公司对知识产权的保护可谓面面俱到，毫无可乘之机。另一方面，中国国内对转基因科研的监管却显示出种种漏洞，导致了属于科研领域的转基因水稻大量流入市场。专利控制和监管不力，已经成为中国转基因生物研发面临的两道难题。而如果这两个问题得不到妥善的解决，中国的转基因水稻之路，也必然继续停断在粮食主权这个绕不过去的问题前：一旦中国批准商业化种植被国外专利技术牵制的转基因水稻，粮食主权必然面临潜在威胁。对此，绿色和平提出建议：

（1）相关政府部门全面调查分析转基因技术专利背后存在的潜在威胁，重新评估转基因专项的巨额投入是否可以真正摆脱专利陷阱，得到独立的自主知识产权

（2）加强涉及转基因技术的监管体系，健全相应的政策法规，有效控制非法转基因产品的种植和流通。

（3）加大生态农业的研究和投入，支持安全，环保和可持续发展的水稻生产模式，避免把转基因技术奉为一劳永逸解决粮食安全问题的圭臬。

注释

[1] 国务院新闻办公室 http://www.scio.gov.cn/xwfbh/xwbfbh/yg/2/201301/t1278047.

[2] 专利持有者有独一无二的经济特权，可以通过操纵转基因种子价格而影响粮食生产成本和农民生计，从而对一国经济和粮食安全产生影响.

[3] 例如在美国和加拿大，专利持有人通常会对违反专利要求的农民采取法律手段.

[4] 发展改革委员会，生物产业发展“十一五”规划. gjss.ndrc.gov.cn/gzdt/P020070429553828541347.doc.

[5] 新华网：“授权发布：2010 年中央一号文件全文”.http://news.xinhuanet.com/politics/2010-01/31/content_12907829.htm.

[6] 例如，仅 US6，017，534 一个专利就衍生出了一系列的相关专利，而这些专利全部都受到延长保护。比如在专利 US7，618，942 中，清晰地表明了专利之间的关系：“The present application is a divisional of application Ser. No. 11/372，065，filed Apr. 10，2006，now U.S. Pat. No. 7，304，206，which is a divisional of application Ser. No. 10/739，482，filed Dec. 18，2003，now U.S. Pat. No. 7，070，982，which is a divisional of application Ser. No. 09/636，746，filed Aug. 11，2000，now U.S. Pat. No. 6，713，063，which is a continuation-in-part of U.S. patent application Ser. No. 09/253，341，filed Feb. 19，1999，now U.S. Pat. No. 6，242，241，which is a continuation of U.S. patent application Ser. No. 08/922，505，filed Sep. 3，1997，now U.S. Pat. No. 6，110，464，which is a continuation-in-part of U.S. patent application Ser. No. 08/754，490，filed Nov. 20，1996，now U.S. Pat. No. 6，017，534; the entire contents of each is herein incorporated by reference.”这就是说专利 US7618942 来自于 US7304206，而 US7304206 来自于专利 US7070982，US7070982 来自于 us6713063，US6713063 来自于 US6242241，US6242241 来自于 US6110464，US6110464 来自于 US6017534.

[7] 舒庆尧，等. 转基因水稻“克螟稻”选育，浙江农业大学学报，1998，24（6）：579-580.

[8] 项友斌，等.农杆菌介导的苏云金杆菌抗虫基因 cry1A（b）和 cry1A（c）在水稻中的遗传转化及蛋白表达，生物工程学报，1999，15（4）：494-500.

[9] Wu gang etc.，Transcriptional silencing and developmental reactivation of cry1ab gene in transgenic rice，Science in China，2002，45（1）：68-78.

[10] Workshop on the Genetic Transformation of Cowpea，Capri，Italy，October 31-November 2，2002. http://www.entm.purdue.edu/NGICA/reports/italy_proceedings.pdf.

[11] 《南方周末》：“漏网之稻——‘高墙深河’、严打整治，仍见转基因违规扩散”，http://www.infzm.com/content/77897.

[12] 中国技术性贸易措施网：http://www.tbt-sps.gov.cn/riskinfo/dataquery/Pages/rasff.aspx，http://www.tbt-sps.gov.cn/riskinfo/dataquery/Pages/rasff.aspx.

[13] 国际商报：“欧盟对中国转基因大米制品采取紧急措施”，http://intl.ce.cn/sjjj/qy/201112/31/t20111231_22965863.shtml 发展改革委员会，生物产业发展“十一五”规划，gjss.ndrc.gov.cn/gzdt/P020070429553828541347.doc.

附件一 孟山都 Cry 相关基因的专利列表

专利名称	专利号	发明人	签发日期	所有人
Hybrid Bacillus thuringiensis. delta.-endotoxins with novel broad-spectrum insecticidal activity	US6，017，534	Malvar；Thomas（Dublin，PA），Gilmer；Amy Jelen（Langhorne，PA）	November 20，1996	Ecogen，Inc.（Langhorne，PA）Monsanto Company（St. Louis，MO）
Polynucleotide compositions encoding Cry1Ac/Cry1F Chimeric O-endotoxins	US6，326，169	Malvar；Thomas（Dublin，PA），Gilmer；Amy Jelen（Langhorne，PA）	March 2，1999	Monsanto Company（St. Louis，MO）
Polynucleotide compositions encoding broad-spectrum .delta. endotoxins	US6，645，497	Malvar；Thomas（Dublin，PA），Gilmer；Amy Jelen（Langhorne，PA	November 30，2001	Monsanto Company（St. Louis，MO）
Hybrid Bacillus thuringiensis. delta.-endotoxins with novel broad-spectrum insecticidal activity	US6，962，705	Malvar；Thomas（Dublin，PA），Gilmer；Amy Jelen（Langhorne，PA）	September 26，2003	Monsanto Company（St. Louis，MO）
Broad-spectrum . delta.-endotoxins	US6110464	Malvar；Thomas（Dublin，PA），Gilmer；Amy Jelen（Langhorne，PA）	September 3，1997	Monsanto Company（St. Louis，MO）
Hybrid Bacillus thuringiensis. delta.-endotoxins with novel broad-spectrum insecticidal activity	US6，156，573	Malvar；Thomas（Dublin，PA），Gilmer；Amy Jelen（Langhorne，PA）	March 2，1999	Monsanto Company（St. Louis，MO）
Chimeric bacillus thuringiensis-endotoxins and host cells expressing same	US6，221，649	Malvar；Thomas（Dublin，PA），Gilmer；Amy Jelen（Langhorne，PA）	March 2，1999	Monsanto Company（St. Louis，MO）
Polynucleotide compositions encoding broad-spectrum . delta.-endotoxins	US6242241	Malvar；Thomas（Dublin，PA），Gilmer；Amy Jelen（Langhorne，PA）	February 19，1999	Monsanto Company（St. Louis，MO）
Broad-spectrum insect resistant transgenic plants	US6281016	Malvar；Thomas（Duglin，PA），Gilmer；Amy Jelen（Langhorne，PA）	February 19，1999	Monsanto Company（St. Louis，MO）
Polynucleotide compositions encoding broad spectrum . delta.-endotoxins	US6521442	Malvar；Thomas（Dublin，PA），Gilmer；Amy Jelen（Langhorne，PA）	July 27，2001	Monsanto Company（St. Louis，MO）
Polynucleotide compositions encoding broad spectrum delta-endotoxins	US6538109	Malvar；Thomas（Dublin，PA），Gilmer；Amy Jelen（Langhorne，PA）	June 4，2001	Monsanto Company（St. Louis，MO）

专利名称	专利号	发明人	签发日期	所有人
Broad-spectrum . delta.-endotoxins	US6713063	Malvar；Thomas（Troy，MO），Mohan；Komarlingam Sukavaneaswaran（Bangalore，IN），Sivasupramaniam；Sakuntala（Chesterfield，MO）	August 11，2000	Monsanto Company（St. Louis，MO）
Antibodies immunologically reactive with broad-spectrum . delta. Endotoxins	US6746871	Malvar；Thomas（Dublin，PA），Gilmer；Amy Jelen（Langhorne，PA）	February 12，2003	Monsanto Company（St. Louis，MO）
Antibodies immunologically reactive with broad-spectrum delta endotoxins	US6951922	Malvar；Thomas（Dublin，PA），Gilmer；Amy Jelen（Langhorne，PA）	April 2，2004	Monsanto Company（St. Louis，MO）
Polynucleotide compositions encoding broad spectrum delta-endotoxins	US7070982	Malvar；Thomas（Troy，MO），Mohan；Komarlingam Sukavancaswaran（Bangalore，IN），Sivasupramaniam；Sakuntala（Chesterfield，MO）	December 18，2003	Monsanto Company（St. Louis，MO）
Hybrid Bacillus thuringiensis Cry1A/Cry1F DNA and plants and host cells transformed with same	US7250501	Malvar；Thomas（Dublin，PA），Gilmer；Amy Jelen（Langhorne，PA）	September 15，2005	Monsanto Company（St. Louis，MO）
Plants transformed with polynucleotides encoding broad-spectrum delta-endotoxins	US7304206	Malvar；Thomas（Troy，MO），Mohan；Komarlingam Sukavaneaswaran（Bangalore，IN），Sivasupramaniam；Sakuntala（Chesterfield，MO）	April 10，2006	Monsanto Company（St. Louis，MO）
Method of detecting Broad-spectrum delta-endotoxins	US7455981	Malvar；Thomas（Dublin，PA），Gilmer；Amy Jelen（Langhorne，PA）	July 23，2007	Monsanto Company（St. Louis，MO）
Method of using broad-spectrum delta-endotoxins	US7618942	Malvar；Thomas（Troy，MO），Mohan；Komarlingam Sukavaneaswaran（Bangalore，IN），Sivasupramaniam；Sakuntala（Chesterfield，MO）	November 28，2007	Monsanto Company（St. Louis，MO）
Broad-spectrum .delta-endotoxins and method and kit for detecting same and for detecting polynucleotides encoding same	US7927598	Malvar；Thomas（Dublin，PA），Gilmer；Amy Jelen（Langhorne，PA）	November 24，2008	Monsanto Company（St. Louis，MO）

附件二　Cry 基因之间 DNA 和蛋白序列对比

专利名称	序列名称	碱基长度	对比序列	差异	氨基酸差异
US6326169	SEQ No. 9	～3500	US6017534 SEQ No. 9	1 个碱基	ATT 变为 ATA，都表达 ILE。 完全一样
US6326169	SEQ No. 11	～3500	US6017534 SEQ No. 11	完全一样	完全一样
US6326169	SEQ No. 13	～3500	US6017534 SEQ No. 13	完全一样	完全一样
US6326169	SEQ No. 25	～3500	US6017534 SEQ No. 25	两个碱基	ASP 变为 HIS THR 变为 ARG 两个氨基酸不同
US6326169	SEQ No. 27	～3500	US6017534 SEQ No. 27	19 个碱基	1.CCA 变为 CCT，都为 PRO，没有变化 2.TTT 变为 TTC，都表达 PHE，没有变化 3.ACG 变为 ATA，THR 变为 ILE 4.CAC 变为 CAT，都为 HIS，没有变化 5.GCA 变为 GCT，都为 ALA，没有变化 6.ACC 变为 GAA，THR 变为 GLU 7.CCT 变为 TTT，PRO 变为 PHE 8.ACA 变为 AAT，THR 变为 ASN 9.ACA 变为 ATA，THR 变为 ILE 10.GAT 变为 GCA，ASN 变为 ALA 11.CCG 变为 TCG，PRO 变为 SER 12.GAG 变为 GAT，GLU 变为 ASP 13.AGG 变为 AGT，ARG 变为 SER
US6326169	SEQ No. 29	～3500	US6017534 SEQ No. 29	完全一样	完全一样
US6645497	SEQ No. 9	～3500	US6326169 SEQ No. 9	13 个碱基	1.GAT 变为 CAT，ASP 变为 HIS 2.CCG 变为 CCC，都表达 PRO 3.GGA 变为 CCA，GLY 变为 PRO 4.CGT 变为 CCT，ARG 变为 PRO 5.AGT 变为 ACT，SER 变为 THR 6.GGA 变为 GCA，GLY 变为 ALA 7.添加了 GAG，即增加了 GLU 8.删减了 GCT，即删减了 ALA 9.GAT 变为 CAT，即 ASP 变为 HIS
US6645497	SEQ No. 11	～3500	US6326169 SEQ No. 11	2 个碱基	1.GAA 变为 CAA，GLU 变为 GLN 2.GAA 变为 CAA，GLU 变为 GLN
US6645497	SEQ No. 13	～3500	US6326169 SEQ No. 13	2 个碱基	1.GAT 变为 GAC，都表达 ASP 2.GGG 变为 CGG，GLY 变为 ARG 仅有一个氨基酸差异

专利名称	序列名称	碱基长度	对比序列	差异	氨基酸差异
US6645497	SEQ No. 27	～3500	US6326169 SEQ No. 27	4 个碱基	1.CCG 变为 CCC，都表达 PRO 2.AGT 变为 ACT，SER 变为 THR 3.AAG 变为 AAC，LYS 变为 ASN 4.TCG 变为 TCC，都表达 SER 仅有两个氨基酸不同
US6645497	SEQ No. 29	～3500	US6326169 SEQ No. 29	5 个碱基	1.GTA 变为 CTA，VAL 变为 LEU 2.GGG 变为 CGG，GLY 变为 ARG 3.AGT 变为 ACT，SER 变为 THR 4.AGT 变为 ACT，SER 变为 THR 5.ATG 变为 ATC，MET 变为 ILE 五个氨基酸不同
US6962705	SEQ No. 13	～3500	US6645497 SEQ No. 13	2 个碱基	1.CAT 变为 GAT，HIS 变为 ASP 2.CGG 变成 GGG，ARG 变为 GLY 两个氨基酸差异
US6962705	SEQ No. 25	～3500	US6326169 SEQ No. 25	14 个碱基	1.GCA 变为 ACA，ALA 变为 THR 2.GGC 变为 CCC，GLY 变为 PRO 3.GAA 变为 CAA，GLU 变为 GLN 4.CAT 变为 GAT，HIS 变为 ASP 5.CGA 变为 GGA，ARG 变为 GLY 6.GGA 变为 CGA，GLY 变为 ARG 7.CCT 变为 GCT，PRO 变为 ALA 8.AAC 变为 CAC，ASN 变为 HIS 9.AAA 变为 AAC，LYS 变为 ASN 10.GCG 改为 GCC，都表达 ALA 11.GCG 改为 GCC，都表达 ALA 12.TCG 变为 TCC，都表达 SER 13.AAC 变为 AAG，ASN 变为 LYS 14.TAC 改为 TAG，终止子

附件三 转基因水稻“克螟稻”的国外专利研究

专利名称	专利号	国家	申请日期	发明者	专利拥有者
Cry1Ab					
Synthetic DNA sequence having enhanced insecticidal activity in maize	US5625136	United States	19920925	Koziel；Michael G.（Cary，NC），Desai；Nalini M.（Cary，NC），Lewis；Kelly S.（Hillsborough，NC），Kramer；Vance C.（Hillsborough，NC），Warren；Gregory W.（Cary，NC），Evola；Stephen V.（Apex，NC），Crossland；Lyle D.（Chapel Hill，NC），Wright；Martha S.（Cary，NC），Merlin；Ellis J.（Raleigh，NC），Launis；Karen L.（Franklinton，NC），Rothstein；Steven J.（Guelph，CA），Bowman；Cindy G.（Cary，NC），Dawson；John L.（Chapel Hill，NC），Dunder；Erik M.（Chapel Hill，NC），Pace；Gary M.（Cary，NC），Suttie；Janet L.（Raleigh，NC）	Ciba-Geigy Corporation（Tarrytown，NY）/NOVARTIS FINANCE CORPORATION/S-YNGENTA INVESTMENT CORPORATION，DELAWARE
	AT221916	Austria	19921005		
	AU2795292	Australia	19920925		
	BG62782	Bulgaria	19920925		
	BG98747	Bulgaria	19940503		
	BR9206578	Brazil	19921005		
	CA2120514	Canada	19921005		
	CZ292953	Czech	19921005		
	DE69232725	Germany	19921005		
	DK0618976	Danmark	19921005		
	EP0618976	EU	19921005		
	EP1209237	EU	19921005		
	EP1213356	EU	19921005		
	ES2181678	Spain	19920925		
	HU220294	Hungary	19921005		
	JPH07500012	Japan	19921005		
	JP2003189888	Japan	19921005		
	RO110263	Romania	19921005		
	RU2202611	Russia	19920925		
	SK37894	Slovakia	19921005		
Synthetic DNA sequence having enhanced insecticidal activity in maize	US6051760	United States	19950602	Koziel；Michael G.（Cary，NC），Desai；Nalini M.（Cary，NC），Lewis；Kelly S.（Hillsborough，NC），Kramer；Vance C.（Hillsborough，NC），Warren；Gregory W.（Cary，NC），Evola；Stephen V.（Apex，NC），Wright；Martha S.（Cary，NC），Launis；Karen L.（Franklinton，NC），Rothstein；Steven J.（Guelph，CA），Bowman；Cindy G.（Cary，NC），Dawson；John L.（Chapel Hill，NC），Dunder；Erik M.（Chapel Hill，NC），Pace；Gary M.（Cary，NC），Suttie；Janet L.（Raleigh，NC）	Novartis Finance Corporation/SYNG-ENTA INVESTMENT CORPORATION
	US6075185	United States	19950602		
	US6320100	United States	20000411		
Synthetic DNA sequence having enhanced activity in maize	US5859336	United States	19950602	Koziel；Michael G.（Cary，NC），Desai；Nalini M.（Cary，NC），Lewis；Kelly S.（Hillborough，NC），Warren；Gregory W.（Cary，NC），Evola；Stephen V.（Apex，NC），Crossland；Lyle D.（Chapel Hill，NC），Wright；Martha S.（Cary，NC），Merlin；Ellis J.（Raleigh，NC），Launis；Karen L.（Franklinton，NC），Bowman；Cindy G.（Cary，NC），Dawson；John L.（Chapel Hill，NC），Dunder；Erik M.（Chapel Hill，NC），Pace；Gary M.（Cary，NC），Suttie；Janet L.（Raleigh，NC）	NOVARTIS CORPORATION/SYNGENTA INVESTMENT CORPORATION

专利名称	专利号	国家	申请日期	发明者	专利拥有者
Synthetic DNA sequences having enhanced expression in monocotyledonous plants and method for preparation thereof	US5689052	United States	19950919	Brown；Sherri Marie（Chesterfield，MO），Dean；Duff Allen（St. Louis，MO），Fromm；Michael Ernest（Chesterfield，MO），Sanders；Patricia Rigden（Chesterfield，MO）	Monsanto Company（St. Louis，MO）
	US6180774	United States	19970805		
Method for producing a plant-optimized nucleic acid coding sequence	US6121014	United States	19950602	Koziel；Michael G.（Cary，NC），Desai；Nalini M.（Cary，NC），Lewis；Kelly S.（Hillsborough，NC）	SYNGENTA INVESTMENT CORPORATION/ NOVARTIS FINANCE CORPORATION
Method of producing transgenic maize using direct transformation of commercially important genotypes	US6403865	United States	19950510	Koziel；Michael G.（Cary，NC），Desai；Nalini M.（Cary，NC），Lewis；Kelly S.（Hillsborough，NC），Kramer；Vance C.（Hillsborough，NC），Warren；Gregory W.（Cary，NC），Evola；Stephen V.（Apex，NC），Crossland；Lyle D.（Chapel Hill，NC），Wright；Martha S.（Cary，NC），Merlin；Ellis J.（Raleigh，NC），Launis；Karen L.（Franklinton，NC），Rothstein；Steven J.（Guelph，CA），Bowman；Cindy G.（Cary，NC），Dawson；John L.（Chapel Hill，NC），Dunder；Erik M.（Chapel Hill，NC），Pace；Gary M.（Cary，NC），Suttie；Janet L.（Raleigh，NC），Carozzi；Nadine（Raleigh，NC），De Framond；Annick（Durham，NC），Linder；James O.（Owatonna，MN），Miller；Robert L.（Cedar Rapids，IA），Skillings；Bruce W.（Innerkip，CA），Mousel；Alan W.（Bluffton，IN），Hornbrook；Albert R.（Bloomington，IL），Clucas；Christopher P.（Washington Court House，OH），Meghji；Moez Rajabali（Bloomington，IL），Tanner；Andreas H.（Plaisance du Touch，FR），Cassagne；Francis E.（Auch，FR），Pollini；Gilles（L'Isle en Dodon，FR），Colbert；Terry Ray（Troy，TN），Cammack；Francis P.（Rochelle，IL）	CIBA-GEIGY CORPORATION/ NOVARTIS FINANCE CORPORATION/ SYNGENTA INVESTMENT CORPORATION

专利名称	专利号	国家	申请日期	发明者	专利拥有者
Expression of the toxic portion of Cry1A in plants	US683349	United States	20030321	BARTON KENNETH A[US]; MILLER MICHAEL J[US]	Monsanto
Agrobacterium-mediated transformation					
Process for the incorporation of foreign dna into the genome of dicotyledonous plants	US4940838	United States	19840223	Schilperoort; Robbert A.（CD Leiden, NL）, Hoekema; Andreas（ET Leiden, NL）, Hooykaas; Paul J. J.（AW Leiden, NL）	SYNGENTA MOGEN B.V., NETHERLANDS
	AT68829	Austra	19840221		
	JP6209779	Japan	19940802		
	JP7036751	Japan	19950426		
	JP2003610	Japan	19951220		
INTEGRATION OF EXTERNAL DNA TO CHROMOSOME OF DICOTYLEDO-NOUS	JP60070080	Japan	19850420	ROBERUTO ADORIAAN SHIRUPERUURU; ANDOREASU HOOKEMA; POORU YAN YAKOBU HOOIKAASU	Unkown
Process for the incorporation of foreign DNA into the genome of dicotyledonous plants	US5464763	United States	19931223	SCHILPEROORT ROBBERT A[NL]; HOEKEMA ANDREAS[NL]	SYNGENTA MOGEN B.V., NETHERLANDS
Shuttle vector comprising a T-DNA Region and RI and RK2 origins of replication	US6, 165, 780	United States		Kawasaki; Shinji（Ibaraki, JP）	National Institute of Agrobiological Resources, Ministry of Agriculture, Forestry and Fisheries（Tsukuba, JP）
New high capacity binary shuttle vector	AU721577	Australia	19980522	KAWASAKI SHINJI	NATIONAL INSTITUTE OF AGROBIOLOGICAL RESOURCES, MINISTRY OF AGR
	AU3922697	Australia	19970925		
	AU6804898	Australia	19980522		
	CA2216596	Canada	19970926		
Methods for agrobacterium-mediated transformation	US5981840	United States	19970124	Zhao; Zuo-Yu（Urbandale, IA）, Gu; Weining（Urbandale, IA）, Cai; Tishu（Urbandale, IA）, Pierce; Dorothy A.（Urbandale, IA）	PIONEER HI BRED INT[US]
	AR011432	Argentina	19980123		
	AU727849	Australia	19980123		
	AU6036898	Australia	19980123		
	CA2278618	Canada	19980123		
	EP0971578	EU	19980123		
	US6822144	United States	19971103		

专利名称	专利号	国家	申请日期	发明者	专利拥有者
GUS					
DNA sequences encoding polypeptides having beta-1，3-glucanase activity	US6632981	United States	20010716	Meins，Jr.；Frederick（Riehen，CH），Shinshi；Hideaki（Tsuchiura，JP），Wenzler；Herman C.（Plano，TX），Hofsteenge；Jan（Reinach，CH），Ryals；John A.（Cary，NC），Sperisen；Christoph（Birmensdorf，CH）	Novartis Finance Corporation（New York，NY）
IDENTIFICATION METHOD OF CHEMICAL REGULATOR AND ANALYZING METHOD OF CHEMICAL REGULATOR	KR100247622	South Korea	19970609	RYALS JOHN[US]；MONTOYA ALICE[US]；HARMS CHRISTIAN[DE]；DUESING JOHN[US]；SPERISEN CHRISTOP[CH]；MEINS FRED[US]；PAYNE GEORGE[US]	NOVARTIS AG[CH]
NptII					
Chimeric genes suitable for expression in plant cells	US6174724	United States	19950504	ROGERS STEPHEN G[US]；FRALEY ROBERT T[US]	Monsanto
Hph					
METHOD FOR SELECTING PLANT CELL AND METHOD FOR REGENERATION OF PLANT BY USING IT	JP8224085	Japan	19951228	KURAIBU UORUDORON	LILLY CO ELI[US]
Selectable marker for development of vectors and transformation systems in plants	US 5668298	United States	19950607	WALDRON CLIVE[US]	LILLY CO ELI[US]/NOVARTIS AG，SWITZERLAND
	US 6048730	United States	19900919		
Hygromycin-resistant transgenic plants	US6365799	United States	20000209	WALDRON CLIVE[US]	NOVARTIS AG，WITZERLAND
Chimeric genes suitable for expression in plant cells	US6174724	United States	19950504	ROGERS STEPHEN G[US]；FRALEY ROBERT T[US]	Monsanto

专利名称	专利号	国家	申请日期	发明者	专利拥有者
CaMV 35s Promoter					
Chimeric genes for transforming plant cells using viral promoters	US 5352605	United States	19931028	FRALEY ROBERT T[US]; HORSCH ROBERT B[US]; ROGERS STEPHEN G[US]	Monsanto
	US5530196	United States	19940902		
	US5858742	United States	19960624		
	US6255560	United States	19990111		
Chimeric genes suitable for expression in plant cells	US6174724	United States	19950504	ROGERS STEPHEN G[US]; FRALEY ROBERT T[US]	Monsanto
Maize Ubiquitin Promoter					
Plant ubiquitin promoter system	US5614399	United States	19950605	QUAIL PETER H[US]; CHRISTENSEN ALAN H[US]; HERSHEY HOWARD P[US]; SHARROCK ROBERT A[US]; SULLIVAN THOMAS D[US]	MYCOGEN PLANT SCIENCE INC[US]
	US5510474	United States	19940825		
	US6020190	United States	19961118		
	US6054574	United States	19980609		
	CA 1339684	Canada	19890516		
	JP11332565	Japan	19990512		
Modified ubiquitin regulatory system	US6878818	United States	20020612	GOLDSBROUGH ANDREW	Monsanto
	AT328098	Austria	20000907		
	AU769567	Australia	20000907		
	AU7515500	Australia	20000907		
	CA2384517	Canada	20000907		
	CZ20020695	Czech	20000907		
	DE60028388	Germany	20000907		
	DK1210446	Danmark	20000907		
	EP1210446	EU	20000907		
	ES2265978	Spain	20000907		
	HU0202852	Hungary	20000907		
	PL354747	Poland	20000907		
	PT1210446	Portugal	20000907		
	ZA200201758	South Africa	20020301		
Novel plant promoter sequences and methods of use for same	US6977325	United States	20020228	JILKA JOSEPH M; HOOD ELIZABETH E; HOWARD JOHN A	PRODIGENE INC
	AU7543301	Australia	20010608		

专利名称	专利号	国家	申请日期	发明者	专利拥有者
Nos terminator and Nos promoter					
Isolated DNA sequence able to function as a termination region in chimeric genes for plant transformation	AU680899B	Australia	19940624	Atanassova Rossitza；Rose Richard De；Freyssinet Georges；Gigot Claude；Lebrun Michel	Rhone Poulenc Agrochimie（FR）/Bayer Sci
	BR9401842	Brazil	19940622		
	CA2126806	Canada	19940627		
	EP0633317	EU	19940623		
	CN1253570C	China	19940627		
	JP7008278	Japan	19940624		
	IL110069	Israel	19940620		
	US6313282	United States	19970718		
PLANT PROMOTERS AND PLANT TERMINATORS	US7202083	United States	19990930	NISHIKAWA SATOMI；OEDA KENJI	SUMITOMO CHEMICAL CO
	AR021842	Argentina	19990930		
	AT345393	Austria	19990928		
	AU5761199	Australia	19990928		
	BR9914130	Brazil	19990928		
	CA2343652	Canada	19990928		
	CN1321198	China	19990928		
	DE69934023	Germany	19990928		
	EP1117812	EU	19990928		
	JP2000166577	Japan	19991001		

专题2　转基因生物相关的社会经济影响

Session 2　Socio-economic Considerations related to GMOs

Socio-Economic Aspects in the Assessment of GMOs

转基因生物的社会经济评估

Dr. Andreas Heissenberger
（Umweltbundesamt – Environment Agency Austria 奥地利环境保护局）

The legal possibility to consider socio economic effects of GMOs in decisions on their authorization is laid down in Article 26 of the Cartagena Protocol on Biosafety. This enables countries which are Parties to the Protocol to take up respective provisions in their national biosafety legislation. As the Article 26 also foresees the compatibility of such provisions with "other international obligations", i.e. mainly WTO obligations, these rules have also to be followed. Socio-economic considerations（SEC）have become more and more important on the agenda of the Cartagena Protocol and have been controversially discussed during the last COP/MOPs. However, despite these controversies the aim to develop guidance on how to take SEC into account has been included in the Strategic Plan of the Protocol in 2010, and the first activities to reach that aim have already started. In a workshop in 2011 and an online forum in March and April 2013 it became quite clear that the viewpoints of different stakeholders and experts are far apart from each other on various issues. First there is the issue of how broad the scope of a socio economic assessment should be（very narrow – economic risk/benefit analysis only, or broad and covering also cultural and social aspects as well）. A second mayor issue is the fear of possible trade restrictions and a delay in the authorization process, if SEC are part of the decision making process. Other points include the rights of indigenous peoples and local communities, challenges regarding ex-ante and ex-post evaluations, and data availability. No agreement has been reached on these points so far and they will be discussed in online conferences and an ad hoc technical expert group later in 2013.

In a project carried out by the Environment Agency Austria a concept on how to identify and assess possible socio economic effects was developed. Based on the well-recognized "pillars of sustainability"（economic, social and environmental aspects）categories were established. For these different categories, e.g. productivity, health and welfare, or ecosystem functioning, assessment criteria were proposed. For each of these criteria the key issues were identified. These key issues should be the basis for the development of assessment tools and the collection of data. The developed concept, which needs to be adapted to the local or regional circumstances and the respective GMO in the authorization process, could serve as a model on how to carry out a comprehensive and standardized assessment of possible socio economic effects.

umweltbundesamt

Socio economic considerations

Andreas Heissenberger

umweltbundesamt

Content

- Cartagena Protocol and national implementation
- Austrian Study
- Conclusions

umweltbundesamt

Regulatory aspects

- Cartagena Protocol on Biosafety, Article 26
- *The Parties, in reaching a decision on import under this Protocol or under its domestic measures implementing the Protocol, may take into account, consistent with their international obligations, socio-economic considerations arising from the impact of living modified organisms on the conservation and sustainable use of biological diversity, especially with regard to the value of biological diversity to indigenous and local communities.*

umweltbundesamt

The Cartagena Protocol on Biosafety

- Agreement under international law
- Currently 166 Parties
- Article 26 was heavily discussed from the beginning
- Activities and discussions only in the context of capacity building
- At COPMOP5 (2010) taken up in the Strategic Plan with the final goal to develop guidance documents by 2020
- At COPMOP6 (2012) decision on activities until 2014

umweltbundesamt

National implementation

- Cartagena protocol does allow to use socio economic considerations in decision making
- National provisions can be taken up in biosafety laws
- Some countries have done so (e.g. Norway, Indonesia, Nigeria)
- National provisions may be stricter and broader than the Cartagena Protocol
- Provisions need to be in-line with other international obligations, e.g. for WTO to be non-discriminatory
- Currently no guidance on assessment criteria and methodology available
- Implementation difficult (mainly due to lack of data)

umweltbundesamt

The Cartagena Protocol - AHTEG

- Planned until COP/MOP7: Ad hoc Technical Expert Group (AHTEG)
- Decision by COPMOP5
- Mandate
  - *Work based on stock-taking by the Secretariat of the Protocol* (consultancy – ongoing)
  - *Based on outcome of online-forum* (finalized, online conferences scheduled for June 2013)
  - *Develop conceptual clarity*
- Conceptual clarity: Definition of the scope of the assessment, e.g.
  - Limited to economic assessment
  - Farm scale or macro-economic
  - Social/cultural aspects
- First meeting scheduled for October 2013

umweltbundesamt

Austrian activities

- Study commissioned by Ministry of Health to provide overview on worldwide situation
  - http://bmg.gv.at/cms/home/attachments/5/0/0/CH1050/CMS1291038713992/assessing_socio-conomic_impacts_of_gmos,_band_2_20101.pdf
- Study commissioned to Environment Agency Austria by Ministry of the Environment: "Socio-economic aspects in the assessment of GMOs - options for action"
  - http://www.umweltbundesamt.at/fileadmin/site/publikationen/REP0354.pdf

umweltbundesamt

Aim of the report

- Analysis of experiences with socio-economic assessments
- Identification of important scientific issues
- Assessment of data availability and limitations
- Identification of socio-economic aspects of major relevance for Austria
  - Focusing on risks and potential negative effects of GMO cultivation
  - Potential benefits not covered (not in mandate by ministry)
- Development of a câtalogue of criteria

umweltbundesamt

Overarching issues

- Important scientific issues
  - Assessment level: Single event vs. plant or trait specific impacts
  - Assessment Unit: Farm level vs. economic sectors vs. political units
  - Assessment method: Assumptions of economic models, data basis, methodology, parameters
  - Regional Assessment: Environmental, economic and socio-cultural conditions of country/region
- Limitations and data availability and quality
  - Data for Europe (developed countries) are rare (most studies from developing countries)
  - Data quality is limited (mainly on farm level)
  - Extrapolation of results is often not possible
  - High regional variation between crops used, geographic conditions, farming systems, environmental conditions and over time
  - Methods influence data quality and results (farmer's interviews, economic models)

umweltbundesamt

Catalogue of criteria

- Structured along the three pillars of sustainability

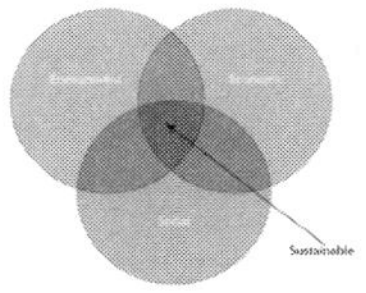

- Additional aspects
  - Ethical issues
  - Third countries

umweltbundesamt

The Cartagena Protocol- Online forum

- Online forum (March/April 2013)
- Controversial discussion
- Agreement that it is responsibility of Parties to decide if they take socio economic considerations into account
- Disagreement on many different issues
- results at:
  http://bch.cbd.int/onlineconferences/portal_art26/se_forum_discussiongroups.shtml

umweltbundesamt

The Cartagena Protocol- Online forum

- Narrow scope, e.g.
  - Economic risk benefit analysis only
  - Main target are adopters of GM technology
  - Limited to direct effects on biological diversity and directly related socio economic effects
- Not compatible with WTO
- Not linked to risk assessment/separate process
- Limited to indigenous peoples
- Limit to hard data (ex post evaluation)
- Enough data and experience available

- Broad scope, e.g.
  - Also social changes and cultural values
  - Assessment needs to cover non-GM farmers and food chain
  - National implementation allows to go far beyond direct effects on biodiversity
- WTO rules allow SECs in decisions
- Linked to risk assessment/parallel process
- Includes also local communities (also in industrialized countries)
- Models need to be included (also ex-ante evaluation)
- Data limited to economic effects and few regions

umweltbundesamt

Catalogue of criteria

- **Categories for each aspect**

Aspect	Category
Economic	Productivity
	Costs
	Tourism
Social	Health and welfare
	Choice
	Cultural aspects
	Social cohesion
Ecological	Ecological limits and ecosystem functioning
	Biodiversity and nature conservation

- **Criteria for each category**

umweltbundesamt

Catalogue of criteria

- Economic criteria
  - Profit (e.g. reduced yield, squeezing out of production of special products like organic products)
  - Prosperity (e.g. reduced employment possibilities)
  - Preservation of the environmental basis (e.g. negative effects on natural resources, ecosystem services)
  - Costs during the whole production chain (e.g. increased costs for coexistence, security & control, image loss)
  - Indirect costs (e.g. welfare system)
  - Tourists' expectations (e.g. loss of attractiveness, landscape changes)
  - Possibility to implement regional policy (e.g. GM-free regions)

Catalogue of criteria

- Social criteria
  - Quality of life (e.g. income, health)
  - Food availability (e.g. GM-free products)
  - Labelling regime (e.g. freedom of choice)
  - Accessibility of alternatives and dependency (e.g. costs for GM-free seed, licence costs)
  - Freedom of research (e.g. patents, patented seed)
  - Preservation of cultural heritage (e.g. traditional production, autonomy)
  - Social changes (e.g. social structure, increased conflicts, burden for future generations)

15

Catalogue of criteria

- Ecological criteria
  - Preservation of resources (e.g. natural resources, energy use)
  - Preservation of environmental quality (e.g. soil, water, air)
  - Preservation of biodiversity in cultivated areas (e.g. agro-biodiversity, seed diversity)
  - Preservation of uncultivated areas (e.g. protected areas, wild relatives)

16

Catalogue of criteria

- Ethics
  - Population's values
  - Solidarity and equality (e.g. regarding weaker groups)
  - Intrinsic value of nature (e.g. species barriers)
- Third countries
  - Safety (legislation ensuring safety assessment)
  - Social cohesion (e.g. distribution of burdens)
  - Minorities' rights (e.g. indigenous people)

17

Catalogue of criteria

- Definition of key-issues for economical criteria (examples)

Category	Criterion	Key issues
Productivity	Profit	Risk of reduced profits in the agricultural sector e. g. by reduced yield, harvest security or price of contaminated products.
		Risk of reduced profits in existing sectors like the production of special products (GMO-free seed, organic products ...). Risk of the squeezing out of existing production that should be maintained.
	Preservation of the environmental basis	Risk of negative effects on environment, energy and other natural resources, especially resulting from economic growth based on GM technology, e. g. regarding genetic resources, seed diversity or gene banks.
		Risk of negative effects on ecosystem services, e. g. pollination.
Costs	Costs incurred during the whole production chain	Risk of increased costs for producers, e. g. by increased herbicide use, resistance management, coexistence management or higher seed prices.
		Specific costs for insurance of GMO-free cultivation could be substantial e. g. for security and control, quality assurance or analysis.
		Legislation regulating liability and redress is necessary to stipulate the costs resulting from e. g. contamination of GM-free products. Besides specific costs e. g. for decontamination there are also non-monetary costs like the image loss of special regions (e. g. GMO-free regions).

18

Catalogue of criteria

- Definition of key-issues for social criteria (examples)

Category	Criterion	Key issues
Health and welfare	Quality of life	Risk of negative effects on people's income especially in rural areas.
		Risk of negative effects on the health of producers and people living in rural areas, e. g. effects of changes in pesticide use.
		Risk of negative effects on recreation abilities e. g. by changes in landscape aesthetics.
Choice	Accessibility of alternatives and dependency	Risk that farmers' freedom of action regarding the cultivation of non-GM crops is reduced e. g. because of cost development and availability of GMO-free seed.
		Risk of increased structural dependency e. g. access to complementary herbicides, contractual obligations, licence costs
		Risk that access to GMO-free products will be reduced e. g. by reduced availability, higher cost, etc.
Cultural aspects	Preservation of cultural heritage	Risk that traditional production techniques or crop varieties, GMO-free/organic production or certain crops will be squeezed out.
		Risk that the autonomy of the local population will be negatively affected e. g. regarding their decisions about GMO-free production.

19

Catalogue of criteria

- Definition of key-issues for environmental criteria (examples)

Category	Criterion	Key issues
Ecological limits and ecosystem functioning	Preservation of environmental quality	Risk of negative effects on soil quality e. g. regarding pollution, nutrient balance, pathogen presence or soil fertility.
		Risk of negative effects on water balance e. g. by increased irrigation or increased water pollution.
		Risk of negative effects on air quality e. g. by increased pollution or emissions of ammonia.
		Risk of increased emissions of greenhouse gases e. g. by changes in the balance of CO_2 emissions and uptake, increased emissions by transport and processing or land use change.
Biodiversity and nature conservation	Preservation of biodiversity in cultivated areas	Risk of negative effects on species diversity e. g. by decrease/disappearance of segetal species or crop diversity including cultivar variety.
		Risk of negative effects on the biodiversity of habitats/ecosystems e. g. by decrease/disappearance of habitats associated with cultivated areas (e. g. boundary structures).
	Preservation of biodiversity in uncultivated areas	Risk of negative effects on biodiversity on various levels (genetic, species, habitats, ecosystems) by e. g. decrease or disappearance.
		Risk of negative effects on wild relatives of GM crops by e. g. decrease/disappearance of gene reservoirs.

20

Conclusions

- To assess sustainability of GMOs comprehensive assessment of socio-economic aspects (3 pillars) is necessary
- The discussion on assessment criteria and methodologies is only at the beginning
- No agreed procedure in place
- Next steps:
  - Definition of scope and goals of socio economic assessment of GMOs
  - Development of assessment framework (criteria, indicators, ...)
  - Development of guidance (framework which allows flexibility)

21

Report available at:

http://www.umweltbundesamt.at/fileadmin/site/publikationen/REP0354.pdf

Contact & Information

Dr. Andreas Heissenberger
andreas.heissenberger@umweltbundesamt.at

Umweltbundesamt
www.umweltbundesamt.at

22

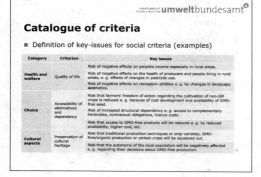

GM Cotton In South Africa

南非的转基因棉花

Mariam Mayet

African Centre for Biosafety 非洲生物安全中心（南非）

Overview

During the early 2000 s，poor black farmers growing GM Cotton in the Makhathini Flats in South Africa became pawns in the 'numbers games' as to whether or not Bt cotton results in increases in yields and savings on pesticide use. The GM machinery，ably assisted by the South African government has peddled the experience of these farmers as a success story，worthy of imitation on the continent. However，beneath the hype lies a tragic tale of oppression and vulnerability，which the introduction of Bt cotton has further exacerbated. The Makhathini farmers have historically been locked into a system of cotton growing due to a range of economic，political and social forces that resulted in chronic indebtedness. Despite cotton growing sliding into sharp decline in the last decade in South Africa，the government and a range of corporate agribusiness actors particularly Monsanto，lured the Makhathini farmers into adopting Bt cotton. This they did by inter alia，providing free production packages，including Bt cottonseeds，duly subsidized with public funds. To date，the South African government has subsidised the Monsanto driven Bt cotton 'success' story with a staggering sum of R30 million from state coffers. Nevertheless，since the arrival of Bt cotton in the Makhathini Flats in 1998 and until 2004 the cumulative arrears of farmers to the Land Bank have amounted to a whopping R22，748，147！

Many reasons may be proffered to explain away the abject failure of the GM project in the Makhathini Flats，however，the central critique must concern itself with the inappropriateness of a development paradigm that seeks to introduce technological solutions to deeply rooted systemic socio-economic problems. Attempts at replicating the Makhathini Flats experience in the rest of Africa，which itself has been caught up in an endless cycle of debt，will undoubtedly yield similar results.

1 Introduction

Cotton is a marginal crop in South Africa. Since the onset of the liberalisation of the agricultural sector in the late 1980 s, cotton's cultivated area has plummeted, from around 180 000 ha to just 6 950 ha in the current (2012/13) cropping season. Production from the present season 2013, is expected to be just under 32 000 lint bales (200 kg). In 2010/11 production was 87 500 lint bales (200 kg). All of the cotton cultivated in South Africa is from Genetically modified (GM) seed, with around 95% being stacked varieties containing both insect resistance and herbicide tolerance traits (which has been the case for the last three seasons). GM cotton accounts for 1% of the total GM crop area in South Africa.[1]

The first GM cotton variety, Monsanto's insect resistant 'Bollgard' (MON 531) was granted regulatory approval in South Africa in 1997, followed by approvals for Roundup Ready (2000) and Bollgard II (2003). In 2005 the first stacked GM cotton variety was approved (Bollgard × Roundup Ready), followed by the approval of Bollgard II × RR flex in 2007. Some sporadic incidences of insect resistance to the early Bollgard varieties had been reported, though these have been withdrawn from the market and replaced with Bollgard II.[2] Monsanto dominates the local cotton seed market; 13 of the 14 GM cotton varieties registered for plant breeder's rights are Monsanto owned. Bayer had one variety registered in 2011, though this contains Monsanto's 'Bollgard' trait.

Bayer, however, have been conducting a spate of field trials for various GM cotton varieties in South Africa since 2008 (see table below), and have been working closely with Cotton SA, the cotton industry association. Though initially reluctant to enter the South African cotton market (owing to its small size), some in the cotton industry believe the prospect of using South Africa as a base to expand into the African cotton market has convinced Bayer to establish itself here.[3] Both Bayer and Monsanto import and export modest volumes of GM cotton seed from South Africa, though in February 2013 Monsanto exported nearly 80 000 kg of BG II × RR Flex to Brazil 'for planting'. Overall, GM cotton activity is marginal when compared to maize and soybeans; of the 135 GMO permits approved so far in 2013[4], only 5 have been for cotton.[5]

The major cotton growing region was formerly in Limpopo, though since the 1990 s many farmers in the province have switched to fruit (particularly citrus) production for export markets. The Northern Cape is now seen as the most important cotton growing area; all cotton production here is under irrigation from the Orange River. However, the current cultivated area is 83% smaller than in 2010/11, due to a higher maize price (maize is also grown under irrigation in the Northern Cape), and climatic constraints.

Small-holder cotton production in South Africa remains minimal, though, due to financial support from the Department of Agriculture, Forestry and Fisheries (DAFF), experienced a noticeable increase from 2011/12—2012/13, particularly in KwaZulu-Natal. There are hopes to re-open the Makhathini Cotton Gin, though to be feasible this will require greater volumes than currently produced, under irrigation. It is not clear from industry representatives if this will happen. Feasibility studies into agro-processing in the area have called the continuing cultivation of cotton 'an appalling waste of good soils'.[6]

There has been a slight increase in small holder cotton production in Mpumalanga over the last 3～4 seasons. Projects to stimulate small-holder cultivation in the Eastern Cape, connected with the Massive Food Production Programme (MFPP), have floundered; there has been no cotton cultivation in the province since 2007/08. There appears little likelihood of this taking off again, as the cotton gin specially constructed for this project has been moved to the Northern Cape, making transport costs prohibitive.[7]

2 Cotton production in SA

Cotton is a marginal crop in South Africa. Since the onset of the liberalisation of the agricultural sector in the late 1980 s, cotton's cultivated area has plummeted, from around 180 000 ha to just 6 950 ha in the current (2012/13) cropping season. All of the cotton cultivated in South Africa is from genetically modified (GM) seed, however, GM cotton accounts for only 1% of the GM crop area in South Africa.[8]Cotton seed must be re-purchased every year.[9]

The South African National Seed Organisation (SANSOR) estimates (based on seed list prices) that the 2012/13 South African cotton seed market was worth R3.9 million. During the same period the monetary sizes of the canola, sunflower and maize seed markets were R34 million, R209 million and R3.5 billion respectively.

Information from the South African National Seed Organisation (SANSOR) indicates that since the 2001/11 cropping season 100% of South Africa's cotton cultivation has been from GM seed, with 95% of this being stacked with both insect resistance and herbicide tolerance.[10]This trend has continued for the 2013 cropping season.[11]

Cotton SA, the "forum and service provider for the South African cotton industry", [12]only recommended GM cotton cultivars for the 2012/13 season: Delta 12 BRF. For non-Bt refuge areas it recommends the herbicide tolerant variety, Delta 18 RF.[13]

GM cotton Statistics

Conditional Environmental Release

Event	Trait	Company	Year approved
Line 531/Bollgard	Insect Resistance (IR)	Monsanto	1997
RR Lines 1445 & 1698	Herbicide Tolerance	Monsanto	1997
Bollgard II, Line 15985	IR	Monsanto	2003
Bollgard × RR	IR × HT	Monsanto	2005
Bollgard II × RR Flex (MON 15985 × MON 88913)	IR × HT	Monsanto	2007

Commodity Clearance

Event	Trait	Company	Year approved
LL Cotton 25	Herbicide tolerant	Bayer	2011

GM cotton field trials，2008—2013

Event	Trait	Company	Year of first trial	Most recent trial
GHB 614 × LL cotton 25		Bayer	2008	-
T304-40		Bayer	2008	-
BGII × LL cotton 25	IR × HT	Bayer	2008	
GHB 119		Bayer	2008	-
GlyTol × LL Cotton 25		Bayer	2009	
BG II × GlyTol × LL Cotton 25	IR × HT	Bayer	2010	
Twinlink × Glytol	IR × HT	Bayer	2012	
GlyTol × TwinLink × COT102	IR × HT	Bayer	2013	

Cotton cultivation in South Africa，2010/11—2012/13（ha）

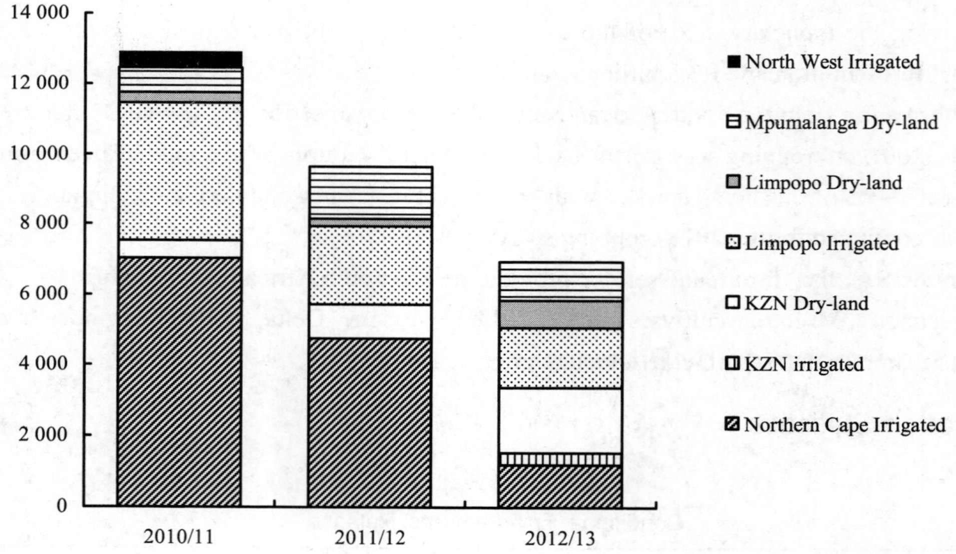

Source：

Agricultural Research Council（ARC）

http://www.arc.agric.za/home.asp？pid=4146

Cotton production in South Africa，2010/11—2012/13（200 kg lint bales）

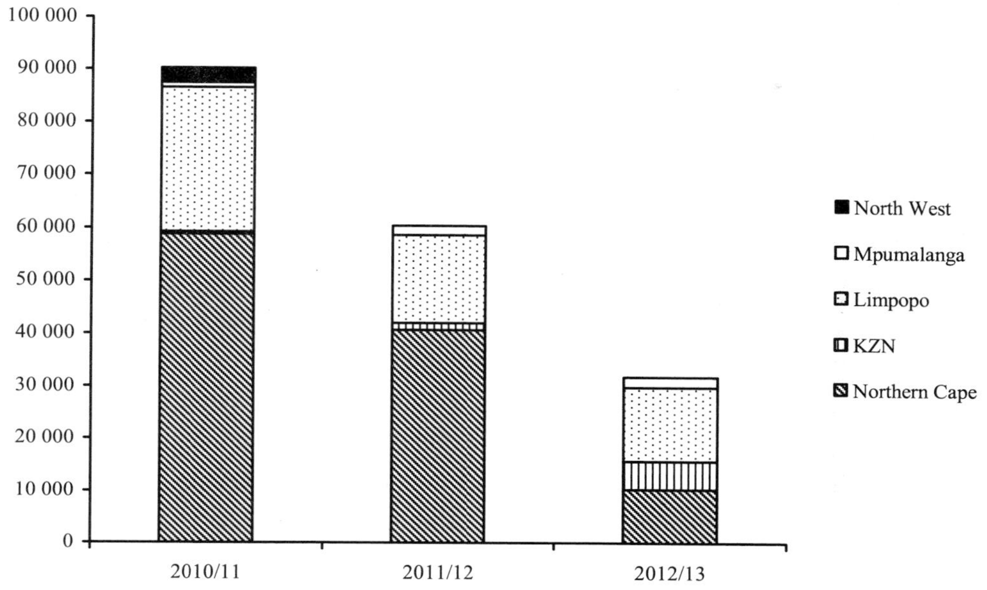

Small holder cotton production

Smalls-scale farmers in Makhathini have consistently identified uneven rain and/or drought as their most significant obstacle to production.[14]Quantitative targets have been set for the upliftment of the sector；35% of local cotton production should come from small-scale farmers by 2014.[15]In 2012/13 they produced 15% of the total crop. This represented a significant increased from 3.3% in the previous year，though that is more attributable to a dramatic overall fall in production than any significant gains in the small holder sector.

3　Cotton Production by Province

3.1　Northern Cape

In 2010/11 the Northern Cape accounted for 55% of cotton ha and 67% of production.

In 2011/12 figures were 47% and 65% respectively.

In 2012/13 figures were 17% and 32% respectively.

During the 2010/11 summer crop season the South African cotton crop was over 150% bigger than the previous season's，the first increase in local production in seven years. This huge spike in cotton production was spurred on by high cotton and low maize prices. The Northern Cape emerged as the leading production area in South Africa during 2010/11.[16]

However，since 2010/11 cotton cultivation in the Northern Cape has fallen precipitously， to 1 199 ha in 2012/13; production has fallen accordingly, from 58 727 to 13 359 over the same period. This is due to a combination of factors, including high maize prices at the time of planting and climatic conditions; for the last two years temperatures in the Northern Cape at planting time have been cooler than usual. There is also general acknowledgment that maize is easier to cultivate and harvest, meaning many farmers would opt for maize if the margins in both crops are similar. Despite the decrease in production, Cotton SA CEO Hennie Bruwer expects cotton cultivation to increase again in the coming season.[17]

3.2 Limpopo

Limpopo province was formerly the centre of cotton cultivation in South Africa. In the 1988/89 growing season over 115 000 ha of cotton was planted, which produced a harvest of 117 000 tons.[18]However, since this high water mark coincided with the onset of the liberalisation of the South African agricultural sector (and the wider economy). In response to cuts in subsidies and increased international competition, many of the border estates switched to producing fruit, particularly citrus, for the export market. Those who did not forsake cotton entirely nevertheless diversified their farming operations considerably. For example, the Grootlplaas estate in Limpopo grow citrus for export, while still remaining (so they claim) to be South Africa's largest cotton planter.[19]By the turn of the century, when Monsanto's Bt cotton had been released onto the market, Limpopo's cotton fields covered just under 33 000 ha. Just over a decade later, and while 100% of cotton plantings may be with GM seed, Limpopo's cotton area has plummeted to around 2500 ha.[20]

In 2010/11 Limpopo accounted for 33% of the nation's cotton area and 31% of its production.

In 2011/12 the figures were 25% and 27% respectively.

In 2012/13 the figures were 35% and 45% respectively.

3.3 Kwazulu-Natal

In 2010/11 KZN accounted for 4% of ha and 1% of production (of national totals).

2011/12 KZN accounted for 10% ha and 2% production.

2012/13 KZN accounted for 31% of national cotton ha and 17% of production.

All of the cotton seed used by farmers in Makhathini is of the stacked (Bt and RR) GM variety.[21]

Since the mid-1990 s(when published records from Cotton SA begin)KZN has consistently accounted for more than 70%, as high as 90% around the turn of the century, of South Africa's small scale cotton farmers. From 2009/10 there was a noticeable decline in small scale cotton farming, in terms of ha cultivated, production and numbers of farmers. The numbers rebounded in 2012/13 though, with the cultivated ha more than doubling, to 2 037 ha, and the re-introduction of irrigated small holder production after a 5 year hiatus; the 200 ha of irrigated land represented 95% of all small holder irrigated cotton during 2012/13.[22]

Cotton has had mixed fortunes and there is a current plan to develop the sugarcane industry in the Makhathini area with the establishment of a sugar mill at Jozini. There has been talk of establishing a sugarmill in the region since 1998.[23]

Small-holder cotton production was severely curtailed by the closure of the Makhathini gin in 2007. There are plans in the offing to re-open the Makhathini Gin. There are also indications that the KZN provincial department of agriculture has set aside funds for Makhathini for the next three seasons. Mr Bruwer is unsure of the feasibility of the facility at Makhathini, stressing that it would require irrigated cotton production to produce the required volumes. Progress has also been slow with national government, with the situation further complicated by land reclamation issues. Around 50% of the land around the Gin is subject to land claims, while the other 50% has been given over to local farmers.[24]

Since the closure of the GIN the farmers' co-op takes care of all the inputs, deducting the costs (of seeds for example) from the final value of the cotton supplied by the farmers.[25]

The DAFF did give some financial support towards cotton production in Makhathini for the last cropping season[26], as well support for the 200 ha under irrigation.[27] It is not widely known if this support will continue for the upcoming season.[28]

DAFF has contributed to the revitalisation of the Makhathini irrigation scheme, covering an area of 3 500 ha. The overall cost for 2010/11, 2011/12 and 2012/13 has been estimated at R71 million.[29]

3.4 Mpumalanga

2010/11 – 5% ha and 1% production (of national totals).

2011/12 – 15% ha and 3% production.

2012/13 – 16% ha and 6% production.

3.5 Eastern Cape

The Cotton Project in the Eastern Cape is a joint initiative by the Eastern Cape Development Corporation and Da Gama Textiles. It involves 500 emerging farmers from Addo, Tyefu, Middeldrift, Keiskammahoek, Qamata, Kat River and the Karoo. A new gin was commissioned in 2005/06 and will eventually jointly be owned by farmers and Da Gama Textiles (the off-take partner of their produce), with small scale farmers having a major share.[30] In the interim, cotton cultivation in the Eastern Cape has collapsed, with no production in the province since 2007/08, while the specially built Gin has been re-located to the Northern Cape.

4 The major players

Monsanto is the dominant player in the South African cotton seed market, owning 13 of the 14 GM cotton seed varieties that have been registered with the DAFF. Ten of these registrations are under the name of Delta & Pineland, who were acquired by Monsanto in 2007.[31] Bayer have one registered variety （Sicot 71B）， though it contains Monsanto's proprietary 'Bollgard' insect resistant trait.[32]

Cotton SA has been working closely with Bayer， who were initially reluctant to invest in the South African cotton seed market， owing to its relatively small size. South African cotton farmers have also been pressurising the company to release commercial varieties in South Africa， as results from field trials so far have indicated that Bayer's varieties may be more suitable for the agronomic conditions in South Africa， particularly the Northern Cape. Industry experts predict that the first Bayer varieties could be commercially available as soon as the 2013/14 cropping season. Bayer appears to have acquiesced， possibly seeing the establishment of operations in the South African cotton seed market as a stepping stone into the rest of the African continent.[33]

5 Analysis/Conclusion

Bt cotton was first approved for cultivation in South Africa in 1997. Its introduction into the highly impoverished Makhathini Flats region of KwaZulu Natal a year later was heralded as a triumph， and proof that， just as it had （appeared） in Makhathini， Bt cotton could lift millions of small scale farmers throughout Africa out of poverty. However， the 'success' of Bt cotton was a façade， kept going by the provision of millions of Rands of agricultural credit to resource poor farmers who had little choice but to accept. In effect， Makhathini was a massive transfer of wealth from the public purse to the private seed provider Monsanto， using some of South Africa's most marginalised people as a conduit. Given the unforgiving environment of Makhathini and its isolation， it was of little surprise when Land Bank had to close its books in Makhathini in 2004， writing off nearly R23 million in the process. Without credit， cotton production plummeted， forcing the closure of the Makhathini cotton gin in 2007.

That Bt cotton largely failed in Makhathini should be of little surprise to anybody with any modicum of the history and geography of the area. Over the twentieth century several government schemes tried and failed to stimulate cotton production in the area. For all the supposed benefits of the Bt trait to cotton farmers， those in Makhathini frequently cite erratic rainfall as their biggest single constraint； something Bt cotton can do nothing to placate. It is for this reason， and not the threat of insect pests， that efforts to re-stimulate cotton production in Makhathini are currently floundering. According to the cotton industry without large scale， irrigated production， it will not be economically feasible to re-open Gin at Makhathini. Though government has provided resources to irrigate 200 ha of cotton fields presently， this is nowhere near enough， and it is not known for how long the Department of Agriculture， Forestry and Fisheries will continue to financially support this. Independent studies have lamented the continuing growing of cotton in the region an 'appalling waste of good soils'.

The wider issue is that， since South Africa's full re-integration into the global economy， cotton has become a marginal crop； from a high of 180000 ha in the late 1980 s， the cotton area has shrunk to around 7 000 ha, with the most important growing areas now being under irrigation in the Northern Cape. The traditional growing regions of South Africa, such as Limpopo, have largely turned their backs on cotton, converting their farmland instead to fruit production, predominantly for export. Even production in the Northern Cape will be largely dictated by the price of maize, which is in direct competition with cotton as a crop in the region.

GM cotton has been introduced in South Africa as a technical solution to a myriad of historical, geographical and economic issues. In itself it will not, and does not attempt, to reverse the centuries of oppression and marginalisation that have been suffered by the peoples of areas such as Makhathini. Neither does it address the vast subsidy regime propping up cotton production elsewhere; since the turn of the century cotton farmers in the United States, for example, have received over $27 billion in subsidies.[34]Small scale production in South Africa is contingent upon the extension of credit, with all the inherent risks of further indebtedness that brings. Irrigation is a pre-requisite to any form of cultivation at all. One has to questions whether, in a water stressed country such as South Africa, where more than 98% of current water supplies already having been allocated, [35]if this is really the most prudent use of this precious resource.

References

[1] USDA (2012). South Africa – agricultural biotechnology annual 2012. US Department of Agriculture Foreign Agricultural Service.

[2] Prof Johnnie van Den Berg, University of the North West. Personal correspondence. 15/05/2013.

[3] Hennie Bruwer, CEO Cotton SA. Personal correspondence. 21/05/2013.

[4] Until 30th April 2013.

[5] DAFF (2013). GMO permit list 2013. Accessed 24/05/2013.

[6] Phipson, J.S (2012). Agricultural and agribusiness status quo assessment: Umkhanyakude district municipality: KwaZulu Natal – 2012. Nemai consulting.

[7] Hennie Bruwer, CEO Cotton SA. Personal correspondence. 21/05/2013.

[8] USDA (2012). South Africa – agricultural biotechnology annual 2012.

[9] Ibid.

[10] SANSOR Annual Reports 2010/11, 2011/12.

[11] Personal communication, Hennie Bruwer.

[12] http://cottonsa.org.za/.

[13] Cotton SA. National Cultivar recommendations for Cotton – 2012/13.

[14] Schnurr, M.A (2012), . Inventing Makhathini: Creating a prototype for the dissemination of genetically modified crops into Africa. *Geoforum* doi: 10.1016/j.geoforum.2012.01.005.

[15] ACB (2012). Hazardous Harvest: Genetically modified crops in South Africa, 2008—2012.

[16] Erasmus, D. Northern Cape drives South Africa' s new cotton boom. 18th March 2011. Farmers Weekly. http://www.farmersweekly.co.za/article.aspx?id=6067&h=Northern-Cape-drives-South-Africa%27s-new-

cotton-boom-.

[17] Bruwer，H. Personal correspondence. 21/05/2013.

[18] Thomas，R. Crop Production in the Limpopo Province.

[19] Bolt，M（2013）. Producing permanence：employment，domesticity and the flexible future on a South African border farm. *Economy and Society*，42：2，197-225.

[20] http://dx.doi.org/10.1080/03085147.2012.733606（accessed 23rd May 2013）.Cotton SA.

[21] Phenias Gumede，Chairman，Ikhwehle co-op/KZN Cotton Growers Association. Personal correspondence. 21/05/2013.

[22] Cotton SA － Small-holder cotton farmer production. http://cottonsa.org.za/Report/GetReport/8（accessed 15th May 2013）.

[23] Gumede，P. Personal correspondence. 21/05/2013.

[24] Bruwer，H. Personal correspondence. 21/05/2013.

[25] Gumede，P. Personal correspondence. 21/05/2013.

[26] Bruwer，H. Personal correspondence. 21/05/2013.

[27] Francina Rakgahla. Directorate：Plant Production（Industrial crops）. Department of Agriculture，Forestry and Fisheries. Personal correspondence. 03/06/2013.

[28] Gumede，P. Personal correspondence. 21/05/2013.

[29] http://www.pmg.org.za/node/32163.

[30] DAFF（2011）. A profile of the South African Cotton market value chain. Directorate：Marketing. Department of Agriculture，Forestry and Fisheries.

[31] http://www.monsanto.com/whoweare/Pages/monsanto-history.aspx.

[32] DAFF（2013）. The SA variety list as maintained by the Registrar of plant improvement － March 2013.

[33] Bruwer，H. Personal correspondence. 21/05/2013.

[34] Environmental Working Group Farm Subsidies Database. http://farm.ewg.org/progdetail.php ? fips=00000&progcode=cotton.

[35] http://www.waterwise.co.za/export/sites/water-wise/industry/tips/downloads/Water_Wise_Tips_for_Your_Business.pdf.

Socio-Economic and Political Factors in the Adoption of GMOs：Comparative Analysis of Various National Choices

转基因生物选择的社会经济和政治因素：不同国家的比较分析

TIBERGHIEN，Yves

（Director，Institute of Asian Research，Associate Professor，

Department of Political Science，University of British Columbia（UBC），Canada）

加拿大不列颠哥伦比亚大学亚洲研究院院长

Why do approval regimes and regulatory regimes of GMOs exhibit such diversity across key countries? Why have some countries chosen to impose strict mandatory labeling and tough regulations，while others have focused primarily on enhancing competitiveness and technological progress?

This presentation argues that pure economic arguments are insufficient to understand this diversity. Rather，the presentation emphasizes the importance of social and political mobilization in the context of institutional limitations. Because GMOs encompass a range of interconnected risks and stakes（health，environment，biodiversity，accountability，intellectual property rights，power relations，cultural identities，and food security），they have often become vectors for discontent against institutional failures. Civil society actors have often acted as conditional catalysts for the crystallization of these issues（including in China）.

The presentation uses evidence from Europe，Switzerland，Japan，Korea，China，and Canada to advance these arguments.

Socio-Economic and Political Factors in the Adoption of GMOs: Comparative Analysis of Various National Choices

The origins of the global divergence in GMO regulations between the US, the EU, Japan, Korea, and China

Yves Tiberghien (肖逸夫)
加拿大-卑诗大学（UBC） 亚洲研究院院长
斯坦福大学 政治学博士
yves.tiberghien@ubc.ca
International Biosafety Forum – the Fifth Workshop
（May 27-28, 2013, Beijing）

A bit of context

A very open and intense debate in China too

GM Rice tested in Hainan

A Battle at Three Levels:

- GLOBAL:
  - Initial Emerging OECD consensus + SPS
  - 1999: EU de facto moratorium + CPB negotiations (2000 Cartagena Agreement)
  - 2003: WTO Panel US, Canada, Argentina vs EU
  - 2000-2005: Competition in Codex, CPB MOP meets
- NATIONAL:
  - Competition between Liberal/Permissive Regime and Precautionary Regime
- REGIONAL:
  - Rise of coalition of GM-free regions

Yves Tiberghien 6

A Case of Global Policy Divergence

- After 1997, major divergence in national regulations
- Polarization with 2 main clusters: US and EU
- Japan, Korea, China de facto joining the EU camp, in form at least
- Brazil and Argentina joining US cluster

Yves Tiberghien 5

GMOs and Global Governance

- A global battle has unfolded since the late 1990s pitting proponents of a liberal and scientific approach (led by the US) against proponents of the precautionary principle (led by the EU). The outcome is polarization and clusterization; internal inconsistence (IPR vs other regulations).
- The battle was unexpected. As late as 1994, there was an emerging consensus (OECD, SPS-WTO)
- The turn against GMOs in some key countries is not primary the result of an economic calculus, but rather the result of political mobilization.

Yves Tiberghien 9

Presentation Outline

- 1. Overall Facts and Stakes
- 2. Analytical Framework
- 3. The EU
- 4. Japan
- 5. Korea
- 6. China
- 7. Conclusion

The Big Puzzles:

- 1. What underpins the diversity of national responses (regulatory polarization) to a new technology with attractive potential for all?
- 2. Why do some countries shift from a permissive to a tight regulatory position despite persistent economic interests and significant investment?

Yves Tiberghien 7

The Big Underlying Questions:

- How should society deal with the introduction of major new technologies and with scientific risk, particularly in biotechnology and life sciences?
- What is clear in many countries: the traditional regulatory system based on interest group primacy and scientific experts is increasingly seen as less legitimate, leading to sub-optimal grassroots rejection.
- GMOs/Biosafety and Global Governance: Difficult coordination at the global level, multipolarity

Special Thanks and Acknowledgements

- Professor Xue Dayuan, the Ministry of Environmental Protection, and Minzu University
- Funding from SSHRC in Canada
- Able research assistantship by Elena Feditchkina and Hyunji Lee.

1. Where GMOS are grown in 2012, ISAAA data

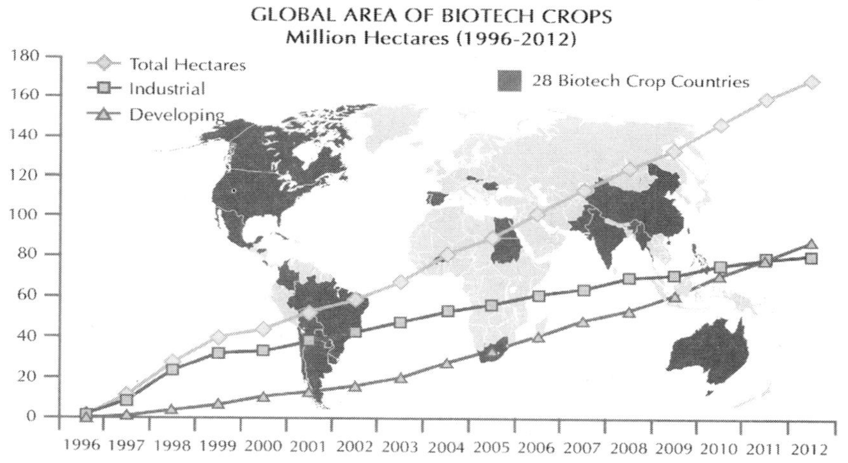

A record 17.3 million farmers, in 28 countries, planted 170.3 million hectares (420 million acres) in 2012, a sustained increase of 6% or 10.3 million hectares (25 million acres) over 2011.

Source: Clive James, 2012.

Biotech Crop Countries and Mega-Countries, 2012

ISAAA

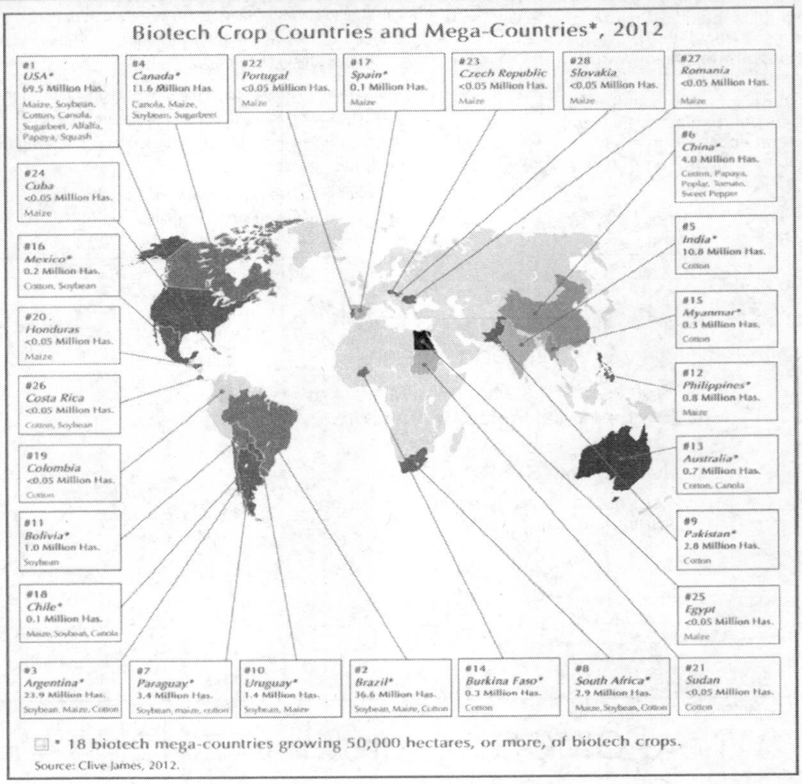

Figure 2: Rate of Adoption of GM Crops as Percentage of Arable Land (source ISAAA 2011 data and FAO data, calculations by Elena Feditchkina)

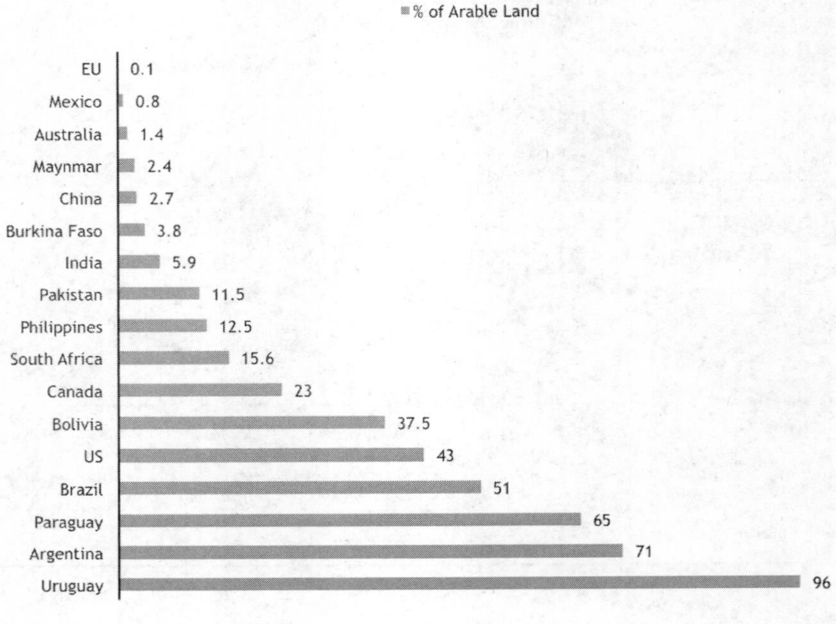

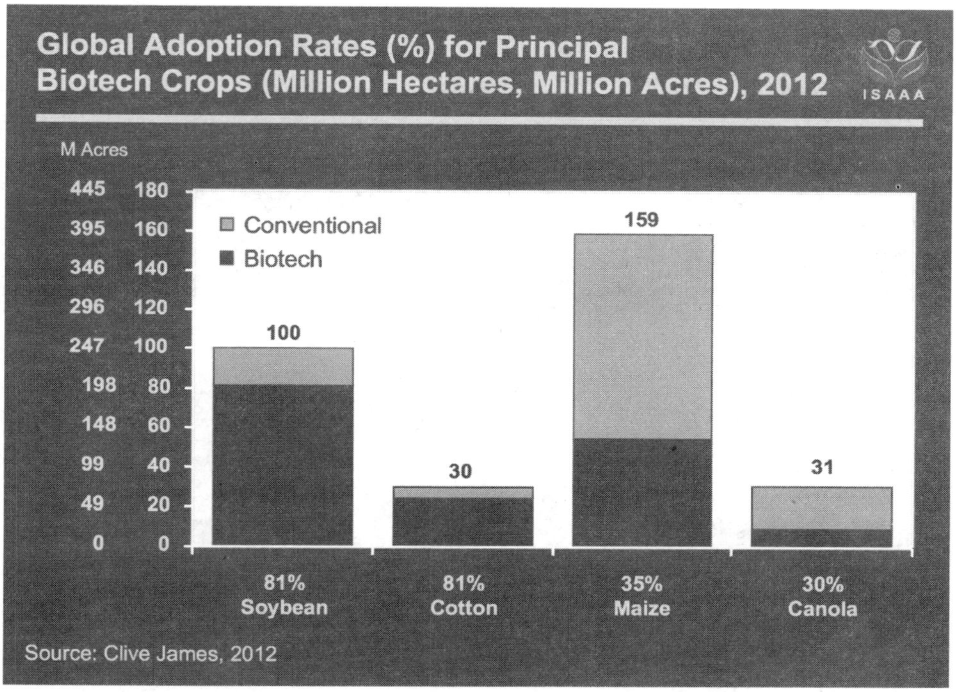

Nine Inter-related Debates on GMOs vs Society

- 1. Food Safety (long-term, holistic, traceable)
- 2. Accountability to Citizens- labeling
- 3. Private Control over Regulation
- 4. Environmental Safety, Biodiversity
- 5. Liability for Unexpected Impacts
- 6. Scientific Progress and Public Interest
- 7. Power Relations between Private Corporations, Farmers, and Citizens
- 8. Society vs Globalization, Cultural Diversity
- 9. Private Ownership of Life (Patents)

Complex Questions Lead to Divergent Regulatory Outcomes

- These 9 issues play out differently in different national settings.
- In every country, the mobilization of interests, political institutions, political leaders, culture, and other variables play as much influence as scientific processes.
- Independence of science in question
- Where lies the public interest on the spectrum between risk and benefits?

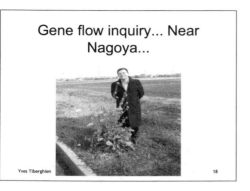

Gene flow inquiry... Near Nagoya...

Yves Tiberghien 18

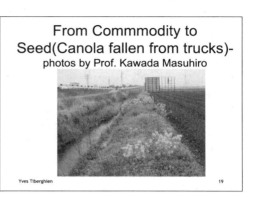

From Commmodity to Seed(Canola fallen from trucks)-

photos by Prof. Kawada Masuhiro

Yves Tiberghien 19

Yves Tiberghien 20

Speaking about Stakes – the next frontier- GM salmon approved

The Next Frontier: cloned animals (approved in US, no labeling)

The Next Battle: Meet one of the first GM goats with human gene

The Overall Lineup

Table 1. Typology of States on the Liberal-Regulatory Spectrum on GMOs

International Level / Domestic Level	Liberal (Pro GMO)	Regulatory (Anti-GMO)
Liberal (Pro-GMO)	**1. Consistent Liberalizers** [USA, Canada, Argentina, Philippines]	**2. Internationalist Regulators** [South Africa, Kenya, Turkey]
Regulatory (Anti-GMO)	**3. Unilateralist Regulators** [Japan until 2003, Australia, New-Zealand, Thailand, Russia, Brazil, Korea, Mexico]	**4. Consistent Regulators** [EU, Switzerland, Norway, China, Japan since 2003, Mexico, Venezuela, India, Malaysia, Indonesia, Kenya]

Paradoxes of Asian Cases

- 1. Developmental State, early support for ag biotech, and US dependence. Yet all experience a partial switch to precaution, Cartagena, mandatory labeling.
- 2. China followed similar path to Japan and Korea. Why?

Arguments

- Japanese and Korean reversals were caused by tensions between a rising civil society and existing bureaucratic structures, leading to political arbitrage in favor of public opinion.
- Chinese policy-making integrated similar tensions through more informal mechanisms.

Theoretical Framework

- Focus on disjunctions (punctuated equilibria?) between traditional gradualist regulatory policies and irruption of external actors with impact on the public.
- Political arbitrage and political entrepreneurship become crucial in that context. Tipover processes.

Explaining Reversals (EU, Japan, Korea)

- 1.GMOs have become proxies for larger debates on the regulation of globalization and on democracy
- 2. Initial shift is triggered by the actions of new civil society actors (conditional catalysts)
- 3. Their success depends on facilitating institutional factors: a/ existing institutional crisis, b/ elite allies, c/ opportunities through local govts
- 4. The actual shape of regulations depends on the balance of power between ministries, as well as between politicians and bureaucrats. The final outcome is often unstable (EU) and another cycle may start

Regulatory Phase: divergent pathways between regulators

- Ideal type 1: EU over-regulation? Large trade impact, large impact on science competitiveness
- Ideal type 2: Japan - compromise regulation, no trade impact thanks to loopholes.
- Middle point - Korea: like Japan with 2 exceptions (context: weak MOE, collapsing NGO networks, MOCIE able to regain control)

Yves Tiberghien　29

Alternative explanations

- Economic realism and interest group models
- International Competition (balancing the hegemon)
- Unintended Events - BSE etc..
- Norm Diffusion

Yves Tiberghien　30

Contrast: Canadian Context

- The GMO regulation regime is potentially unstable: gap with public opinion on labeling, critique by Auditor General (2004), regional debates (PEI, Quebec, now BC), ad hoc situation with patents (Percy Schmeiser case 5-4)
- Emblematic topic: a/ underlines cultural differences between Quebec-Canada, b/ tension between democratic accountability and trade integration with the US (big issue of GM wheat in 2004)
- A case of democratic deficit? (lack of public accountability, lack of mandatory labeling despite public preferences, and lack of debate)

2. The EU as Trouble-Maker in Chief

Meanwhile in France

And the A Surprise Out of Germany

But Pig Farmers Use Mon 810 corn and support it

The EU as Leader of Precaution – an unpredicted event

- GMO governance has shifted from national level to EU level since 1990, particularly after 1996
- Since 1997-1999, the EU has emerged as the generator of new international norms and tight regulations. It was not pre-determined.
- Global advocate of precautionary principle
- Tight Safety + environmental assessment process (2001/18, updated 1829/2003)
- Tight labeling and traceability: process-based, 0.9% threshold
- Yet also, continuation of support of research and some GMOs crops grown (corn, Spain, FR)

Agenda Capture in the EU

- A stable regulatory game with a set number of insider players was disrupted by the irruption of new civil society actors (environmental NGOs, anti-globalization NGOs, organic farmers).
- Civil society actors were successful due to three institutional opportunities:
  - a/ crisis of EU governance / democratic deficit and tilt toward neo-liberal agenda
  - b/ existence of institutional allies: EP, small states (Austria, Denmark, Greece)
  - c/inability of big states to come up with unified positions (France, Germany, UK)

Yves Tiberghien 38

Political Mechanisms in EU:

- 1. Strong initial pro-biotech position of Commission and big States (hence SPS agreement, R&D..)
- 2. Equilibrium disrupted by entry of civil society actors with political relays: small states (Austria, Dk, Greece), European Parliament, and regions (then big states)
- 3. Context of changing institutional environment: EP power, new DGs (Sanco, Environnement), enlargement (Poland, Hungary, etc..)
- 4. Positive loop-back response by public opinion to regulations in UK, Sweden, Netherlands, and Dk

Rise of New Actors in EU galaxy

- 1. Civil Society Networks (since 1996): Environmental groups, Anti-globalization groups, Smaller Farmer unions, and consumer groups
- 2. Impact of Small states: Denmark, Greece, Austria (seizing agenda)
- 3. Rise of GM-free Regional Network
- 4. Uninvited Swiss Input: Nov 2005 referendum as catalyst for further mobilization in EU

Yves Tiberghien and Marko Papic 40

3. Japan: an Unusual Reversal

- Context: low food sufficiency (wheat 13%, soy 5%, corn 4%, canola 0%) - 40% overall
- Initial Situation: government promotion of biotech (hundreds of field tests), public funding, GMO tests since 1989, substantial equivalence.
- 1999: government shift and move toward mandatory labeling (effective 2001)
- 2003: tightening with Cartagena ratification and creation of independent FSC
- 2005: Hokkaido Govt passes a law making GM production impossible and restricting research after a pioneering deliberative process.

Yves Tiberghien 41

Civil Society Actors

- 1996: Creation of NGO federation:"No GMO! Campaign" (Consumers) (Yasuda Setsuko, Amagasa Keisuke)
- Alliance with Seikatsu Club Consumers' Cooperative Union (22 Million members) - common focus on labeling demand.
- Numerous new groups: Soybeans Trust, Rice Trust, Slow Food Cafes, etc..
- After 2001: alliances with other groups: environment, organic farmers (key in Hokkaido 2005 = alliance consumers and farmers)

Yves Tiberghien 42

Yves Tiberghien 43

Agenda-Capture: Key Political Mechanisms

- 1. Massive Petitions that capture the public attention (and MAFF): 1997-1998: petition signed by 2 million citizens requesting labeling.
- 2. Resolutions by 50% of local governments (1600).
- 3. Activities in Diet: Benkyokai and relay through Key Committee [MAFF sandwiched]
- 4. Impact of EU processes: ideas, models, blueprints
- 5. Facilitating Factor: general institutional crisis (bureaucracy under attack)

Yves Tiberghien 44

Key Diet Committee:

- 消費者問題等に関する特別委員会、遺伝子組み換え食品の表示問題等に関する小委員会
- Created in 1973 during oil shock, low visibility
- Took the GMO issue in Dec 1997
- 1999: issued strong report demanding labeling
- Key driving force: coalition of women MPs from opposition parties (DPJ, Komeito, JCP, Shaminto) led by Ishige Eiko (DPJ), gaining support of pivotal LDP MP: Kono Taro
- Window closed in 2001: committee dissolved

Yves Tiberghien 45

In sum - a tipping point

- Encounter between 3 forces:
- Issue Entrepreneurs (civil society)
- Political opening: party politics in flux, increase in role of urban politics, weakening bureaucracy
- Decentralization and changing balance of power: role of local governments

Yves Tiberghien 46

Shape of Actual Regulations in Japan

- Mandatory labeling after 2001 with high threshold (5%) and excluding key products (oil, soy sauce, papayas). Not very trade disruptive.
- Assessment process rigorous, but no power for civil society groups (purely scientific).
- Number of GM products approved for import = high (relative to EU).

Yves Tiberghien 47

4.Korea – some elements

- An initial position in favor of GMOs (eg when joining OECD): possibility as future developer, strong biotech industry
- Major actions by civic groups after 1998: PSPD, KFEM, Soshimo, organic farmers,etc..
- 2 NGO coalitions split until 2003: Anjeonyunli Shiminsahoi-danche Yondae Moim (Park Byung-sang) and GMO bandae Saengmyung Yunli Yondae
- Big issue in trade negotiations with US (2002 amical agreement, impact on Cartagena ratification, 2007 FTA)
- Labeling: major role by MOA Minister Kim Sung Hoon

5.The Chinese GMO Puzzle

- China is increasingly seen as the crucial battlefield in the global conflict over GMOs. Where China goes, global governance will go.
- China has a strong technological and economic interest in developing GMOs.
- China does not have the free civil society and democratic institutions that channel public voice and have led the EU or Japan to take strong regulatory stances.
- Yet, China has adopted a strong precautionary regulatory position, including mandatory labeling.
- Why?
- Secondary HK puzzle: why nothing here?

A Large Stake in GMOs

- Second largest research base in the world after the US (over $120M of public funding per year)
- Over 100 different crops had been tested in open field tests by 2006.
- Pioneer in GM rice
- High stakes – food sufficiency, rising prices, etc..

GMOs in China

- GMOs approved for production in China: cotton essentially
- Rice and Corn just approved late Nov09, but not OK'd for commercialization
- But also: tomatoes, sweet peppers, chilli peppers, and petunia
- Productive testing approved: oilseed rape, and papaya
- Imports: GM corn and soy (from US), canola (from Canada)

Food Sufficiency as Key Issue

- Since joining the WTO in 2001, China lost its overall situation of self-sufficiency in agriculture.
- Exports of agricultural products increased from US$10 billion a year in 2000 to about $30 Billion in 2008.
- Meanwhile, imports shot up from about $10 Billion to nearly $60 Billion in 2008.
- China was a net exporter of soybeans and corn until 1995. Now, China is the world's largest importer of soybeans and has retained only a barely positive position on corn.

Yet, also strong EU-like Regulations

- 2001 State Council Law imposing strict food safety assessment and mandatory labeling with 0% threshold
- Limited labeling coverage: soy, corn, canola, cotton, tomato
- Process-based labeling (strong than Japan), but only primary process
- 2005 ratification of Cartagena biosafety protocol (faster than Korea)

Hypothesis for Chinese Governance Outcome:

- Representation without Institutions
- China shows significant incorporation of public voice (public good formation), due to several access points
- Outcome is surprisingly more like EU and Japan than the US

Public Opinion of Consumers

- Greenpeace survey on urban consumers
- 65% of the consumers shows a preference for non-GE in all the four categories that are included in the question (soy oil, rice, other food products of a plant origin, food products from an animal origin), specifically:
  - 79% choose non-GE soy oil over GE soy oil
  - 77% choose non-GE rice over GE rice
  - 85% choose non-GE food from an animal origin over their GE counterparts
- 80% of the consumers oppose the use of GE ingredients in baby food

Channels for Public Voice in Chinese GMO Governance

- 1. Active NGOs
- 2. Public opinion, media
- 3. Intellectuals (ex Prof Xue – MOE)
- 4. Consumer Council: broad base
- 5. International Norms (EU, Japan)
- 6. Bureaucratic competition, MOE-MOA-MOST
- 7. State Council Keeping Balance

Consumer Court Battle Against Corporation

Conclusion

- The global confrontation and unstable regulatory environment around GMOs is emblematic of the difficulties of our traditional regulatory models to deal with novel risks brought upon by modern science and technology:
- How to balance progress/prosperity with a reasonable risk management across several key dimensions?
- How to ensure a healthy governance process that represents the public good and not just specific interests

Social and Economic Impacts of GMOs: Global and Local

转基因生物的社会和经济影响：全球和地区

Jerry McBeath，Professor of Political Science，
and Jenifer H. McBeath，Professor of Plant Pathology/Biotechnology，
（*University of Alaska Fairbanks*，*USA*）

Abstract: Genetically-modified organisms(GMOs)remain problematical on several scales，from global to local. This article reviews both the promise and perceived perils of GMOs in the evolution of agriculture. It discusses the two global protocols—the Cartageña Protocol and the Nagoya-Kuala Lumpur Supplementary Protocol—instituted for the purpose of improving safety in use of GMOs. Then the article compares the way two large but quite different agricultural powers，the United States and China，have developed and regulated GMOs.

Introduction

For many years，farmers and agricultural scientists have sought improvements in the disease-，pests-，weeds- and drought-resistance of crop species through breeding programs. They have methodically selected seeds from crops that have produced the most sought-after characteristics. Their efforts have produced hybrid species through grafting superior elements of one variety onto another.

While these developments have reflected scientific rigor，the much-acclaimed scientific revolution in agriculture accommodates more invasive methods utilizing technological requirements. These entail penetration within the seed possessing the essential property（for example insect resistance chemical，herbicide resistance and disease resistance），and extraction of the relevant gene（from bacteria，silkworm or other sources），which is then inserted into the crop species. This requires the equipment of a modern research laboratory；only those with requisite knowledge—for example geneticists and agricultural scientists—are able to do the work.

Most studies refer to this new biotechnological approach as genetically-modified organisms（GMOs or more simply GMs）. Less frequently，they are called genetically-engineered species

(GEs) or the name used in the Cartagena Protocol, living-modified organisms (LMOs). Another form of reference is transgenic (*Zhuanjiyi*) species.

In this article we treat four large topics. First, we consider the original promise of GMOs—what the experts said they would accomplish in the global evolution of agriculture. Second, we treat global resistance to and concerns about GMOs. Then we turn to development and change of global protocols on safety of GMOs. The final substantive section compares responses in two large agricultural powers—the U.S. and China—to several dimensions of GMOs. We include knowledge and awareness of GMOs, transparency in the decision-making process concerning GMOs, labeling of GMOs, roadblocks to the commercialization of GMOs and finally attitude change toward GMOs.①

1 The Global Promise of GMOs

The promise of GMOs arises from what they may provide to improve global food security, reduction in use of chemical pesticides and insecticides, improved health of consumers, reduction in damage to farmers' health and improvement of farmers' income. We discuss each in turn.

Improvement of Food Security. Feeding a global population expected to reach 8 billion by the year 2030 and eliminating hunger will require large advances in food production and distribution that enhance food supplies, without damaging the environment. Agricultural biotechnology is a tool with great promise for the alleviation of hunger and poverty.

Already in the first generation of GMOs, the technology has brought changes to farmers' fields. In 2009, GM crops were being grown on 10 percent of the world's arable land.② The International Service for the Acquisition of Agri-biotech Applications (ISAAA) reports that at the global level, GM soybeans (Roundup Ready® soybean) occupied 41.4 million hectares (about 60 percent of the global GM area). GM corn (Bt-corn, Roundup Ready ®corn) was planted on 15.5 million hectares (or slightly less than a quarter of the global GMO area. Transgenic cotton (Bt-cotton, Roundup Ready® cotton) was grown on 7.2 million hectares (nearly 11 percent of the global area). Finally, GM canola (Roundup Ready® canola) occupied 3.6 million hectares, or about 5 percent of the total area.③

A number of technological changes can be made to plants, which will increase their productivity on a per hectare and per capita basis. For example, their drought resistance can be enhanced, and they can be made resistant to diseases and pests through genetic manipulation.

---

① See Jerry McBeath and Jenifer Huang McBeath, "The Socio-Political Environment of Biosafety Regulation in China: The Case of GMOs," in Dayuan Xue, editor, *Risk Assessment and Regulation of Genetically Modified Organisms: Proceedings of the International Biosafety Forum—Workshop 4, Beijing*. Beijing: China Environmental Science Press, 2011, pp. 104-34.

② Matin Qaim, "The benefits of genetically modified crops—and the costs of inefficient regulation," *Weekly Policy Commentary*, Resources for the Future, April 2, 2010.

③ Chi-Chung Chen and Wei-Chun Tseng, "Do humans need GMOs? —A view from a global trade market," *Journal of American Academy of Business*, Vol. 8 (1), 147.

Other attributes of GMOs include herbicide resistance, delayed ripening, improved sweetness, miniature size, cold-resistance, high starch content, improved fiber properties, and faster growth.①

In the first generation, GMOs have been broadly advertised as not only increasing agricultural yields but also promoting more efficient land use. The latter aspect is of great importance because of upward pressures of increased human population leading to urbanization and economic development. Climate change too is expected to increase pressures on available arable land. Thus, GMOs that increase plant yields will have the effect of lessening the upward pressure on food prices.

Reduction of Pesticide Use. Many experts herald the ability of GM crops to develop issue-specific solutions to new pest and disease problems. Genetic engineering can be specifically applied to insects, bacteria and other non-food life forms. Too, they can be used to reduce populations of insect pests whose damage to crops is quite severe .

Monsanto created Roundup Ready soybeans, which can withstand application of the strong herbicide glyphosphate (the trade name of which is Roundup) . The herbicide helps GM soybeans flourish after Roundup application in order to control weeds; it also reduces the costs of spraying of less effective herbicides.②

In addition, agricultural GMOs, i.e. Bt-crops, can minimize pesticide use (particularly of organophosphates and pyrethoid insecticides), which are very toxic. Field trials in China showed that use of two kinds of insect-resistant GM rice reduced farm-level pesticide use by 80 percent.

Agricultural pesticides play a perverse role in species extinction, and reduction in their use would have significant and positive environmental effects. While there is considerable variation in the overall effect on pesticide/insecticide use of the introduction of GMOs, it appears to be the case that when there is high disease pressure and the population of weeds is high, specific GM crops, perform better than conventional crops.③

Improved Health of Consumers. The biology of gene splicing allows for the opportunity to create plants that will produce food that is more nutrient rich. For example, scientists have developed a new strain of rice called "golden rice." This species promises to boost vitamin A levels and reduce blindness in children living in lesser-developed countries (LDCs). (Perhaps as many as one million people consume insufficient amounts of the A vitamins, including a majority who survive on rice-dominant diets.) Rice has greater advantages than maize or soybeans (a focus of GMO development and broadly traded), because it is produced extensively in developing countries, where rates of malnutrition are high, and is primarily consumed in the

---

① See U. of Michigan, "GM crops: Costs and benefits," (http://sitemaker.umich.edu/see006grouop5/gm_food), retrieved5/17/2013.

② David G. Victor, "Trade, science and genetically modified, " Council on Foreign Relations, March 13, 2001 (retrieved 5/21/2013, http://www.cfr.org/genetically-modified-organisms/trade-science-gen) Victor repeats that about half of the surplus gained from the innovation is returned to Monsanto, which sells both the seeds and the roundup; one quarter goes to farmers; and the remainder to consumers.

③ Flannery, M-L., Fiona Thorne, Paul Kelly and Ewen Mullins, "An economic cost-benefit analysis of GM crop cultivation, " *AgBioForum*, Vol. 7, No. 4 (2001), 37.

country where grown.① Another example is sweet potatoes. Field trials have been conducted in Kenya for a variety of sweet potatoes that resists the sweet potato virus (which kills up to 80 percent of the crop and significantly reduces yield).②

Although scientists have long advertised the positive health effects of GM crops, the first (current) generation had not lived up to expectations. Still, there may be hope for the next generation of GMOs as development of genetically-modified foods that have vital vaccines and nutrients is advancing.

Improvement of Farmers' Health. In lesser-developed countries such as China, farmers' (many of whom are women and children) use of pesticides, especially insecticides, has endangered lives, because often farmers are not aware of how to apply them correctly. Also, farmers do not always wear protective apparel when using the chemical pesticides in their fields. This is particularly the case regarding controlling cotton boll-weevils in China. Contrary to the modern varieties, traditional Chinese cotton grows from tall stalks, and is difficult to spray without showering the sprayer with insecticide.

Farmers in economically developed countries (EDCs) are more likely to be aware of the risks in use of chemical agents, and their countries invariably have adopted higher standards (than LDCs) for use of pesticides. Farmers in EDCs report greater convenience in the use of GM crops.

Improvement of Farmers' Income. A large number of studies have been conducted to assess the benefits of GMOs to farmers. Surveys typically point to increased yields for farmers through improved pest control and reduction in pesticide costs. Studies in the U.S. indicate that this is the primary reason for adoption of a majority of farmers.③ This counterbalances the higher costs for GM seeds.

Returns to land and labor have been marginally greater for GMO than for non-GMO crops, albeit there are significant differences across crops. Other results of surveys include increased flexibility in planting and adoption of more environmentally-friendly practices.

A group of studies has examined "welfare effects," of GMO crops, by which is meant a drop in prices and greater export sales. The relevant example is Bt cotton, where the introduction of the GMO is associated with a decline in cotton prices.

Other Benefits. The focus on climate change has brought attention to the importance of carbon sequestration. GM trees possibly can provide significant carbon storage capacities (but only if not harvested).④ Too, genetically-engineered bacteria have been approved for

---

① Nielssen, Chantal P. and Kym Anderson, "Genetically modified foods, trade and developing countries: Is golden rice special?" *AgBioWorld* (Retrieved, 5/21/13, http: //www.agbioworld.org/biotech-info/topics/goldenrice/specialgol..) See also, McKie, Robin, "After 30 years, is a GM food breakthrough finally here?," *The Guardian*, February 2, 2013.

② Ibid., p. 6.

③ Director-General for Agriculture, "Economic impacts of genetically modified crops on the agri-food sector," European Commission (retrieved, 5/17/13, http: //ec.europa.eu/agriculture/publi/gmo/fullrep/ch3.htm). Also see Duffy, M., "Does planting GMO seed boost farmers' profits?, *Leopold Letter* (retrieved 5/20/13, http: //www.leopold.iastate.edu/news/leopold-letter/1999/fall/does-pl...).

④ See, among others, Rachel Asante-Owusu, *GM Technology in the Forest Sector*, Gland, Switzerland: WWF International, 1999.

agricultural use in the U.S. in order to increase nitrogen-fixing properties of agricultural crops. The objective is for GM bacteria to replace naturally occurring species.①

An important but at this stage potential benefit of GMOs is the concept of edible vaccines. Currently undergoing testing, edible vaccines could eliminate the needs for injection of vaccines via needles as well as the cold storage of vaccines. Thus, the vaccines would be more readily available, particularly for infectious diseases such as cholera and HIV/AIDS.②

It should be kept in mind that a critical stream of studies challenges the benefits of GMOs. For example, Wolfenbarger and Phifer (2000) claim that statistical and other supported documentation of benefits is extremely limited.③

2 Global Concerns about GMOs: Toxicity, Allergenicity, Reduced Nutritional Value, Biodiversity Threats

More than 39 million hectares (and some observers believe nearly twice this amount) of genetically modified crops have been grown worldwide. Although there is little clear evidence of damage caused to humans by ingestion of GM products, concerns remain about their toxicity and allergenecity. Too, critics allege that GMOs have less nutritional value than conventional crops, and that they threaten global biodiversity. We examine each of these complaints in turn.④

Toxicity. Although no evidence has been found of human health problems associated with consumption of GM food products, evidence has been presented on one case regarding non-human animals. In this instance, mice fed with Bt potatoes were alleged to have toxins in their systems. However, the study was not completed, raising questions about the allegations.⑤ This notwithstanding, GMO critics allege a strong biological parallel between genetically engineered food and adverse health effects in humans.

The larger issue is that trials and tests of GMOs have been insufficiently long and thorough to detect human health effects. This leads to frequent calls for the use by national regulatory authorities of the precautionary principle, discussed below.

Allergenecity. A second concern is that GM crops cause allergic reactions in consumers. The Food & Agriculture Organization (FAO) has evaluated protocols for tests for GM foods. It found no allergic effects related to GM foods then on the market. However, recent testing has focused on a new variant of Bt StarLink," which includes a protein (cry9 c) that allegedly is unsafe for human consumption.⑥

---

① D. VanAken, "Genetically engineered bacteria: U.S. lets bad gene out of the bottle," *Greenpeace Report*, January 2000.

② See C. Amtzen, "Oral immunization with a recombinant bacterial antigen produced in transgenic plants," *Science*, 266 (1995), 716.

③ Wolfenbarger, L.L. and P. R. Phifer, "The ecological risks and benefits of genetically modified plants," *Science* (September 15, 2000), 2093.

④ For overall assessments of food safety effects, see Zhang Tao and Zhou Shudong, "The economic and social impact of GMOs in China," *China Perspectives*, 47 (May-June 2003).

⑤ Randerson, James, "Arpad Pusztai: Biological divide," *The Guardian*, November 10, 2010.

⑥ D. Magnus and A. Caplan, "Food for thought: The primacy of the moral in the GMO debate," in Ruse & Castle, eds., *Genetically Modified Foods*, New York: Prometheus Books, 2002.

Several critics of GMOs point to the associations between rise in allergic reactions of humans and the development of the GM industry, particularly in the United States. For example, the publication "GMO Risks/GMO Awareness" notes that "Over the last 10 years (during which GMOs have become widespread in the U.S.), 1 out of every 17 children in the U.S. developed a food allergy, and hospital emergency rooms across the nation experienced a 265% increase in food allergy emergencies."① This is typical of GMO critics who correlate association with causation, without providing evidence of causation.

Reduced Nutritional Value. Critics of GMOs allege that they tend to have lower nutritional value than non-GM crops, and in the main have lower protein levels. The genetic modification in fruit and vegetables might lead to changes in metabolic pathways and to variation of the molecular pattern, with particular attention to antioxidant compounds that are not well-covered in the literature. At this stage, insufficient research has done on this issue, but one recent study by Venneria et al indicates that "GM events are nutritionally similar to conventional varieties of wheat, corn, and tomato" in the market.②

Threats to Global Biodiversity. A major concern regarding GMOs is the potential consequences of gene flow from GM to non-GM individuals of the same species and to the possibility of unpredictable crosses with other species. This cross-pollination with wild types of GM species might lead to genetic contamination of the wild types, which could alter local ecosystems.③

Second, some opponents of GMO development claim that crops modified to be tolerant to herbicides could foster the development of "super weeds." Initially, the introduction of herbicide tolerant plants decreased the use of herbicides, but afterwards this increased the usage and scope. Weeds have become more resistant to herbicides. In the United States, studies indicate that the combined total of pesticide and herbicide use has increased about 10 percent.④

Roundup resistant weeds have surfaced after many years of selection under Roundup use. This is similar to the insects, pathogenic resistance to insecticides and fungicide use. Untransformed soybeans are very susceptible to Roundup. Roundup Ready® soybeans are highly resistant to glyphosate (Roundup), hence Roundup can be used in the weed control scheme of soybean fields. The purpose is for ease of management. A recent report indicated that glyphosate-resistant weeds has expanded to 61.2 million acres in 2012 ⑤. Weed species resistant

① Retrieved 5/18/13 from: http: //gmo-awareness.com/all-about-gmos/gmo-risks.

② Venneria, Eugenia, Simone Fanasca, Giovanni Monstra, Enrico Finotti, Roberto Ambra, ElenaAzzini, Alessandra Durazzo, Maria Stella Foddai and Giuseppi Malani, "Assessment of the nutritional values of genetically modified wheat, corn, and tomato crops," *The Journal of Agricultural and Food Chemistry, Vol. 56, no. 9 (2008)*.

③ Chevre, A.m., F. Eber, A. Baranger and M. Renard, "Gene flow from transgenic crops," *Nature*, 389 (1997), 924. See also G. Persley and James Siedow, "Application of biotechnology to crops," in Ruse and Castle, eds., *op.cit.*

④ Cherry, B., "GM crops increase herbicide use in the United States," Institute of Science and Technology (retrieved 11/10/10 from: http: //www.i-sis.org.uk? GMcropsIncreasedHerbicide.; php.).

⑤ *Farm Industry News*, "Glyphosate-resistant weed problem extend s to more species, more firms," January 1, 2013 (retrieved 9/5/2013 from http: //farmindustrynews.com/print/herbicides/glyphosate-resistant.w...).

to glyphosate also increased from goosegrass (1997) and rigid rye grass (1998) to a total of 11 weed species in the U.S. and the world.[①]

Third, concerns have been raised about secondary pests that have become major scourges. Bt crops in India occasioned the increase in mealy bugs; in China, mirid bugs, once presenting little threat to agriculture, progressively grew in number following the introduction of Bt cotton.[②]

A related concern is that GMOs could threaten the world's biological diversity and lead to excessive dependence on just a few crop varieties, which would increase the vulnerability of crops to diseases. This has led some critics to allege that GMO trials may "pollute the genetic commons."[③] Also, critics allege that GMOs increase risk of ecosystem damage or destruction.

3 Global Agreements on Regulation of GMOs

Protocols to regulate GMOs internationally developed after the adoption of the Convention on Biodiversity in 1992. Negotiators formed two protocols, the Cartageña Protocol on Biosafety (named after the city at which it was signed) developed in 2000 and the Nagoya-Kuala Lumpur Supplementary Protocol on Liability and Redress, developed in 2010. We discuss each.

Cartageña Protocol on Biosafety. The biosafety protocol is the world's first international treaty regulating the transboundary movement of GMOs. Specifically, it requires advanced informed agreement of an importing country prior to trade in certain GMOs. It is hailed by its supporters as a key global environmental agreement to institutionalize a precautionary approach to the governance of risk.[④] Adopted in 2000 and entering into force in 2003, currently it is being implemented in some of the world's largest agricultural import markets.

The Cartageña Protocol was created at the instigation of most developing countries and the European Union (EU) members, who demanded strict rules on GMO governance. In opposition were a group of countries led by the U.S., which objected to the trade-intrusive nature of international biosafety regulation. Its focus is transboundary movement, transit, handling and use of Living Modified Organisms (LMOs) that may have an adverse effect on the conservation and sustainable use of biological diversity, also taking into account the risks to human health.

International leadership played a critical role in countering U.S. opposition to binding biosafety rules. Notwithstanding the importance of the international regulatory interest to multi-national corporations, nonstate actors with an emphasis on environmental and consumer associations played a more significant role than did industry actors in negotiation. A number of

① Boerboom, Chris and Michael Owen, "Facts about glyphyosate resistant weeds," The Glyphosate, Weeds, and Crops Series, Purdue Extension Service, 2006.

② Xia, Bing et al., "Mirid bug outbreaks in multiple crops correlated with wide-scale adoption of Bt cotton in China," *Science*, 328 (May 13, 2010), 1154.

③ See B. Johnson and A. Hope, "GM crops and equivocal environmental benefits," in Ruse and Castle, eds., op.cit.

④ Gupta, Aarti and Robert Falkner, "The influence of the Cartageña Protocol on Biosafety," *Global Environmental Politics*, Vol. 6, No. 4 (November 2006), 24.

scholars have focused on the influence of environmental NGOs such as Greenpeace，Friends of the Earth，Third World Network and others in influencing the contents of the Biosafety Protocol. They lobbied government delegates，cooperated with developing countries by mobilizing public pressure，and were particularly active in the pre-negotiation phase（where participation procedures were lenient）.①

One of the more interesting aspects of Cartageña is its incompatibility with World Trade Organization（WTO）rules，and whether it will take precedence in trade disputes with the WTO on biodiversity issues. Critics of the design and implementation of the biosafety protocol suggest that "trade policy（following Cartagena）is open to exploitation by protectionist interests that have nothing to do with the maintenance of biodiversity."②

4 The Nagoya-Kuala Lumpur Supplementary Protocol on Liability and Redress

Adopted in 2010，this protocol provides international rules and procedure on liability and redress for damage to biodiversity resulting from living modified organisms.（When agreement on rules could not be reached at Cartageña，the first meeting of the Conference of the Parties[COP-MOP] established an open-ended ad hoc working group of experts to fulfill the mandate of adding a supplemental protocol on liability and redress.）A small group of government negotiators resolved contentious issues. Other decisions included adoption of a 10-year strategic plan for implementation of the protocol，a program of work on public awareness，education and participation.

At the negotiations of the Cartageña Protocol，business nonstate actors were active agents，but were unsuccessful in attaining their objectives（to avoid a treaty at all or to significantly modify it）. At follow-up negotiations for the purpose of specifying how liability and redress could be arranged for environmental damage caused by GMOs，six leading agro-biotechnology multi-nationals（Monsanto，BASF，Syngenta，DowAgroSciences，Dupont/Pioneer and Bayer Cropscience）presented a private，international agreement. Governments in attendance rejected this proposal and instead adopted a binding supplementary liability and redress protocol to Cartageña. No content was transferred from the business proposal.③

---

① Arts，Bas and Sandra Mack，"Environmental NGOs and the biosafety protocol：A case study on political influence，" *European Environment*，Vol. 13（2003），33. This applied to the supplemental protocol，where organizations such as IUCN—The World Conservation Union—emphatically articulated the "safety first initiative" to "anticipate and resolve safety issues as far upstream of commercialization as possible." See IUCN Law Center，"Genetically Modified Organisms and Biosafety，" IUCN，August 2004.

② Hobbs，Anna，Jill Hobbs，and William Kerr，"The Biosafety Protocol：Multilateral agreement on protecting the environment or protectionist club？" *Journal of World Trade*，Vol. 39，No. 2（2005），297. Also see Simonetta Zarrilli，"International trade in GMOs and GM products，" Policy Issues in International Trade and Commodities Study Series No. 29，UNCTAD，New York：United Nations，2005.

③ For an analysis of different business groupings' strategies in the negotiations，see Amandine Bled，"Global environmental politics，regulation to the benefit or against the private？ The negotiations of the Cartagena Protocol on Biosafety，" *Political Perspectives*，Vol. 1，No. 5（2007）.

Scholars attribute this failure of business to the fracturing of business interests. The six proposing firms were not agreed among themselves, and they did not enjoy the support of other business interests, not to speak of non-governmental organizations with interests in biosafety.① A second explanation for this case of business failure in the development of environmental regulations was the lack in development of sophisticated networking capabilities. They failed to include NGOs with food safety interests, while unifying the business position.②

Another notable factor of negotiations concerning the supplementary protocol (and the Cartagena Protocol as well) was the failure of the United States to dominate, which expresses the limitations of hegemonic stability theory. This theory posits that the superpower (currently the U.S.) seeks to construct the world in her image and to create rules and institutions both reducing global conflict (which otherwise would need to be addressed by additional demands on the resources of the superpower) and establishing her values and practices as globally supreme.③

To the present, the United States and most other GMO exporting nations have failed to ratify the protocol. They remain skeptical of and hostile to many of its provisions. Yet it remains unclear whether this case supports the claim that international cooperation in opposition to the current world hegemon is in place. Although the U.S. and a number of other GMO producing countries are not signatories to the protocols, their exporters will be faced with import regulations based on the protocols. Whether these will have a significant impact on trade volumes—bilateral or multilateral—is still being studied.④

In the absence of a robust international regime regulating GMOs, safety of consumers in different countries is dependent on the strength of national regulation. As we see below, there is great variation in national law and regulations concerning GMOs.

5 Comparative Responses of U.S. and China to GMOs

Both China and the United States are large countries (the world's third and fourth largest respectively) with large populations—1.314 million Chinese and 310 million Americans. Yet the U.S. is considered an economically developed country (EDC), while China—notwithstanding

---

① Orsini, Amandine, "Business as a regulatory leader for risk governance? The compact initiative for liability and redress under the Cartagena Protocol on Biosafety," *Environmental Politics*, Vol. 21, No. 6 (November 2012), 960-68.

② Ibid., 974.

③ See Robert Falkner, "American hegemony and the global environment," *International Studies Review*, vol. 7, no. 4 (2005), 596; and his "Non-hegemonic regime-building: Negotiating the Cartagena Protocol on Biosafety" in Falkner, ed., *The International Politics of Genetically Modified Food: Diplomacy, International Trade and Environmental Law*, New York: Palgrave, 2008.

④ For example, see R. Colitt, "Washington takes the battle over future of genetically modified crops to Brazil," *The Financial Times*, June 20, 2003; Falkner's discussion of the U.S. objection to new Chinese safety regulations for GMO imports, which led it to extract major concessions from the Chinese government in the form of interim safety certificates for US GM soybean exports (R. Falkner, "International Sources of Domestic Environmental Policy Change in China: The Case of Genetically-Modified Food," *Pacific Review* 19 (2008); and Xue Dayuan and Clem Tisdell, "Effects of the Cartagena biosafety protocol on trade in GM0s," Working Papers on Economics, Ecology and Environment series 49, School of Economics, University of Queensland, 2002 (which argues that the protocol will have a profound effect on China's trade).

many shared characteristics with EDCs, remains in the category of developing countries. Both countries have established leadership positions regarding production of GMOs, and both countries' government have invested heavily in agro- biotechnology.

Because of the critical difference of the two powers in economic development (and in the resulting relationship between economic and social groups and the state), we would expect there to be more dissimilarities than similarities between the two nation's biosafety regimes regarding GMOs. In this exploratory comparison, we review five aspects of the regulatory process: knowledge/awareness of GMOs, transparency in decisions about GMOs, labeling of GMOs, commercialization roadblocks, and attitude change toward acceptance of GMOs. We consider data from China and the United States with respect to each point.

Knowledge/Awareness of National Populations regarding GMOs. In the U.S., studies point to a relatively greater acceptance than rejection of GMOs. Some 33 percent of a national sample approved GMOs in the marketplace as compared to an outright rejection rate of 5 percent. However, overall 23 percent of the sample was not aware that GMOs had reached the market.

In China, governmental public outreach efforts have focused on consumer education and explained benefits and development of agricultural biotechnology products. Still, public knowledge of GMOs is limited. A national survey reported on by Zhang and Zhou indicated that 32.1 percent had no knowledge at all about GMO products; 47.3 percent did not know whether GMO products could influence human health; and 50.4 percent did not know whether GMO products had an adverse impact on the environment.[①]

Transparency in Decisions about GMOs. Some scholars argue that transparency of biosafety and regulation is becoming a "transformative force" in global biosafety governance.[②] The U.S. regulatory process regarding GMOs is unlike the European adoption of the precautionary principle. For example, the U.S. views GM corn portions in food as additives, and it does not require the Food and Drug Administration (FDA) to approve them. In the European Union (EU) on the other hand, GM products are viewed as a potential hazard to human health.

In the U.S., regulators have made relatively minor modifications to existing food and pesticide regulatory regimes as GMOs entered the marketplace. However, overall, the U.S. food and pesticide regulatory regimes are transparent. In the U.S. increasingly, the concern over transparency has been reflected in scrutiny of the membership of all panels having to do with regulatory authority, and their connections with GMO-producing firms, such as Monsanto. For example, a dietitian working on a panel charged with setting policy on GMO foods for the National Academy of Nutrition and Dietetics contended she was removed for pointing out that two of its members had ties to Monsanto.[③]

China's legal framework of GMO biosafety consists of what are called the "four pillars"

---

① Zhang and Zhou, op. cit., p. 7.

② See Aarti Gupta, "Transparency as contested political terrain: Who knows what about the global GMO trade and why does it matter? , *Global Environmental Politics*, Vol. 10, No. 3 (August 2010), 32.

③ Stephanie Strom, "Food politics creates rift in panel on labeling," *New York Times*, April 10, 2013.

of agro-GMO biosafety protection. These include 1) the environmental protection law, 2) legislation specifically addressing agro-GMO biosafety issues such as the "Regulations on Agro-GMO Biosafety Management (2001) or "Implementation regulations on the safety of import of agricultural genetically modified organisms (2002) ①," 3) regulations addressing biosafety issues in other relevant fields, for example in forestry biological safety, and 4) technical standards for agro-GMO biosafety. Summarizing this overall system, Yu and Wang attest that "China's legal framework is currently insufficient to properly manage and supervise agro-GMO biosafety." ② They point to lack of a comprehensive biosafety law (for example lack of specific legislation in key fields, such as forestry), ambiguities in the administrative system and various implementation deficits.

Labeling of GMOs. In the U.S., the lack in labeling of GMOs in food products creates confusion as to what foods consumed are genetically modified. Too, there is some concern in the U.S. that labels will not be informative to consumers, and some label designs may introduce bias against GM technologies.

Industry attention has focused on the threshold that determines what is "GM free." Moreover, the argument is made that labeling requirements are appropriate only when there is clear evidence that products are harmful to human health.

The United States is a federal system in which states have the opportunity to make laws—such as environmental regulatory laws—that are more demanding than national laws. In the environmental area, we often see conflict at the state level when the national government has not responded to the progressive trend line. In fall 2012, California voters had the opportunity to vote for Proposition 37, which required that food with GMOs be so labeled.③ The measure would have made California the first state in the U.S. to require labels on some fresh produce and processed foods, such as corn, soybeans and beet sugar, whose DNA has been altered by scientists. Opponents said it would cost households from $350 to $400 million if enacted, that it was bureaucratic and full of illogical loopholes for certain foods, such as meat, dairy products, eggs and alcoholic beverages. Opponents prevailed in the election, with 51.4 percent of voters casting ballots against it. A significant factor in the campaign was the amount spent by opponents. Monsanto and other agribusinesses and food companies (for example PepsiCo and Nestle) spent $45 million on advertising and lobbying, compared with about $8 million spent by advocates. Another factor was problematic wording of the proposition, which brought support

① Adopted at the 5th Executive Meeting of the Ministry of Agriculture on July 11, 2001, promulgated by Decree No. 9 of the Ministry of Agriculture of the People's Republic of China on January 5, 2002.

② Yu Wenxuan & Canfa Wang, "Agro-GMO biosafety legislation in China: Current situation, challenges, and solutions," *Vermont Journal of Environmental Law*, Vol. 13 (2012), 873. See also, Joshua Lagos and Ma Jie, "China-Peoples Republic of, Agricultural Biotechnology Annual," USDA, FAS, Gain Report, 7/13/2012.

③ For analyses of Prop 37, see, among others, Alex Philippidis, "Weighing the costs of GMO labeling," *Genetic Engineering & Biotechnology News*, October 11, 2012; and David Zilberman, "Why labeling of GMOLs is actually bad for people and the environment," (retrieved 5/17/2013 from http://blogs.berkeley.edu7/2012/06/06/why-labeling-of-gmos-is-actual...).

for the anti-Prop 37 campaign of the *Los Angeles Times* and the San *Francisco Chronicle*.①

Some 50 countries, including China (and Mexico) have labeling laws.

Commercialization Roadblocks. The idea that commercialization of GMOs should occur easily in countries is based on the liberal trade premise that any product for which there is demand and is safe for consumers should be allowed free entry into the market.② The U.S. regulatory system, as mentioned, does not adhere to the precautionary principle, which means that in the absence of documented evidence that GMOs cause harm to humans, there are no roadblocks to commercialization. The national attitude toward commercialization of GMOs is thus among the most laissez-faire in the world, making the U.S. the target of criticism by anti-GMO NGOs.③

In China, the situation is more complex, as its economy has both free and controlled aspects. For example, studies of both Chinese and Americans indicate that the beta-carotene of golden rice was easily taken into the blood stream. However, Greenpeace objected to tests on Chinese children without full information being revealed to parents, leading to halts in testing. GreenpeaceChina raised concerns over the ecological and health risks of GMOs. It publishes consumer guides for GM foods, has investigated several allegedly illegal sales of unapproved GM rice, and collaborated with Chinese scientists to register concern over adverse ecological impacts of GMOs.④ This is symptomatic of often carefully orchestrated campaigns by environmental interest groups opposed to GMOs. Reactions of bureaucrats and testing requirements for golden rice commercialization in China far exceed those for standard crops.

Globally, there lacks standardization of the ways in which genetic modification is managed. In China, regulations such as those requiring asynchronous approval delay marketing and trading new biotech products already approved elsewhere. Further, re-registration of expired biosafety certificates is cumbersome, with many redundant requirements. Additionally, there are problems regarding variety-based registration, import testing done by the state quarantine administration (AQSIQ), and treatment of proprietary information.

Attitude Change toward GMOs. In the U.S. attitudes toward GMOs have shown more propensity to buy GM foods than reluctance, but support has declined. Public opinion plays a major role in U.S. regulatory decision-making and recent protests against Monsanto and GMOs likely will cause a change in attitudes of the general public. Demonstrations were held in about 300 cities in more than 44 countries in late May, 2013, the first such global protest, which was

---

① See Anna Almendrala, "Prop 37 Defeated: California Voters Reject Mandatory GMO-Labeling," *The Huffington Post*, November 8, 2012 and Adam Vaughan, "Prop 37: Californian voters reject GM food labeling," *The Guardian*, November 7, 2012.

② Using a deconstructionist paradigm, Richard Hindmark and Anne Parkinson argue in "The public inquiry as a contested political technology: GM crop moratorium reviews in Australia,"(*Environmental Politics*, Vol. 22, No. 1 [March 2013], 293-311) that review panels were "stacked" with pro-agriobiotechnology interests, eager to facilitate commercial introduction of GMOs.

③ See, for example, Lynne Peeples, "GMO debate heats up: Crtitics say biotech industry manipulating genes, and science, *Huffington Post*, September 24, 2012.

④ See Ma Tienjie, "Wielding the double-edged sword: The Chinese experience with agricultural genetically modified organisms," *China Environmental Forum*, May 13, 2008.

the product of a call to action through social media such as Facebook and Twitter. In the U.S., protesters distributed information about health and political concerns regarding GMOs with the intent of raising awareness.①

The Chinese government regards biotechnology as an indication of the scientific advancement of the state, and take pride in its growth. Recent studies indicate that Chinese media, the voice of the government, are largely supportive of GMOs, and this strongly influences public attitudes. Print media such as *People's Daily* and *Guangming Daily* omit negative references to GMOs: 48.1 percent of news articles were largely supportive of GMO research and development and adoption of GM cotton; and 51.9 percent were neutral in describing the Chinese government as actively pursuing national GMO research and development programs.② To the present, discussion of risks associated with GMOs is minimal in Chinese media reports, for reasons mentioned above.

However, in China, consumer opinion toward GMOs may have become more critical. A rumor mill of NGOs and some opposing scholars challenge government scientists who praise biotech advances. Some scholars believe that popular attitudes in China reflect an important change in government policy, from strong promotion of agricultural biotechnology to a more precautionary approach.③

6 Conclusions

GMOs are still a relatively young product in the global market place. They did not reach the marketplace in large numbers until the mid-1990 s and did so in select product areas—such as cotton, soybeans and corn. In these product areas, while profitability increases in some areas, it did not in others. For example, the herbicide tolerant soybeans reduced cost of weed management but did not increase yields over conventional varieties, nor reduce herbicide use. Better assessments still are needed of the first generation GMOs.

Increasingly, social and economic concerns about GM products have dinned out positive news about their promise in the global arena. It is likely that critical concern will increase as GMOs enter their second generation of life in the global food community. Development of the second generation products—for example fruits and vegetables that deliver vaccines and crops leading directly to significant boosts in yields—will doubtless be influenced by controversies over the first generation ones.

Two protocols specifically regulate GMOs, the Cartageña Protocol, which went into force in 2003 and the Nagoya-Kuala Lumpur Supplemental Protocol on Liability and Redress. Developing countries strongly supported these protocols, as did most NGOs with interests in

---

① Xia, Rosanna, "Hundreds in L.A. march in global protests against Monsanto, GMOs," *Los Angeles Times*, May 25, 2013.

② Li Du and Christen Rachul, "Chinese newspaper coverage of genetically modified organisms," *BMC Public Health*, 12 (2012), 326.

③ Falkner, Robert, "International sources of environmental policy change in China: The case of genetically modified food," *The Pacific Review*, Vol. 19, No. 4 (2006).

food security and safety. The protocols are interesting analytically because they do not reflect the interests of the large GMO producing multinational firms, such as Monsanto, BASF and Syngenta. In fact division of the global business community on the GMO issue reduced the influence of business. A second factor of interest is that both protocols were developed notwithstanding the opposition of the United States, the world's remaining superpower.

It is too early to say whether the supplementary protocol will improve the global biosafety regime.Certainly, given the heterogeneous forces in global food production, and increasing conflict over time among nonstate actors (not only multinational business firms but also consumer-oriented NGOs and food safety NGOs), entrance into both the Cartageña and Nagoya-Kuala Lumpur protocols has been slow.

Although biosafety is now an area covered by international conventions, it is national law and practice which continue to shape the GM market. China and the U.S. remain different in their approach and strategy to development of this agricultural technology.

Socio-economic Considerations in GMOs Decision-making

转基因生物决策的社会经济考虑

Georgina Catacora-Vargas

Introduction

The inclusion of socio-economic considerations in biosafety decision-making is a widely debated issue at international，regional and national levels. Despite significant experience and acceptance of the inclusion of social and economic aspects in environmental decision-making（Freudenburg 1986；Bereano 2012），the recognition of the eco-social inter relationship and its practical implementation in regulation related to genetically modified organisms（GMOs）have been difficult and contentious（Secretariat of the CBD 2003；MacKenzie et al. 2003）.

The arguments both in favour and against the inclusion of socio-economic considerations in biosafety decision-making are diverse. Points of view in favour acknowledge the relevance of socio-economic considerations in risk assessment and management of GMOs due to their potential impacts on biological diversity that may in turn jeopardise rural livelihoods，indigenous knowledge，market opportunities and even national economies，among others. These concerns have been more force fully raised by governments and institutions in countries that are centres of origin and genetic diversity（MacKenzie et al. 2003；Khwaja 2002；Secretariat of the CBD 2011；Pavone 2011）. In contrast，opinions against consider socio-economic considerations of limited relevance in GMO regulation. Moreover，it is argued that their inclusion could delay the process of adoption of new technologies and increase the cost of compliance with biosafety policy（Falk-Zepeda and Zambrano 2011；Falk-Zepeda 2009，Secretariat of the CBD 2011；Secretariat of the CBD 2003）.

Nevertheless，several countries have been - and are in the process of - including socio-economic provisions in their national biosafety frameworks，including countries that are not Parties to the Cartagena Protocol（Spök 2010；Bereano 2012）. The Cartagena Protocol on Biosafety is the multilateral environmental agreement that sets international rules and procedures for the safe transfer，handling and use of GMOs in order to prevent ‘adverse effects on the conservation and sustainable use of biological diversity，taking also into account risks to human health'（Article 1）（Secretariat of the CBD 2000：3）.

Based on the current experience related to impacts of GMOs at the socio-economic level, and the need for greater conceptual clarity on its utility, the following sections provide some elements on the basic questions of what, why, when and how to include socio-economic considerations in GMO decision-making.

1 What are socio-economic considerations related to GMOs?

There is not yet a clear and agreed definition on what socio-economic considerations entail in the context of biosafety regulation, despite more mature use of the concept in other fields of environmental decision-making（e.g. Sadler and McCabe 2002）. In order to advance some conceptual clarity and for the purpose of this paper, the definition of social impacts given by Sadler and McCabe（2002）in United Nations Environmental Programme training manuals could be adapted to preliminarily describe socio-economic considerations related to GMOs as the set of the intertwined social and economic consequences resulting from the changes arising from the introdution of GMOs into the environment, which need to be taken into account in the biosafety decision-making processes.

Three aspects need to be pointed out from this proposed description:

（1）The core of the analysis is the consequences or impacts of changes rather than only the changes. This is because some changes may not result in impacts（Vanclay 2002）, or more importantly, may overshadow the real relevant effects（see Box 1 for further discussion on this point）.

（2）The socio-economic considerations embrace two general types of impacts:（i）tangible and mostly quantitatively-measured impacts, such as changes and resulting outcomes in income generation, trading opportunities, forms of livelihoods, work generations local organisation, access to food, food quality, health status, gender equity, etc., and（ii）intangible and mostly qualitatively-measured impacts such as cultural and psychological changes and related impacts, e.g. alteration in values, attitudes, perceptions, communities, visions for the future, etc.（Sadler and McCabe 2002）.

（3）Since social and environmental contexts vary from place to place, socio-economic impacts and therefore socio-economic considerations will vary from community to community and even from group to group（Vanclay 2002）. This brings methodological challenges as discussed below.

2 Why socio-economic considerations in decision-making related to GMOs?

The 'Nature-Society co-evolution' perspective in development, i.e. the process of development from the mutual influence between the environment and social systems（Norgaard and Sikor 1999）recognises that all interventions（e.g. projects and technology）have implications for both the environment and society（Pavone et al. 2011）. This gives the rationale for why socio-economic considerations are relevant in environmental decision-making processes，such as the introduction of GMOs into the ecosystems. In addition to this evident mutual relationship Barrow（2002）adds two other reasons for the consideration of socio-economic aspects in decision-making：One is the growing demand for social responsibility by markets and regulations（exemplified by the rising demand for fair trade and socially-responsible products）; and two，the global necessity of advancing sustainable development objectives.

3 When should socio-economic considerations be considered?

The debate on when to consider the socio-economic impacts of GMOs in the decision-making process is another unresolved issue in the biosafety discussions.

Socio-economic assessment can be performed either before（ex-ante）or after（ex-post）the GMO introduction. Both are different in purpose and information provided.

Ex-ante assessments are anticipatory，in other words，they aim to determine the potential impacts and risks of GMO，information that is relevant during the decision-making process over applications of introduction of GMOs. These kinds of assessments are precautionary and have the potential to better contribute to sustainable development efforts（Barrow 2002）. The Cartagena Protocol on Biosafety highlights the ex-ante consideration of socio-economic impacts. Article 26.1 of the Protocol mentions that for Parties who choose to include socio-economic considerations in their biosafety procedures，these are applicable in the process of reaching a decision on import of GMOs（Secretariat of the CBD 2000）.

Conversely，ex-post assessments focus on the monitoring of the risks identified in the ex-ante evaluation，and detecting any potential or real unforeseen adverse effects from either approved or illegally introduced GMOs. Ex-post assessments are relevant for identifying and taking preventive or corrective measures in the case of risk or damage，respectively，from GMOs.

Based on the differentiated aims and types of information provided，these two sorts of assessments are not inter-changeable. This means that one cannot replace the other since they fulfill different purposes and supply information for particular decision-making processes.

4 How to include socio-economic considerations?

Generally speaking，socio-economic aspects and impacts related to GMOs are complex for diverse reasons：（i）they vary along time and across space，this variation may occur even over short time periods or within locations geographically close to each other;（ii）multiple factors may influence social systems simultaneously，highlighting the importance of their inclusion in the socio-economic analysis（e.g. social，economic，cultural，political，ethical and other factors）; and（iii）societies are embedded in the natural environment（a more complex system in itself） giving place to another set of socio-economic considerations arising from the Nature-Society relationship（Barrow 2002；Norgaard and Sikor 1999）.

These various features provide the rationale for the inclusion of the following methodological and decision-making approaches related to socio-economic considerations：

— Integrated and complementary assessment to environmental risk assessment. As mentioned above，ecological and socio-economic factors are intertwined and influenced mutually. This is clear in the example given in Box 1 where a socio-economic change i.e. the introduction of a GM glyphosate-tolerant variety and the inherent intense use of the specific herbicide that this variety is tolerant to，resulted in ecological changes（such as appearance of glyphosate-tolerant weeds）that at the same time gives place to a new set of eco-social implications：the need for other herbicides to combat weed resistance that further pollute the agro-ecosystem，increase in the production costs and in the risk to public health.

— Holistic by including direct and indirect as well as cumulative and combinatorial effects. Changes and their consequences rarely occur in a linear or isolated manner in nature or societies. Since both systems are complex，changes result in direct and indirect，combinatorial and cumulative，and hence are often unforeseen impacts（Stabinsky 2001；Cardinale et al. 2012），out of which some may be undesirable. This justifies the need for monitoring the performance of GMOs if introduced into the environment. Following the example given above and from Box 1，the increased use of glyphosate is a direct impact from cultivating GM glyphosate-tolerant varieties. A reported indirect impact is the use of more toxic herbicides（e.g. paraquat）to control glyphosate-resistant weeds that appear over time. This，combined with the need for larger investments to purchase such herbicides and the higher risks to the health of ecosystems and human populations，the example of Box 1 points to a potentially unsustainable production system in the long term at ecological，social and economic levels.

Box 1 Consideration of the impacts of changes，rather than only the changes themselves

Genetically modified(GM)glyphosate-tolerant soybean is promoted under the claim that its adoption will contribute to reducing the use of toxic herbicides. Since the provisional approval (in 2003) of this GM soybean in Brazil，the use of glyphosate has increased considerably，from 62.5 thousand kilogrammes of active ingredient applied in 2003 to approximately 300 thousand in 2009（Meyer and Cederberg 2010）. This change in volume equals a rise of 380% in the use of glyphosate as active ingredient. This increase results mainly from two processes the expansion in the area planted with GM soybean tolerant to glyphosate(Catacora-Vargas et al. 2012），and the loss of efficacy of this herbicide in controlling weeds（Waltz 2010）due to the appearance of glyphosate-resistant weeds（Cerdeira et al. 2011）. In order to control them，herbicides more toxic than glyphosate are used，such as paraquat. Although paraquat was banned in Europe in 2007 due to its implications for neurological and reproductive disorders（Wright 2007；Frazier 2007），its import and use is increasing in the largest(GM)soybean-producing states of Brazil(Meyer and Cederberg 2011). In 2009 alone，3.32 million litres of paraquat were applied in the country（Catacora-Vargas et al. 2012）.

This case shows two changes in the production systems of soybean in Brazil: The first related to the introduction of GM varieties and the second, to increased glyphosate use. Since there is a wide controversy on the safety of GM crops and glyphosate, probably these changes may not say much. However, the consequences are the core of the socio-economic impact analysis, such as the development of glyphosate-resistant weeds that results in increased use of highly toxic herbicides, which at the same time are linked to other impacts: higher production costs as a result of the purchase of additional herbicides other than glyphosate and larger health risks. If the socio-economic assessment focuses only on the changes (e.g. introduction of GM glyphosate-tolerant soybean as a means of reducing the use of other more toxic herbicides) and not on the effects of those changes, there is the danger of overlooking the related consequences and, as a result, failure to consider these aspects in the GMO decision-making process.

— Multi- and transdisciplinary approaches. The complexity of socio-economic issues, particularly the ones related to the environment, requires an assessment and decision-making process that includes different disciplines for a more comprehensive exchange of knowledge and information. In the case summarised in Box 1, ecological, health and social sciences are needed to adequately understand and estimate the corresponding risks taking place: alterations in weed populations, exposure to different herbicides, and changes to local livelihoods that may result from introductions of GMOs. Also highly relevant yet often ignored areas, such as ethics, play an important role. For instance, the ethical considerations of increasing export and use of pesticides banned in some regions (such as paraquat) and its impacts on the welfare of local ecological and social systems.

— Methodologically pluralistic. Based on the above, an expected conclusion is the application of different research and decision-making approaches utilising diverse fields of knowledge; also necessitating the broader inclusion of questions to be answered and concerns from actors to be involved or impacted. The application of not only quantitative but qualitative (including participatory) methods is essential in socio-economic assessments. The participation of an informed public is crucial for achieving societal-relevant outcomes in both GMO research and biosafety decision-making.

— Context specific. As stated earlier, the eco-social interrelationship varies at temporal and geographical scales. This requires a case-by-case and regularly updated assessment of the socio-economic impacts of GMOs according to the social and ecological context where they are introduced.

— Long-term oriented. Only long-term assessments will provide proper information on the socio-economic impacts of GMOs and their consequences on sustainability. The indirect, combinatorial and cumulative effects of GMOs introductions in complex systems such as nature and society will not be appropriately captured or assessed in short-term scenarios.

5 Final comments

Socio-economic impacts（positive or negative，predicted or unforeseen）are an inherent part of technology introduction and adoption. This points to the need for including socio-economic considerations in biosafety decision-making related to GMOs.

The Nature-Society interface defines the complexity of the socio-economic dimension of any intervention（e.g. projects or technologies）and calls for a thoughtful and comprehensive methodological approach characterised by：a holistic view，integrative with environmental risk assessments，multi- and transdisciplinarity，methodological pluralism，context specificity and long-term orientation. In other words，proper socio-economic assessments will require going beyond the common practice of mostly economic assessments，but aiming towards sustainable-development-relevant appraisals. To this end，precautionary or anticipatory（also called ex-ante）assessments are needed，complemented with regular monitoring（or ex-post evaluations）.

The challenges ahead for the appropriate assessment of socio-economic implications related to GMOs and their inclusion in environmental decision-making processes are significant. However，equally significant is their relevance，particularly in light of sustainable development. Hence，failing in adequately including the socio-economic dimension biosafety processes may jeopardise nature's and society's welfare.

Georgina Catacora-Vargas is an agricultural engineer specialising in environmental sciences，agroecology and sustainable development. She has practical experience（as practitioner，policy adviser and researcher）in sustainable rural development，agroecology and biosafety regulation. She has worked in these fields in different，mostly developing，countries. From February 2009 to August 2013 she was part of the Society，Ecology and Ethics Department（SEED）of GenØk — Centre for Biosafety，Norway. Currently she does research on socio-economic aspects of agroecology and genetically modified crops in Brazil，in collaboration with the Federal University of Santa Catarina，and Bolivia，as well as research on international regulations on biosafety.

References

[1] Barrow C.J. 2002. Evaluating the Social Impacts of Environmental Change and the Environmental Impacts of Social Change：An Introductory Review of Social Impact Assessment. International Journal of Environmental Studies，59（2）：185-195.

[2] Bereano P. 2012. Why the US should support full implantation of Article 26，the consideration of socio-economic consequences of LMOs. ECO（43），CBD Alliance.

[3] Catacora-V G.；Galeano G.；Agapito-Tenfen S.；Aranda D.；Palau T.；Nodari R.（2012）. Producción de Soya en las Américas：Actualización Sobre el Uso de Tierras y Pesticidas. GenØk/UFSC/REDES-AT/BASE-Is，Cochabamba，43.

[4] Cardinale B.J.，Duffy J.E.，Gonzalez A.，Hooper D.U.，Perrings C.，Venail P.，Narwani A.，Mace G.M.，Tilman D.，Wardle A.A.，Kinzig A.P.，Daily G.C.，Loreau M.，Grace J.B.，Larigaude A.，Srivastava D.S.，Naeem S. 2012. Biodiversity Loss and Its Impact on Humanity. Nature 486：59-61.

[5] Cerdeira A.L.；Gazziero D.L.P；Duke S.O.；Matallo M.B. 2011. Agricultural Impacts of Glyphosate-Resistant Soybean Cultivation in South America. J. Agric. Food Chem. 59：5799-5807.

[6] Freudenburg W.R. 1986. Social Impact Assessment. Annual Review of Sociology，12： 451-478.

[7] Falk-Zepeda J.B. 2009. Socio-economic Considerations，Article 26.1 of the Cartagena Protocol on Biosafety：What are the Issues and What is at Stake？. AgBioForum 12（1）： 90-107.

[8] Falk-Zepeda J.B.，Zambrano P. 2011. Socio-economic Considerations in Biosafety and Biotechnology Decision Making：The Cartagena Protocol and National Biosafety Frameworks. Review of Policy Research 28（2）：171-195.

[9] Frazier L. 2007. Reproductive Disorders Associated with Pesticide Exposure. Journal of Agromedicine 12（1）：27-37.

[10] Khwaja R.H. 2002. Socio-Economic Considerations In C. Bail，R. Falkner and H. Marquard（eds），The Cartagena Protocol on Biosafety：Reconciling Trade in Biotechnology with Environment and Development?，362-365.

[11] MacKenzie R.，Burhenne-Guilmin F.，La Viña A.G.M. Werksman J.D. 2003. An Explanatory Guide to the Cartagena Protocol on Biosafety. IUCN，Cambridge，295.

[12] Meyer D.；Cederberg C.（2010）. Pesticide use and glyphosate-resistant weeds - a case study of Brazilian soybean production. SIK-Rapport Nr 809，55.

[13] Norgaard R.；Sikor T.O. 1999. Metodología y Práctica de la Agroecología In M.A. Altieri，Agroecología. Bases Científicas para una Agricultura Sustentable. Nordan-Comunidad，Montevideo，337.

[14] Pavone V.，Goven J.，Guarino R. 2011. 'From Risk Assessment to In-context Trajectory Evaluation-GMOs and Their Social Implications'. Environmental Science Europe 23（3）：1-13.

[15] Sadler B.；McCabe M.（Eds）. 2002. Environmental Impact Assessment Training Resource Manual. UNEP，Geneva，561.

[16] Secretariat of the CBD（Convention on Biological Diversity）. 2000. Cartagena Protocol on Biosafety to the Convention on Biological Diversity：Text and Annexes. Secretariat of the Convention on Biological Diversity，Montreal，20.

[17] Secretariat of the CBD（Convention on Biological Diversity）. 2003. The Cartagena Protocol on Biosafety. A Record of the Negotiations. Secretariat of the Convention on Biological Diversity，Montreal，140.

[18] Secretariat of the CBD（Convention on Biological Diversity）. 2011a. 'Socio-Economic Considerations：Summary of Submissions Received from Parties，Other Governments and Relevant Organizations'. Retrieved on 31 July 2012 from：http://www.cbd.int/doc/meetings/bs/bsws-sec-01/information/bsws-sec-01-bsregconf-sec-ap-01-inf-01-en.pdf.

[19] Spök A. 2010. Assessing Socio-Economic Impacts of GMOs. Issues to Consider for Policy Development. Lebensministerium/Bundensministerium für Gesundheit，Vienna，123.

[20] Stabinsky D. 2001. Bringing Social Analysis into a Multilateral Environmental Agreement：Social Impact Assessment and the Biosafety Protocol. Journal of Environment and Development 9（3）：260-283.

[21] Vaclay F. 2002. Conceptualizing Social Impacts. Environmental Impact Assessment Review 22：183-211.

[22] Waltz E. 2010. Glyphosate resistance threatens Roundup hegemony. Nature Biotechnology 28（6）：537-538.

[23] Wright G. 2007. Paraquat：The Red Herring of Parkinson’s Disease Research. Toxicological Sciences 100（1）：1-2.

中国六省转基因抗虫棉种植对棉花害虫的影响调查

Investigation on Impacts of GM Cotton's Plantation on Cotton Pests in the Six Provinces of China

汪文蓉[1]　陈晨[2]　薛达元[2、3]　王艳杰[2]

（1. 上海有机道文化发展有限公司，上海；2. 中央民族大学生命与环境科学学院，北京，100081；
3. 环境保护部南京环境科学研究所，南京，210042）

WANG Wen-Rong[1]　CHEN Chen[2]　XUE Da-Yuan[2,3]　WANG Yan-Jie[2]

（1. Shanghai Organic Food Culture Development Company；
2. College of Life and Environmental Science，Minzu University of China，Beijing，100081；
3. Nanjing Institute of Environmental Science，Ministry of Environmental Protection，Nanjing，210042）

摘　要：中国种植转基因抗虫棉已经超过 15 年，为了解多年种植抗虫棉产生的影响，对中国黄河流域的河北省、河南省、山东省以及长江流域的湖北省、安徽省和江苏省进行田间调查以及农户访谈。调查发现 *Bt* 棉花对靶标害虫棉铃虫和红铃虫仍旧有效，但是很多地区种植抗虫棉 5～8 年后引起次生害虫以及棉花病害的爆发，这成为困扰农户的重要问题。调查还发现因为棉田周围存在大量天然庇护所，靶标害虫棉铃虫，红铃虫并未对抗虫棉的 *Bt* 基因产生明显抗性。但是近年一些非靶标刺吸式口器害虫逐渐发展成为新的棉花主要害虫，主要是棉蚜、盲蝽蟓和棉蓟马。然而各地区仍存在一定差异：例如长江流域的湖北省和安徽省受到斜纹夜蛾和甜菜夜蛾的威胁；黄河流域的河北省和山东省遭受盲蝽蟓和红蜘蛛的严重危害；而河南省近 3～5 年开始遭受粉虱类害虫的困扰。

关键词：抗虫棉，棉铃虫，非靶标害虫，影响

Abstract: *Bt* cotton has been cultivated in China for more than 15 years. In order to the impacts of multiple years' plantation of *Bt* cotton，the field surveys and interviews were conducted in the six provinces in China – Hubei，Anhui，Jiangsu，Henan，Shangdong and Hebei. The investigation revealed that the *Bt* cotton is generally effective to cotton bollworm（*Helicoverpaarmigera*）and pink bollworm（*Pectinophoragossypiella*），but，in many places，the *Bt* cotton has led to the outbreak of the secondary pests as well as several cotton diseases，after 5-8 years' plantation. It has plagued cotton farmers as a serious problem. Based on the survey，it is found that the target pests of *Bt* cotton like cotton bollworm haven't developed significant resistance to *Bacillus thuringiensis* because there exist a lot of natural sanctuaries around *Bt* cotton fields，but some non-target piercing-sucking insects such as cotton aphid（*Aphis gossypii*），mirids（*Hemipteramiridae*）and cotton thrips become the new dominating cotton pests in recent years. For example，the *Bt* cotton in Hubei and Anhui province of Yangtze

River Valley area are threatened by the cotton leaf worms and the beet armyworms; while in Hebei and Shangdong provinces of Yellow River Valley area，the mirids andspider mite are the main pests; however in Henan Province the whitefly became the main insect 3-5 years ago.

Key words: *Bt*cotton，cotton bollworm，secondary pests，impacts

棉花是我国重要的经济作物和纤维作物，其生长过程中遭受多种害虫危害，其中最严重的是棉铃虫、棉红铃虫、棉蚜及棉叶螨等。20 世纪 90 年代，我国棉花主产区棉铃虫连续爆发，造成严重的经济损失。自 1997 年引进美国转基因抗虫棉品种 33B 和 99B 至今，转基因抗虫棉在我国的种植历史已超过 15 年，并成为我国长江流域棉区和黄河流域棉区的主要品种。2012 年，我国超过 700 万农户种植了约 400 万 hm^2 的转 *Bt* 基因棉，占棉花种植总面积的 80%左右。转基因抗虫棉的大量推广，有效地降低了棉铃虫和红铃虫等靶标害虫的数量[1]，并缓解了棉花生产中棉铃虫抗药性带来的压力，减少了化学杀虫剂对环境的影响[2]。但是转 *Bt* 基因棉花在防治靶标害虫的同时，会对非靶标害虫的生物学特性、种群动态和群落结构等产生相应的影响[3,4]。在一些棉区，刺吸式口器害虫，如棉蚜、棉盲蝽、棉叶螨、白粉虱、棉蓟马等种群呈现上升趋势，甚至出现种群爆发的情况。因此，有必要调查和研究转 *Bt* 基因抗虫棉对非靶标害虫的影响。

实验室以及田间试验都认为，*Bt* 抗虫棉的推广能有效控制靶标害虫——棉铃虫和红铃虫的种群爆发。崔金杰等 1997 年[5]的研究认为，转基因棉花对鳞翅目害虫具有良好的防治效果，只是在抗虫棉生长后期根据需要使用农药。2007 年黄季焜等人[6]对全国四省抗虫棉田农药施用量的跟踪调查发现，*Bt* 抗虫棉的多年推广，棉铃虫种群数量大大降低，使农户减少了对化学杀虫剂的依赖。吴孔明[7]等对 *Bt* 抗虫棉和常规棉田中棉铃虫的种群动态进行了连续 10 年的监测，结果表明，虽然 *Bt* 棉花与常规棉花的棉铃虫落卵量没有明显差异，但 *Bt* 棉花的棉铃虫幼虫发生数量显著减少；同时随着 *Bt* 棉的连续种植，棉铃虫落卵量和幼虫发生数量逐年下降，棉铃虫种群发生世代数量明显减少。

随着转基因抗虫棉在我国商业化种植推广，对非鳞翅目生物可能产生的影响存在很多争议。吴孔明等经过多年的研究证明，种植转 *Bt* 基因棉花田块中的瓢虫类、草蛉类和蜘蛛类捕食性天敌的数量较常规棉田大幅度增加[8]。另外多个研究认为，转 *Bt* 基因抗虫棉仅对鳞翅目害虫具有良好的防治效果，但对棉蚜、棉叶螨、棉蓟马、白粉虱等刺吸性害虫没有效果[5,9]。其他的研究也表明，转 *Bt* 基因棉花虽然够有效地控制棉铃虫和红铃虫，但原本处于次要地位的害虫——盲蝽蟓，其种群急剧增加，逐渐取代棉铃虫，成为主要害虫[10,11]，并造成严重影响。

然而，上述研究只是限定于科研人员的小面积实验地，尚没有大范围棉田的调查和统计数据，特别是缺少农户对种植转基因抗虫棉的认识和经验。本研究希望通过在主要棉区进行大量的农户面访，调查农户在过去 10 多年种植转基因抗虫棉的过程中，对转基因抗虫棉在防治棉铃虫等靶标害虫的效果、棉铃虫对转基因棉的抗性发展，以及非靶标害虫种群变化等方面表现的认知和作出的判断，进而证实与科学家们的小面积科学实验结果的一致性。

1　研究方法与研究内容

本研究选择对我国棉花种植的六个大省进行实地调查，分别为黄河流域的河北省、山东省、河南省和长江流域的湖北省、安徽省以及江苏省，共 25 个县市，65 个村。调查时间为 2011 年 7 月至 2012 年 8 月。调查样本为随机选择，在随机选取调查的村庄后，调查人员从当地镇政府和行政村村委会提供的植棉农户中，随机选择 15～20 个农户（根据村庄规模大小，并且选择种植过常规棉的农户）。经过培训的调查员采用访谈式问卷调查法，与每个农户面谈 30～40 min。每省完成问卷 200 份，六省共完成问卷 1 200 份，其中有效问卷 1 167 份。调查内容主要是了解种植常规棉与转基因抗虫棉 15 年后，靶标害虫棉铃虫和红铃虫的生态位变化，盲蝽蟓、红蜘蛛等非靶标害虫种群变化，以及靶标害虫抗性发展情况等及其原因。调查结果见表 1。

表 1　棉农对转基因抗虫棉田其虫害严重程度评价及平均得分

棉区	害虫名称	种植抗虫棉前							种植抗虫棉后						
		无	轻	偏轻	中度	偏重	重度	平均得分	无	轻	偏轻	中度	偏重	重度	平均得分
黄河棉区	棉铃虫**	0	12	20	61	356	136	4.98	12	260	213	65	29	5	2.74
	棉蚜**	0	109	94	185	155	39	3.83	1	83	171	186	105	34	3.67
	盲蝽蟓**	213	186	68	51	27	22	2.15	4	54	138	220	125	33	3.80
	斜纹夜蛾**	59	161	75	57	18	11	1.71	38	186	102	47	16	3	1.73
	甜菜夜蛾**	41	183	85	75	34	13	2.09	17	207	130	61	31	2	2.12
	棉蓟马	76	90	51	21	15	0	1.01	60	94	61	39	19	0	1.21
	红蜘蛛**	7	201	121	122	100	11	3.12	6	82	185	188	100	9	3.47
	地老虎**	15	179	129	114	68	18	2.85	22	225	144	104	35	5	2.61
	粉虱**	50	55	15	15	12	12	1.37	3	31	22	31	69	13	2.35
	蜗牛	1	2		1	0	0	.05	0	0	0	0	2	1	0.08
长江棉区	棉铃虫**	3	18	55	124	240	138	4.71	25	320	156	45	28	6	2.57
	棉蚜**	3	62	171	153	162	20	3.80	8	198	179	117	66	7	3.09
	盲蝽蟓**	89	241	108	51	63	10	2.57	1	24	97	188	228	38	4.26
	斜纹夜蛾**	21	170	185	59	89	15	3.06	16	133	121	143	118	19	3.47
	甜菜夜蛾*	21	145	195	52	51	2	2.84	25	179	138	80	53	4	2.91
	棉蓟马**	67	104	167	50	45	4	2.72	40	164	127	72	47	3	2.91
	红蜘蛛**	6	88	204	95	146	25	3.59	3	196	181	108	74	11	3.14
	地老虎**	10	97	239	100	95	4	3.25	10	239	168	65	67	13	2.96
	粉虱**	17	14	9	3	6	1	2.40	2	12	5	3	19	10	4.00
	蜗牛	0	3	6	1	1	0	3.00	0	3	2	0	7	4	4.56

注：1. *显著；**极显著；2. 表中数字为接受访谈的农户数量；3. 平均得分的计算：将害虫严重程度从轻至偏重以 1～6 赋值，计算其平均分。

2　种植抗虫棉前后棉田主要害虫种群变化情况

对表 1 中种植抗虫棉前后虫害变化的分析结果表明，种植转基因抗虫棉能够明显抑制两大棉区的棉铃虫以及地老虎等害虫的危害，且种植前后危害程度差异极显著。值得注意的是，盲蝽蟓、斜纹夜蛾及粉虱的危害呈显著增加的趋势。棉蚜虽是主要害虫之一，但其危害程度显著降低。

黄河地区种植常规棉品种时，主要棉花害虫为棉铃虫、棉蚜、红蜘蛛；次要害虫为斜纹夜蛾、甜菜夜蛾、地老虎等；而种植转基因抗虫棉后，棉铃虫、棉蚜、地老虎的危害明显减轻，棉铃虫造成的危害仅为轻和偏轻两个等级。但是盲蝽蟓取代棉铃虫成为危害棉花的主要害虫，成为中度危害。同时棉蚜、红蜘蛛仍旧是威胁抗虫棉的主要害虫，为中度危害和偏轻两个等级。

长江棉区在种植 *Bt* 棉后，棉铃虫危害程度明显降低，而盲蝽蟓的危害则比黄河棉区更加严重，达到中度和偏重的程度；棉蚜、红蜘蛛的危害程度虽然相比种植常规棉降低，但是仍是威胁棉花的主要害虫。值得注意的是，粉虱在黄河棉区为轻度危害，但是在长江棉区则达到中度危害；棉蓟马是种植常规棉品种时的主要害虫之一，对棉花造成中度危害，但是种植抗虫棉多年后，则为轻度和偏轻的等级，不再是主要害虫；斜纹夜蛾在种植转基因抗虫棉后仍是主要害虫，而且危害程度显著提高；甜菜夜蛾危害也有增加的趋势，但并不显著。

从表 1 可以看出，斜纹夜蛾、甜菜夜蛾在黄河棉区的危害为轻度等级，明显低于长江棉区的中度和偏轻的等级。调查发现，长江棉区超过 45.15%的棉农认为种植转基因抗虫棉后，棉铃虫种群减少，但是次生害虫种群增加，而在黄河棉区只有 38.8%的农户支持这种观点；而在这两个棉区持反对意见的则分别为 23.8%和 24.7%，比较接近。此外，棉农认为造成棉田害虫种群变化的主要因素是气温变化或者降雨量变化。

3 靶标害虫的抗性发展情况

虽然两大棉区种植抗虫棉已有 15 年的历史，但是靶标害虫棉铃虫和红铃虫并未对转 *Bt* 基因棉花产生明显的抗性，转基因抗虫棉对棉铃虫和红铃虫的防治作用仍旧比较明显。长江棉区以及除河北省外的黄河棉区，更多的棉农认为种植抗虫棉后，棉铃虫、红铃虫种群数量递减（表 2）；黄河棉区中，超过 57.4%的棉农对递减的趋势表示认可，仅仅 14.0%持反对态度，认为棉铃虫种群数量逐年递增，同时 20.3%的棉农认为呈现波动变化趋势；长江棉区 69.0%的农户认为抗虫棉对棉铃虫、红铃虫控制效果非常明显，仅仅 8.1%的农户持反对态度。

表 2　农户对种植 *Bt* 棉后棉铃虫数量变化的评价　　单位：%

棉区	省份	递增趋势	递减趋势	基本不变	波动趋势	不清楚
黄河棉区	河北省	15.3	36.2	5.6	42.9	0.0
	河南省	15.6	66.7	12.0	4.2	1.5
	山东省	10.1	70.2	3.5	14.6	1.5
长江棉区	安徽省	13.6	59.2	13.6	12.5	1.1
	湖北省	5.6	71.1	14.7	8.1	0.5
	江苏省	5.5	76.0	13.0	5.5	0.0

棉农对于该现象的解释，主要有以下理由：①棉田周边存在大量的天然庇护所；②抗虫棉种籽质量好；③抗虫效果好，未产生退化；④农药对害虫的防治效果越来越好。而实际调查结果也支持棉农的认知。

4 结语

本研究的结果表明，在中国抗虫棉种植的主要地区，由于天然庇护所的存在，靶标害虫棉铃虫和红铃虫并未对 *Bt* 基因产生明显的抗性，在历经 10 多年的种植后，转基因抗虫棉的防治作用仍较为明显；但是，近年来一些非靶标害虫如盲蝽蟓、棉蚜、红蜘蛛等逐渐取代棉铃虫成为危害棉花的主要害虫，而白粉虱种群在两大棉区均有增长的趋势。而且，省份之间还表现出一定的特殊性：长江流域的湖北和安徽省受到棉蚜虫和甜菜夜蛾的威胁；黄河流域的河北和山东省受到甜菜夜蛾和棉蓟马的危害；河南省在近 3～5 年中，白粉虱逐渐成为主要害虫。

参考文献

[1] 刘晨曦，李云河，高玉林，等. 棉铃虫对转 *Bt* 基因抗虫棉花的抗性机制及治理，中国科学：生命科学，2010，10：3.

[2] Wu，K.-M.，*et al*.. Suppression of cotton bollworm in multiple crops in China in areas with Bt toxin-containing cotton. *Science*，2008，321（5896）：1676-1678.

[3] Henneberry，T.，L.F. Jech，and T. De la Torre. Effects of transgenic cotton on mortality and development of pink bollworm（Lepidoptera：Gelechiidae）larvae. *Southwestern Entomologist*，2001，26（2）：115-128.

[4] WILSON，D.F.，*et al*.. Resistance of cotton lines containing a Bacillus thuringiensis toxin to pink bollworm（Lepidoptera：Gelechiidae）. *Journal of Economic Entomology*，1992，85（4）：1516-1521.

[5] 崔金杰，夏敬源. 转 *Bt* 基因棉对棉田主要害虫及其天敌种群消长的影响. 河南农业大学学报，1997，31（4）：351-356.

[6] 黄季焜，米建伟，林海. 中国 10 年抗虫棉大田生产：*Bt* 抗虫棉技术采用的直接效应和间接外部效应评估. 中国科学，2010，（40）：260-272.

[7] 吴孔明，陆宴辉，姜玉英. 在中国种植含 Bt 毒素棉花的地区，棉铃虫在多种作物中受到抑制. 分子植物育种，2008，6（6）：1106-1106.

[8] 李进步，方丽平，张亚楠. 类型品种棉花上棉蚜适生性及种群动态. 昆虫学报，2007，50（10）：1027-1033.

[9] 崔金杰，夏敬源. 麦套夏播转 *Bt* 基因棉田主要害虫及其天敌的发生规律. 棉花学报，1998，10（5）：255-262.

[10] 陆宴辉，吴孔明，姜玉英. 棉花盲蝽的发生趋势与防控对策. 植物保护，2010，36（2）：150-153.

[11] Wang，Z.-J.，*et al*.. Bt Cotton in China：Are Secondary Insect Infestations Offsetting the Benefits in Farmer Fields？ *Agricultural Sciences of China*，2009，8（1）：83-90.

发达国家与发展中国家转基因生物认知的比较研究

Comparative Research of the Public Cognition to GMOS between Developed and Developing Countries

曲瑛德，叶凌风，陈源泉，康定明
（中国农业大学 农学与生物技术学院，北京，100193）
QU Ying-De，YE Ling-Feng，CHEN Yuan-Quan，KANG Ding-Ming
（Agronomy and biotechnology college，China Agricultural University，Beijing，100193）

摘　要：转基因生物技术是引领和解决新世纪农业与环境等诸多问题的新技术，但也引起了公众对转基因生物安全的关心和担心。重视转基因生物安全交流认知，做到明明白白转基因，能够推动转基因生物技术的发展，使其为我国社会经济发展做出更大贡献。论文分析引述比较了美国公众对转基因生物安全的认知状况，发现美国公众对生物技术与转基因的了解大众普及的程度不一，公众对转基因食品消费与安全的认同也有差异。对比了转基因产品在发达国家美国、英国、日本以及发展中国家哥伦比亚和我国公众的认知与影响，发达国家的消费者愿意购买转基因食品的意愿低于那些发展中国家消费者家里没有适宜的或高质量食品可消费的比例。分析了影响公众转基因产品认知的因素主要有对政府的信任度、对科学的态度和媒体宣传等，以及公众判定转基因产品利益与风险在不同经济发展水平国家的差异，总结了引起这种差异的可能因素，以及最后发达国家和发展中国家转基因标签与转基因认知的相互影响。

关键词：转基因生物；认知；发达国家；发展中国家；差异

Abstract: Transgenic biotechnology is one of frontiers for resolving numerous agricultural and environmental dilemmas happened in a new century，but it is invoked several biosafety problems cared and worried by the public. Smart cognition and communication with the public takes an important role in facilitating development of transgenic biotechnology so as to allow the public to know the reasons and principle of transgenic biotechnology，and more，to make great contribution to China social and economic development. This paper cited and commented the transgenic biotechnology biosafety cognition of the American public and compared the cognition and its impacting factors to transgenic products between developed countries，such as the US，UK，Japan and developing countries including Columbia and China. We also analyzed the different impact factors of the public cognition to transgenic products and the varied elements in defining benefit and risk of transgenic products among the developed and developing countries. Eventually，we collected those possible candidate factors leading to the abovementioned phenomena and summarized the interaction between the public cognition to transgenic products and tagging in transgenic product among the developed and developing countries.

Key words: Transgenic organism; Cognition; Developed country; Developing country; Difference

现代生物技术起始于 20 世纪 50 年代初华特森和克里克发现 DNA 双螺旋结构。60 年代完成的 DNA 重组技术和 80 年代开创的遗传转化技术，使人类利用现代生物技术在不同生物之间转移利用优质基因成为可能，90 年代中期首创的包含有主要粮食作物的禾本科作物遗传转化技术，对于利用现代生物技术特别是基因工程技术，也即转基因技术，推动和加快了粮食作物优良品种的培育进程，特别是给利用不同物种之间人类所需要的优良基因带来可能与应用。

由于转基因技术的极大进步与突破，克隆人包括可用于患病器官和丧失功能器官医疗替换的主要功能性器官的克隆，克隆动物特别是与人类日常生活密切相关的大动物，如猪、牛、羊等。转基因作物，包括人类主要粮食作物小麦，水稻和玉米的转基因品种已有应用。全面遗传重组化的微生物，如与人体健康和食品密切相关的，广泛应用于现代生物技术制药的重组疫苗以及干扰素，以及酿造发酵技术中的新型重组菌株等的应用。由此可见，转基因技术的快速发展，已经全面渗透到我们现代生产生活的各个层面。由于任何技术发展都存在自身的限制和应用条件与环境的约束，而且人类对新技术的意识和心理的认识，也有一个不断深入的过程，这个过程应该是伴随新技术的发展而开展。事实上，任何技术都有自身的局限和潜在的隐患，特别是涉及人类自身繁衍和食品安全，影响范围大，涉及后代子孙的转基因技术。

转基因技术自产生和应用的那一刻起，在其显示自身强大功能和影响的同时，就昭示了可能的安全隐患和风险。目前，我国还没有开放转基因主要粮食作物的种植，实施了转基因产品标识的管理政策，农业部也建立颁布了严格的转基因作物管理政策和跨七个部委的国家转基因生物安全管理委员会。据最新统计，2011 年我国转基因棉花种植达 390 万 hm^2，种植比例高达 71.5%，2012 年达 400 万 hm^2。2012 年，我国进口转基因大豆 5 838 万 t。“十二五”期间国家在科技项目上投入了 200 亿元作为国家重大项目，以推动我国动植物转基因研究与应用。与此同时，在环境恶化带来的生活困境，假、冒、伪、劣、毒食品和药品事件时有发生的社会情势下，一些公众对转基因生物的管理政策和依据，以及目前的科学研究水平对转基因生物的安全认知与风险缺乏科学认知。因为我国目前还没有专门从事转基因生物安全风险交流的部门、预算和人员，沿用的是计划经济的模式实施管理。少数具有开展风险交流功能的民间机构也几乎少有从事转基因生物安全风险交流。从事转基因生物有关的企事业单位，更没有进行针对其产品和研究成果与社会公众进行风险交流的意识和责任。由此，导致了媒体和社会公众似乎越加关注转基因生物安全，但同时又缺乏科学和权威的证据与声音，把这种关注看作吸引眼球的新闻炒作，而使得在这类关注国际民生的大问题上，形成政府无法主导，安全观点由少数不负责任媒体主导，随风而动，以群体事件掩盖科学研究的事实与结论，造成谣言伺起的情势，给政府的公信力和转基因技术和产业在我国的发展带来极大的伤害和被动。

本文试图通过引述转基因技术在美国、英国、日本等发达国家，以及哥伦比亚以及我国等发展中国家的公众关注主题和内容的变化过程，及其影响因素，以期在我国进行转基因技术推广和转基因生物安全交流中，对照世界其他国家公众在转基因生物安全认知上的变化，根据我国的社会经济发展水平对不同层次公众生活水平与心理的影响，制订我国的转基因生物安全交流方案，选择和借鉴最为影响我国公众对转基因技术生物安全态度的主题、内容和交流方式，以及利益认知，有效开展我国转基因生物安全交流工作，推动我国转基因生物安全在政府、企业、专家和公众之间的良性互动，确保转基因技术为我国的社会经济发展做出更大贡献。

1 发达国家公众对转基因生物安全的认知状况

美国生物安全风险交流学者 Hallman 的研究报告（Hallman et al.，2003）[1]采用问卷调查的形式研究美国人对转基因农业生物技术知识了解的基本情况、认识的深度和广度，以及美国公众的这些观察和态度随时间变化趋势。调查了美国人对转基因食品的态度，对未来可能的转基因食品标准的态度，以及涉及直接和间接消费者的利益是如何认识的。另外，该报告还评估了美国人在生活中实际认识到的科学、技术和食品生产的知识，以及他们对食品的总体态度和行为，探讨了他们对转基因食品的相关意见，最后也测试了美国人对转基因食品标签的想法，以及转基因标签制度是如何影响他们的购买行为的。

报告对调查对象信息采集了性别、年龄、种族、种族特点、教育水平、收入水平、雇佣类别、自由或保守型、购物的义务等方面，同时对上述情况进行归类，年龄段以 18～24、25～34、35～44、45～54、55～64、65 以上等各段。种族分为白人、黑人和西班牙、葡萄牙裔以及其他等类别，教育分为低于高中教育、高中教育以上、大学教育以上。地域分为东北部、中西部、南部和西部等。

报告指出，美国货架上含有转基因成分的食品，尽管无法精确估算，大都认为对于加工食品应该占 60%～70%的比例。这是因为大部分大豆和菜籽以及 1/3 的玉米在美国和加拿大都来自转基因品种（2002 年）[1]，因为作物产品的组分或多或少地都混合于大部分食品中。

调查结果发现，43%的人听过或者读过但很少或者几乎不知道遗传工程和生物技术。45%的人听过或读过一些，12%的人听过或读过很多。美国人每天读报的人有 39%，从来不看的占 22%，每天关注国内新闻的人有 54%，从来不看的有 8%，每天关注当地新闻的人有 58%，从来不看的 6%，每天听新闻的 22%，不听的 49%，每天收听广播的 30%，从不听的 38%，每天听新闻的 28%，从不听的 38%，每天读新闻的 8%，从不读的 51%，每天读网络新闻的 19%，从不读的 55%。尽管现在市场有丰富的转基因食品存在，但有 52%的受访者说根本就没有意识到这个问题，25%的受访者不相信有转基因食品销售，23%的人说不清楚。只有 41% 的人说知道市场有转基因食品销售。美国人根本没有花很多时间来讨论和关注转基因食品，调查发现 62%的人从没有讨论过这个问题。只有 38%的人说他们至少在一次交谈中提过这个话题。89%的人说偶尔或者只有 1～2 次说到这个问题。只有 11%的人市场谈论这个问题。20%的人有超过 1～2 次谈论这个话题。2003 年上述数据比 2001 年有少许增加。但是，2003 年有 62%的人从不谈论生物技术，2001 年有 68% 的人从不谈论生物技术。美国人中也只有 10%的人表示非常了解生物技术和基因工程知识，而 55%的人表示知道的很少。但是，总体看有关转基因食品基本认识方面的 11 个问题，如普通番茄不含基因，转基因番茄含基因等，美国人的回答正确的人员比例要比欧洲人高出10～20 个百分点。在回答有关食品方面的 7 个知识问题时，调查者有 50%以上的正确性，显示出美国人对本国食品行业的认识与自信。但是对于遗传工程、遗传修饰和生物技术的区别与认识 1/3 的人表示不知道，只知道科学知识的只有 14%，知道负面报道的有 15%，了解正面知识的有 6%。美国人完全同意转基因的 2001 年有 16%，2003 年有 12%，有一定程度同意的 2001 年有 42%，2003 年有 37%。美国人经常阅读食品标签的人有 31%，总是阅读的有 23%，有时阅读的有 30%，很少阅读的有 10%，从不阅读的有 7%。在食品标签的内容影响购买意愿上，有 38%的人说没有差别不影响，30%的人说有很少的影响，1%的人说非常影响，3%的人说有一定影响，6%的人说不知道。上述调查数据说明，美国公众对生物技术与转基因的了解也没有做到大众普及化，公众对转基因食品消费与安全的认同也有差异。

2　发达国家和发展中国家公众对转基因食品的认知与影响因素

2.1　发达国家的状况

转基因成分在食品中的应用已经引发了高度争议，欧盟的许多国家以及日本的消费者对转基因食品持强烈的否定态度，这些国家的消费者对转基因食品持否定态度，一般都归因于转基因产品所产生的无法预知的环境和人体健康的后果。这些可能的结果包括但不只限定在非预期的过敏反应，害虫抗性的流行，或者野生植物上产生杂草抗性，以及野生物种上产生的不可逆转毒性等。

在欧盟和日本的研究提供了强有力的证据，消费者只有在转基因产品明显便宜于非转基因产品的条件下，他们才愿意承担未知的风险而消费转基因食品。Grimsrud，McClusky，Loureiro 和 Wahl（2003）[2]在挪威的一项研究，总结了在挪威的消费者的平均观点，即如果这个面包比非转基因的面包便宜 49.5%，消费者才愿意购买用转基因小麦制作的面包。Burton，Rigby，Young 和 James（2001）[2]在英国进行的一项消费者对转基因食品的态度研究中总结到，男性购物者愿意付额外的 26%，以购买非转基因产品。而女性购物者则愿意付 49.3%，McClusky、Grimsrud、Ouchi 和 Wahl [3]发现，日本消费者在他们的样本中，平均来看在购买由转基因成分制作的面条，要比非转基因面条便宜 50%以上才愿意购买。

研究发现美国的消费者比欧盟和日本的消费者更愿意接受转基因食品，例如，由 Lusk 等（2002）[4]的研究发现 70%的应答者不愿意去付更多的钱购买更优质的非转基因的炸薯片。加拿大的研究（国际农业科学与技术中心，1995）发现，消费者只有在有一些折扣价格时才愿意购买转基因的马铃薯。

在欧盟各国特别是英国，在疯牛病后转基因食品迅速成为一个公众关注的主题。关键问题是在这种情况下，释放含有转基因成分的食品到英国市场，是否会进一步损害消费者对英国食品供应链的信心，影响消费者选择购买食品的一些类型，而又如何来评估这种可能的损害程度？

当前预测消费者选择转基因食品的影响因素是困难的。当然，意识到转基因是一个食品问题的消费者比例正在日益增加，但却几乎没有人被很全面地告知了转基因的实际内容和意义是什么，转基因对食品安全、环境保护和动物福利等方面是如何影响的。对消费者的调查说明，消费者在选择他们的食品过程中，其实是忽视是否是含有转基因成分的，在面对可选或不选含有转基因的食品时，通常只有小部分消费者会明确表示他们的态度，即坚决选择或不选择含有转基因的食品。调查表明，目前消费者在市场购买食品时，还没有面对必须做出决定时选择转基因或非转基因食品。所以，是很难预测英国消费者对转基因食品是广泛接受或者是排斥的态度。只有在一种情况下，如存在转基因和非转基因的相同罐装的面食用番茄酱，存放在相邻货架上，如果这时转基因的产品是便宜的，消费者购买转基因番茄酱的意愿则高于非转基因的。

对英国消费者对转基因食品的态度调查结果表明（见图 1）[5]，42%的被调查者表示可能愿意尝试转基因食品，只有 2%的被调查者表示坚决不尝试，可能尝试和可能不尝试的被调查者分别为 27%和 20%，由此可见，在一定条件下有可能尝试的高达 89%（42%+27%+20%），而这个条件，是对转基因食品的了解和安全管理责任方的信任。所以，影响消费者对转基因食品态度的接受程度的重要因素是“意识”，因为如果这期间消费者在关于转基因方面的知识与信息上有充分的交流，就能使消费者有更多可能的机会去认识转基因的缺点和优点，这说明转基因食品的供应商应该开展更多的积极主动地与消费者的交流活动，如通过产品标签提供更多的产品信息等。然而，也应该注意，如果供应的转基因产品质优价廉，价格便宜、风味好等，那么消费者将只购买转基因产品，他们不会去评价转基因的一些宣传的所谓转基因产品优点等。例如，番茄能够保存在冰箱 2 个月等，因为这样的番茄的一些特性不能被认为还新鲜。

当然，也应该注意，社会公众也可能表现出片面化地拒绝转基因产品，尽管供应这样的产品是优惠的，就如 Seymour Cooke 建议：年纪大的，较少受教育的消费者可能总是对生物技术食品有更多的意识，而瑞士和德国人比美国人则更少地去购买这样的产品，而啤酒和面包类和果汁中如果含有转基因可能会引起消费者更多的反感。这说明当含有转基因的食品比例增加时，对非转基因食品将会有一个更活跃的需求，也将会有更优质的市场。

前面的讨论指出，消费者的选择在影响英国生产转基因食品的接受程度中是处于第二位的因素，已有清晰的证据表明，伴随释放 Round-up 抗除草剂大豆进入欧洲市场，消费者希望有一种选择，即是否可以消费或者不消费转基因食品。这里强调了消费的需求和精确的产品标签。

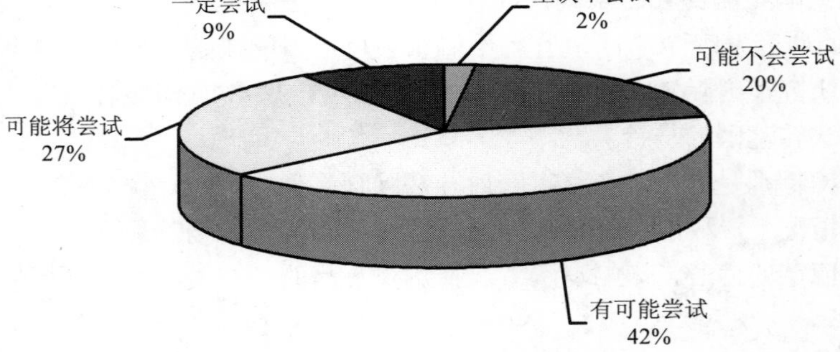

图 1 英国消费者尝试转基因食品的意愿（1995 年）（资料来源：英国食品与饮品联盟）[5]

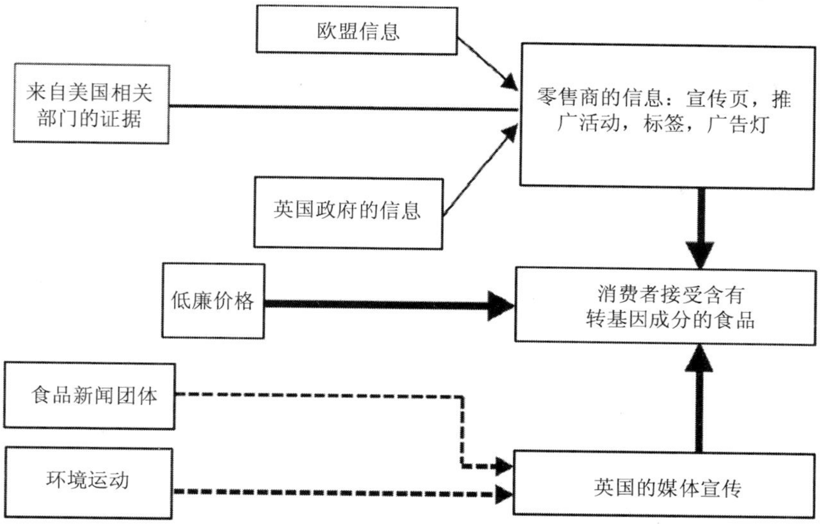

图 2　影响英国消费者对转基因产品态度的因素

图 2 对英国消费者针对转基因食品态度的影响因素分析。示意图中的粗箭头知名各类信息来源的相对重要性。这个强调了各个食品零售商在提供信息的重要性。示意图说明来自美国的证据可能会有疑问的。效应对消费者的信任（某种程度的怀疑）来自英国政府或欧盟委员会信息的有效性。框架之间的点线说明了潜在的负面影响。

下一阶段欧盟在这个问题上将怎样发展，是需要做出许多规划的，以便使更广泛的消费者接受。显然在一个近年来经受了高风险食品的地区，由于其社会主体和消费者已经开始关注食品安全，因此任何与转基因产品有关的风险所带来的威胁绝对都将是全面的，并在阻止转基因产品中反映。更多状况是，一旦如果依照于政府或欧盟要求的水平来决定，产品获得了释放证书和标签，欧盟的一大部分消费者可能将会以价格的更便宜，或者与他们有更充分满意的交流，才接受转基因产品。

2.2　发展中国家的状况

尽管也有一些研究关于发展中国家消费者对转基因食品的态度，在中国和哥伦比亚的研究也表现相似的结果。Li、Curtis、McClusky 和 Wahl（2003）[6]研究表明，平均来看中国消费者更愿意购买相比非转基因产品优惠 16%的转基因大豆食用油和 38%的转基因大米。Curtis（2003）[2]还发现中国消费者更愿意购买便宜了 35%的转基因马铃薯制作的油炸薯条、薯片和土豆泥。Pachico 和 Wolf（2002）[2]发现在哥伦比亚 66%的调查响应者愿意去尝试转基因食品。愿意购买转基因食品的意愿高于那些他们在家里没有适宜的或高质量食品可消费的比例。

3　影响消费者转基因生物安全认知的因素

由上节研究调查发现，世界各国的消费者对转基因食品的态度的差异是很明显的，从折扣价格超过 50%到价格优惠需要达到 38%。近期的研究集中在解释美国和欧盟消费者态度上的这种差异。Nelson（2001）[2]总结认为欧盟消费者一般是关注和转基因产品相关的不知道的风险，而不是优点与利益。而美国消费者通常即关注风险也关注利益。Nelson 用Margolis（1996）[2]的风险矩阵来表示，见表 1。由此争论认为欧盟消费者应落在框架 2 中，即“注重安全杜绝危害”，因为他们对待转基因产品的潜在危害是持肯定态度的，所以他们在所有的成本环节应尽可能避免这种危害产生。这种注重安全杜绝危害的方法基本是一种“预先警戒原理”的做法。这一原理支撑着欧洲转基因的标签政策。预先警戒原理呼吁，当一个活动产生威胁时，即使在还没有被科学证实会直接引起一个效应关系时，就应该要采取防御的方法。

表 1　风险矩阵

		机会	
		是	否
风险	是否	1. 利益与危害兼有（支付成本和获取利益）	2. 注重安全杜绝危害（只有成本）
		3. 不要成本不要危害（只有利益）	4. 无差异（没有成本也没有利益）

注：来自 Margolis 1996.

用这个方便和明显的形式，来表现那些潜在风险所应该具有的单一举证责任，甚至即使那些潜在的危害还没有表现出来之时（Sunstein；2002）[2]就必须要有调控行动。

Nelson 认为将美国的消费者归属于第 4 框架“无差异”，因为美国人典型感到转基因食品是与其他食品没有差别的，而且是用相等标准来评估的。然而，当一个人对于源自是否对转基因食品权衡利益和感受风险时，信任政府管理者，以及针对科学发现和媒体影响的态度，这是值得高度赞赏的。美国和欧盟的消费者思考转基因食品的利益和潜在成本，是有区别地认知风险，这种方法解释了在他们不同态度背后的原因与根据。用相同方法，你也可以总结出发展中国家的消费者也适用转基因食品的利益和潜在成本评价。发展中国家对食品的供给，以及食品营养成分和用于支配购买食品的收入更加重视，如果可以降低食品成本，那么发展中国家开发转基因食品是有潜力的。

4　发达国家和发展中国家的公众对转基因生物风险与利益的认知比较

4.1　发达国家公众对转基因生物风险与利益认知

在美国和欧洲的转基因产品是通过更多有效生产而产生的利益，源自减少投入成本。因为在美国和欧洲都不缺少有营养食物供应。在这些国家，消费者从转基因食品中感受不到这些在发展中国家所具有的利益。在日本和欧洲，食品威胁和丑闻已经影响了消费者的信任。例如，在日本不景气的经济政府丑闻，政府对发现疯牛病的处理，可能会引起消费者，对政府安全措施和转基因食品具有更少的信任。

进一步看，欧洲和日本文化对自身传统方式生产有自豪感，并不迫切需要用科学发展来改善生活，欧洲人常常怀疑新发现，而且总有一个“假设”，即如果现状不需要被打破的话，我们为什么要完善食品系统的态度。正是因为这样的原因，所以欧洲人和日本人，将消费转基因食品划归为高概率的潜在风险。2000 年和 2001 年的欧洲晴雨表调查发现，59.4%的欧洲人认为，转基因可能对环境有负面影响，70.9%的人无论任何情况下都不愿意看到转基因的活生物。高潜在风险的认知，结合转基因食品微弱的利益，通过成本利益分析的结果，给在欧洲反对转基因食品的认知提供了强烈的争议。结果欧盟已经对一些含有转基因成分的食品要求强制标签。在 1999 年 10 月，对有关管理法规[7]，欧盟已经给出了最初的认可，即要求标签所有含有超过 1%量转基因成分的食品。在日本管理部门已经针对 29 类可能含有任何转基因成分的食品颁布了强制标签的法令。在美国，消费者对转基因食品的认识是有限的，只有一部分活跃分子公开强烈反对转基因技术，也只有很少的媒体关注转基因技术。另外，消费者相信政府管理部门管理食品安全的标准是相对高的，由于密布管理部门机构，食品药物管理局（FDA），环境局（EPA）和美国农业部（USDA），所有部门都开展安全评价和管控转基因生物对食品和环境，以及公共健康的风险。所以，对美国消费者来说，来自转基因食品的微小的利益和微小的潜在成本无法使媒体有更多的关注，也是对科学创新持积极态度的结果，相信政府管理部门是负责任地监控食品安全的，对没有大的价格优惠和折扣的转基因食品一般是接受的。

4.2　发展中国家对转基因生物风险与利益的认知

发展中国家可以从转基因技术中获得极大利益，当这些利益与相对小的与转基因食品有关的认知成本比较时，由于消费者的风险认知是来自对政府管理者的信任，对媒体关注的正面反应，以及一种对科学发现的积极正面的对待。要理解在发展中国家的消费者会全面接受转基因食物产品通过成本利益分析。

关于转基因作物的利益，由此说明转基因作物主要是生产成本降低，能使产量增高。而发展中国家正是关注的是满足粮食的供给，以及食物的营养和经济性等优点。由于发展中国家短缺和需要的就是粮食，所以调查参与者的 40%在 Pachico 和 Wolf（2002）[2]在哥伦比亚的研究调查中都感到，他们的家庭没有合适充足的食物。而中国有几乎 13 亿人口，在 2050 年可能超越 14 亿（人口普查，2002），中国认识到如果国家要继续养活这样众多的人口，那么就必须发现和采用更有效的农业生产方法。另外，已有很多转基因作物已经被批准在中国开发（Huang，Rozelle，Pray & Wang，2001）[6]。期望运用转基因技术能够增加作物产量，以满足发展中国家对粮食短缺的需求。James 和 Krattiger（1999）[7]估计，在未来 10 年里，转基因技术可增加亚洲水稻生产的 10%～20%。

发展中国家的第二个主要问题是营养缺陷，特别是维生素 A 缺乏（VAD）。据估计，每年每百万人中有 1/4 或一半的维生素 A 缺陷儿童将失明（Zimmerman 和 Qian，2002）[8]，因为水稻是发展中国家广泛被消费的粮食，所以运用转基因技术的黄金大米能够提供维生素 A，而由此减少 VAD 的病患者。Zimmerman 和 Qian（2002）[8]估计黄金大米能够减少健康维护成本，仅在菲律宾就几乎高达 32%，而且每年有 2 200 到 10 200 的病例成为盲人。发展中国家的消费者的营养吸收正在受到关注。Li et al.（2003）[6]发现，与转基因大豆的食用油相比，中国的消费者是愿意在转基因水稻上付更多钱，因为在水稻产品中含有额外的维生素含量。

发展中国家面临的第三个问题是经济优先。为了在世界市场上获取竞争力，生产者必须找到成本投入有效的生产方法。转基因大豆也被称作 Roundup Ready（RR）大豆已给阿根廷在大豆生产中的所有增产因素提高了 10%，由于成本节约（Qian 和 Traxler，2002）[8]，一项由 Kirsten Gouse 和 Jenkins（2002）进行的研究发现，由于较高产量并节约了杀虫剂的使用费用，无论是大规模还是小规模的 Bt（Bacillus thuringienis）棉花对比于传统棉种的使用，即使产生较高的种子成本费和技术费用，南非的棉花种植者都获得了净收入。在中国由于推广应用 *Bt* 抗虫棉减少了 14%～33%的成本费用（Pray，2000）[6]，因此转基因 *Bt* 棉花也受到棉农的欢迎。

5 影响消费者转基因生物风险与利益认知的因素

一般地，理性消费者对不确定事件做决定前，对其不确定结果是通过事件的概率来决定的。分析一个行为的潜在成本和利益，正如消费转基因食品，针对这样的行为将产生一个期望的效果。

消费者衡量预期的利益和成本依赖于他们对风险的耐受性。科学的一致性证明，转基因食品并不置任何风险给消费者。但是，在科学评价风险和感知风险之间存在明显的区别，公众对风险的理解和专家的理解是非常不同的。（在环境风险文献中，见 Jenkins-Smith 和 Bassett，1994；Lindel 和 Earle，1983；McClelland，Schulze 和 Hurd，1990）[9]。这些认知的风险，被消费者看作为将来的成本，而且是由消费者来划归这个可能性存在的概率大小。由此可见这种划归是主观的。我们认为消费者划归每个潜在成本和风险，主要源自 3 个原因：①对食品供应安全负责的政府管理者的信任程度。②对科学发现所持的态度。③媒体宣传的影响（表 2）。

表 2　与转基因食品相关的风险认知的影响

代表国家	政府管理	媒体宣传	对科学态度
中国/哥伦比亚等“发展中国家”	+	+	+
美国/加拿大	+	+/-	+
欧洲/日本	-	-	-

注：+正面影响；-负面影响；=/-含糊的.

5.1 对政府的信任度

我们调查的（见 Li et al.，2003；Curtis，2003）[6] 599 名中国北京的消费者都支持这个理论，发现调查响应者是信任政府关于供应食品的安全性的管控，对科学也有非常正面的积极态度，包括在农业中应用生物技术，当被问到他们为什么愿意对转基因食品予以惠顾的问题时，许多调查响应者均反应相信科学，愿意尝试新产品，或者说价格变化并不影响他们的购买。

5.2 对科学的态度

欧洲各国和日本逐渐发展的现代资本社会，在保护文化传统方面倾注了极大的关注和骄傲。

在哥伦比亚调查，Pachico 和 Wolf（2002）[2]发现，对科学创新存在一种积极的倾向，被证明拥有很高的一致性，表现出有 68%的调查响应者，他们认为科学改善了生活的质量。另外，有 75%的调查响应者同意或强烈同意他们的政府对他们的食品供应商所提供的适度安全水准。

5.3 媒体宣传

来自媒体的讯息，有可能媒体是私营公司，他们提供的讯息要受自己媒体利润最大化动机的影响来主导他们的报道内容与观点，而来自公共资源的媒体，就能够有正常的目标追求和提供客观信息。但是，管理者和政府可能有他们自己的（私利的）动机，而使信息有意出偏。（Swinnen McCluskey 和 Francken，2003）[6]研究调查认为这个问题和消费者对生物技术的认识特别有关。超过 90%的消费者收到关于食品和生物技术的信息，主要是通过大众出版物和电视（Hoban 和 Kendall，1993）[9]研究认为广泛的媒体宣传能够强化风险认知，例如媒体对疯牛病的报道在欧洲就使大家都知道了“使牛发疯的疾病”，包括在日本和现在的加拿大，结果就产生了牛排需求的下滑。Verbek、Ward 和 Viaene（2000）[6]研究发现，电视在肉制品安全问题上的宣传，对疯牛病在比利时暴发后的红色肉品的需求具有负面效应。

年轻人和拥有小孩子的主妇对这样的负面媒体宣传特别敏感。Frewer、Howard 和 Shepherd（1998）[6]研究认为，电视收音机和报纸节目后接着的讨论是主要的信息渠道，使人们对生物技术做决定的基础。

在中国政府管理媒体关注转基因作物是很正面的，中国在世界上是转基因作物的第六大生产国，在努力满足自身粮食供给的政策下，持续支持生物技术研究。所以，并不惊讶在我们的中国研究调查中，只有 9.3%的调查响应者有某种程度的负面或很负面的关注来自食品中应用转基因成分。54%的调查响应者声称对转基因产品没有任何知识，只有 7.8%的认为转基因食品和高风险有关。另外，64.6%的调查响应者认为广告在他们的食品选择中起决定作用。在哥伦比亚，Pachio 和 Wolf[2]的调查指出，电视是他们有关转基因食品信息的主要来源，包括其中的讨论。收音机和广播处于第二、三位。近 75%的消费者调查同意转基因食品存在一些风险，但几乎许多人都愿意在任何情况下尝试转基因食品。

6 转基因食品标签制度在发达国家和发展中国家的认知

6.1 发达国家

当前在欧盟推行的零售商制度，是针对零售商自己标签产品的制度。但是，随着标签制度的要求与水平的提高，市场份额的增加，食品供应中需要更多的垂直协调，这样产生的后果是，大多数主要的各类食品零售商的市场力量，特别是像 TESCO（乐购）和

SAINSBURY（圣果）已经明显增加了，这样的企业被日益看作为在英国食品市场中主流趋势和关键的驱动力。由于众多食品零售商与食品消费者建立起了相当的信任关系，消费者的跟踪调查显示，消费者信任零售商胜过其他团体，甚至胜过政府。然而，这样的结果就会容易产生在某些情况和事件中，零售商的极端风险逆转，而正是这些事件可能会引发消费者的信任损失。例如，他们对自己供应的自己标签的食品开发了完善的质量保险系统，零售商对自己供应给消费者的食品问题提供信息持积极的态度。

质量保险和跟踪这个方面已经导致了零售商仔细考虑他们对可能含有转基因产品推广的反应，特别是对含有来自转基因大豆成分，因为他们已经知道他们不可能同时囤积转基因食品和非转基因食品，大多数主要食品零售商，已经选择精确地标签他们的产品，严格控制和说明如果存在可能含有转基因成分的有任何机会，这个已经导致建立了全英国的食品零售商团体条例，以此来管理和标准化转基因食品的标签。

主要食品零售商 Iceland 已达成共同协议[9]，将推行全面详细的标签，即他们自己所有的可能含有转基因大豆的产品都给予标签“含有转基因大豆”的字样，这个决定是建立在消费者研究的基础上，消费者更接受确定的“含有”字意，而不是较少肯定的（尽管更精确的字意如“可能含有”）。

食品零售商 Iceland 是一个综合类食品零售商（只有相对小的份额，有 770 个店，只占全英国食品店的 1.6%）已经接受更激进的政策，在 1997 年 3 月 18 日已宣布[6]，遵循全英食品零售团体的推荐，冰岛将在 1997 年 5 月 1 日后的商品中使用非转基因原料在他们自己标签的食品中。而自 11 月开始，美国生产商将不再被要求区分转基因和非转基因种类。

其他一些小的零售商，将受他们所处的市场地位和营业点分布的影响。预算紧的零售商可能会感到如果转基因产品能够让他们建立起以价格为基础的竞争优势而接受标签，因为这样的产品会更适合一定的市场品味，另一方面，有些零售商可能会盯着品优质好的市场。这种零售商可能也会把这种看作为一种市场机会，而确定非转基因产品的地位。

关于对转基因产品是否需要立法方面，欧盟内部仍然存在相当的分歧。英国的标签倡议可能代表了一种预先空白的努力，对部分英国食品企业，根据预期中的欧盟的计划，针对转基因产品的标签，当前立法限定在 1997 年 5 月前经过批准进入的食品不需要标签，但是针对那些可能含有可育的转基因原料在这个日期之后进入欧盟必须标签。尽管在欧盟目前内部不一致已经拖延了这个统一的计划。

当前的建议范围是相对适度的，要求标签种子和含有转基因大豆蛋白的食品（因为这中间转基因原料仍然存在）。针对所有经营家禽家畜饲喂者的极端情况，如对于未确定成分的肉制品，也作为转基因饲料喂养生产的肉予以标签，因为仍然感到才从疯牛病危机解脱出来的那种紧张的影响。由于这样的原因，可能吸引政策制定者努力去避免潜在的公共健康方面的风险（哪怕是很小的），进而失去消费者的信任。

欧盟标签制度的性质（基于新食品条例下），强调了围绕这个问题的一些混淆，条例允许使用这个用语，即“可能含有转基因原料”，这种英国政府授权的用语。然而这个并不要求在生产和市场对转基因的大豆及产品的检验和隔离。但是在欧盟内部也有一些（显然包括了农业委员会和法国渔业）对名词“可能含有”不满意，争辩说这或许会导致消费者的更加混淆。这是英国内部食品零售商所持的看法。

6.2 发展中国家

尽管发展中国家对转基因食品的接受程度是远高于发达国家，近期的调查显示，消费者希望知道哪种食品含有转基因成分。Pachico 和 Wolf（2002）[2]的调查响应者显示，90.7%在哥伦比亚认为强制标签转基因食品是非常重要或者某种程度的重要。但是，只有 64%的响应者说他们非常关注或者有些重视阅读标签内容。Curtis（2003）[2]发现在中国 89.8%的调查响应者认为标签有转基因成分食品时很重要或某种程度重要。在这种关注的反应中，中国已经要求，自 2001 年 6 月始，所有转基因产品进入中国无论是用于研究或是生产、或是加工，均需要有来自农业部颁发的安全证书，以保证他们对人类消费、动物和环境等是安全的。中国也要求所有列入清单的转基因产品都必须标签。

结语

综上所述，针对转基因生物安全的认知，与美国、欧洲和日本等发达国家比较，在欠发达国家，更加关注可用的食品及可从食品吸收的营养。增加作物产量和食物的供应通过转基因食品将获得巨大的利益，根据欠发达国家存在的食品短缺和营养需求。而在欠发达国家，转基因生物的潜在风险认知却是相似的，这些潜在的利益伴随低风险认知，通常对产生转基因食品持正面态度是有贡献的，因此仍需要对风险进行小心评估。Rissler 和 Mellon（1996）[6]认为联合国应该开发国际生物安全监控方案，以确保发展中国家防止转基因作物对栽培作物多样性的风险。Nelson（2001）[6]总结到，在考虑成本和利益的同时，特别是对公共健康保护的关注，公共评价转基因生物是在未来开发转基因生物之前必需的环节。

关注发达国家的转基因生物安全认知，尤其是那些依靠食品出口到发达国家地区，像欧洲和日本是潜在的市场损失，由于需要标签的新规定和所有含有转基因成分食品的可跟踪性要求，在保证非转基因产品出口的努力中，对于种植转基因作物欠发达国家可能遭受打击，由于区分转基因和非转基因作物所带来的额外支出。这种策略在巴西近年来的收购玉米，以优惠于美国转基因玉米每吨 6～7 美元卖给西班牙和日本，由于他们是非转基因玉米的优势而显现出来。世界贸易组织的 Codex 委员会正在努力协调标准，解决由于要求食品标签而带来的贸易争端，促进在保护消费者健康前提下的公平贸易。由 Nielson、Robinson 和 Thierfelder（2001）[6]完成的实验调查表明，全球市场将调整区分转基因和非转基因食品，已减缓食品行业必须标签“非转基因”而带来的成本与检测的压力，这样将可以使生产转基因食品原料的南美和低收入的亚洲国家获取更多的利益。

参考文献

[1] Hallman，W. K.，Hebden，W. C.，Aquino，H.L.，Cuite，C.L. and Lang，J.T.. 2003. Public Perceptions of Ge-netically Modified Foods：A National Study of American Knowledge and Opinion.（Publication number RR-1003-004）. New Brunswick，New Jersey；Food Policy Institute，Cook College，Rutgers - The State University of New Jersey.

[2] Curtis R.Kynda，McCluskey J. Jill，Wahl I.Thomas. Consumer acceptance of genetically modified food products in the developing world. AgBioForum[J]. 2004，7（1&2）：70-75.

[3] McCluskey J. Jill，Ouchi Hiromi，Grimsrud M.Kristine and Wahl I. Thomas. Consumer response to genetically modified food products in Japan. 2001，（in press）.

[4] Lusk J. L.，Sullivan P. Consumer Acceptance of Genetically Modified Foods Consumer acceptance of and willingness to eat genetically modified foods depend on the reason for the modification and other factors. Food Technology[J]. 2002，56（10）：32-37.

[5] Loader Rupert and Henson Spencer. A view of GMOs from the UK. AgBioForum[J]. 1998，1（1）：31-34.

[6] Li Quan，Curtis R.Kynda，McCluskey J.Jill，and Wahl I.Thomas. Consumer Attitudes Toward Genetically Modifi ed Foods in Beijing，China. AgBioForum[J]. 2002，5（4）：145-152.

[7] Gaskell George，Bauer W.Martin，Durant John，Allum C.Nicholas. Worlds apart？ The reception of genetically modified foods in Europe and the U.S. Science[J]. 1999，285：384-387.

[8] Frewer J. Lynn，Miles Susan，and Marsh Roy. The media and genetically modified foods evidence in support of social amplification of risk. Risk analysis[J]. 2004. 22（4）：701-711.

[9] Nisbet Matt，Lewenstein V. Bruce. A Comparison of U.S. Media Coverage of Biotechnology with Public Perceptions of Genetic Engineering 1995-1999. Biotechnology，the Media，& Public Perceptions.（Paper Presented to the 2001 International Public Communication of Science and Technology Conference，Geneva，Switzerland，February 1-3.）.

生态农业的公众认知、态度和需求

Public Cognition, Attitude and Requirement to Eco-Farming

王知凡，晁文庆，俞江丽，张菁，廖萍，高玉珍，齐英
（清华大学媒介调查实验室，绿色和平）
WANG Zhi-Fan，ZHAO Wen-Qing，YU Jiang-Li，ZHANG Jing，LIAN Ping，GAO Yu-Zhen，QI Ying
（Tsinghua Media Survey Lab，Greenpeace）

摘　要： 随着化学农业弊端的显现，探索更加稳定和可持续的农业生产方式成为全球共同的议题，其中生态农业的生产方式被认为是粮食可持续供给的解决方案之一。在食品安全问题频发，耕地在化学耕作模式下污染日益严重的今天，我国政府也开始探索发展生态农业的可行方案。本研究旨在了解公众和专家对生态农业的认知、对生态农产品的需求和购买意愿，以及对目前经费投入等相关政策的评价。我们选择北京、上海作为调查区域，通过问卷调查、深度访问、小组座谈会展开调查，先后走访及电访了农业、环保、食品安全相关领域的 8 位专家、官员，5 位生态农业从业者，以及两个城市共计 1 597 位居民。问卷结果显示公众对于生态农业的认知度为 51.4%；95.1%的受访者认为政府应该增加对生态农业的投入；如果价格下降，82.7%的受访者会增加对生态产品的购买。对于如何发展生态农业，受访专家认为政府应该充分发挥监督指导职能，增加在生态农业技术研究推广以及对认证过程的监测方面的投入。同时，本研究也对转基因作物的研发和审批的公众关注进行了调查，在 500 位受访者中，84.5%的受访者希望了解国家对于转基因作物安全证书审批和发放的流程；53.8%的受访者认为目前的转基因审批过程不是公开透明的；90.4%的受访者认为所有的转基因食品都需要标识。

关键词： 生态农业，公众认知与需求，转基因，专家建议，调查问卷

1　研究背景与意义

现代农业依靠大量石油资源的化学品和机械的投入，维持着较高的生产效率，对缓解人口压力和摆脱饥饿与营养不良做出了巨大的贡献。但是与此同时，现代农业也带来许多难以克服的问题，其中最主要的是能量消耗大，其次是对生态环境造成了严重的污染，再加上由此带来的食品安全方面的问题，相互交织，相互作用，使农业陷入难以自拔的恶性循环中。为了加强环境保护以拯救我们赖以生存的地球，确保人们的生活质量和经济健康发展，政府也开始探索有利于环境和食品安全的生态农业的发展道路。

为了了解公众对发展生态农业的认知、倾向、对经费投入等相关政策的评价以及对生态农产品的认知、需求和购买意愿等，我们通过问卷调查、深度访问、小组座谈会展开调查，最大化地运用调查结果分析当前公众对发展生态农业的支持度，审视国家对发展生态农业的投入是否符合公众、专家意愿，并试图从市场需求的角度反映投入的必要性，根据调查判断生态农业是否是同时解决食品安全和应对粮食危机的最佳途径。

2 研究方法

2.1 定性研究

在本次调研中，清华大学媒介调查实验室研究人员先后走访及电访了农业、环保、食品安全相关领域的 8 位专家、官员，5 位生态农业从业者，通过深度访谈的形式了解其对生态农业的发展现状、发展面临的机遇和挑战以及未来发展趋势等方面的看法。

除深度访谈以外，清华大学媒介调查实验室还组织了公众小组座谈会，深入洞察社会公众对生态农业及农产品的认知、态度、需求以及支付意愿等，以获得启发性观点。

2.2 定量研究

项目组选择北京、上海作为调查区域，根据统计学原理，有效样本量计算公式为：$N=Z2\sigma^2/d^2$，Z 为置信区间 Z 统计量，为保证准确度，本次调查取置信度 95%，对应 Z 值为 1.96；σ 为总体标准差，一般取 0.5；d 为抽样误差范围，本次调查取 3%，以保证调查精确度。此次调查共访问北京、上海两个城市的 1 597 位居民，访问方式分为深度访谈、座谈会和问卷调查。深度访谈选择北京、上海的公众各 10 位；每个城市男女比例均为 1∶1，年龄 19～35 岁、36～59 岁、60 岁以上，每个年龄段至少 3 位，政府机构、事业单位、白领员工和蓝领员工至少各 2 位。座谈会，北京、上海各 1 场，每次座谈会 6～8 位，年龄包括 18～35 岁、36～45 岁、46～59 岁、60 岁及以上，每个年龄段至少 1 位、政府机构、事业单位、白领员工和蓝领员工至少各 1 位。深度访谈和座谈会样本是通过清华大学媒介调查实验室在线样本库发布消息进行征集，并对报名者按照配额条件进行筛选。问卷调查通过在北京、上海各区县进行街头拦访进行，其中北京样本量为 733，其中城区样本 630，农村样本 103；上海样本量为 864，其中城区样本 661，农村样本 203；区县拦访配额按照人口比例进行分配。

职业分布为：27.0%的人为公司/企业一般职员/职工，退休人员占 21.5%，个体劳动者/自由职业者、农民/工人/服务人员、学生、教学/科研/医生/律师等专业技术人员、公司/企业领导/管理人员、家庭主妇、下岗/失业/无业人员、兼职工作、机关/事业单位干部/公务员以及其他的比例分别为 16.3%、12.8%、7.5%、3.9%、3.9%、2.8%、1.6%、1.3%、0.9%、0.5%。

收入分布，44.6%的人选择了 1 500～3 000 元，其次是 3 000～6 000 元，选择比例为 32.6%，选择 1 500 元以下、6 000～8 000 元、8 000～10 000 元的比例分别为 13.4%、5.4%、2.4%。另外有 1.5%的人选择 10 000 元以上。

学历方面，42.3%的人选择了初中及以下，其次是高中/中专/职高，选择比例为

26.4%，选择大专、大学本科的比例分别为 16.0%、13.9%。另外有 1.4%的人选择研究生及以上。

3 问卷调查结果

3.1 生态农业的认知度分析

在国外，生态农业又称自然农业、有机农业和生物农业等；其生产的食品称生态食品、健康食品、自然食品、有机食品等。

各国对生态农业提出了各自的定义。例如，美国农业部的定义是：生态农业是一种完全不用或基本不用人工合成的化肥、农药、动植物生长调节剂和饲料添加剂的生产体系。生态农业在可行范围内尽量依靠作物轮作、秸秆、牲畜粪肥、豆科作物、绿肥、场外有机废料、含有矿物养分的矿石补偿养分，利用生物和人工技术防治病虫草害[1]。

德国对生态农业提出了以下条件：①不使用化学合成的除虫剂、除草剂，使用有益天敌或机械除草方法；②不使用易溶的化学肥料，而是有机肥或长效肥；③利用腐殖质保持土壤肥力；④采用轮作或间作等方式种植；⑤不使用化学合成的植物生长调节剂；⑥控制牧场载畜量；⑦动物饲养采用天然饲料；⑧不使用抗生素；⑨不使用转基因技术。另外，德国生态农业协会（AGOEL）还规定其成员企业生产的产品必须 95%以上的附加料是生态的，才能被称作生态产品[1]。

尽管各国对生态产品的叫法不同，但宗旨和目的是一致的，这就是：在洁净的土地上，用洁净的生产方式生产洁净的食品，提高人们的健康水平，促进农业的可持续发展。

我国生态农业的基本内涵是：按照生态学原理和生态经济规律，因地制宜地设计、组装、调整和管理农业生产和农村经济的系统工程体系。它要求把发展粮食与多种经济作物生产，发展大田种植与林、牧、副、渔业，发展大农业与第二、三产业结合起来，利用传统农业精华和现代科技成果，通过人工设计生态工程、协调发展与环境之间、资源利用与保护之间的矛盾，形成生态上与经济上两个良性循环，经济、生态、社会三大效益的统一。生态农业是在现有条件下兼顾经济、生态、社会效益，实现生态环境保护、资源培育和高效利用的成功模式，是解决我国农村人口、经济发展与资源、环境之间矛盾的有效途径，是我国农业和农村经济可持续发展的必然选择。

2002 年，国家农业部从全国征集到的 370 种生态农业模式或技术体系中，选择了十大类典型模式和配套技术，加以推广。分别是："四位一体"、"猪—沼—果"、平原农林牧复合、草地生态恢复与持续利用、生态种植、生态畜牧业生产、生态渔业、丘陵山区小流域综合治理、设施生态农业、观光生态农业模式及配套技术[2]。

3.1.1 生态农业的概念认知度分析

我们的调查问卷显示，关于生态农业的概念，49.5%的受访者了解一些，48.6%的人不了解，只有 1.9%的人选择了解很多。

关于生态农业的具体概念，最多的人认为生态农业产品也就是绿色食品（60%）、生态农业有助于保护环境（59%），其次是生态农业不使用现代化学制剂，只运用传统农业优秀的经验方法进行生产（56.1%）、生态农业可充分利用空间和时间，把粮食生产与多种经济作物生产相结合，发展种植业与林、牧、渔业相结合，发展农业与第二、三产业相结合（44.3%）。

这说明公众对于生态农业概念所包含内容的认识存在一些误区，比如认为生态农业产品就是绿色食品，以及生态农业只运用传统农业经验。

3.1.2　生态农业的价值认知度分析

关于生态农业最重大的意义，50.0%的人选择了改善生态环境，其次是加强食品安全，选择比例为 32.0%，选择提高农民收入、避免资源浪费的比例分别为 12.0%、3.0%。另外有 2.0%的人选择应对粮食危机。

3.1.3　生态农业的政策认知度分析

关于生态农业，我国尚未见到一个对生态农业的总体目标、发展规划、保障机制有完备规定的纲领性文件。目前已有的相关法规例如：《全国生态示范区建设规划纲要》《基本农田保护条例》《农业生态建设技术规范》等。

在调查问卷中关于生态农业的政策部分，67.7%的人选择了不了解，其次是了解一些，选择比例为 31.4%。另外有 0.9%的人选择了解很多。从地区来看，上海对生态农业政策的了解程度明显高于北京。从职业来看，对于生态农业选择“了解一些”的比例，机关/事业单位干部/公务员、农民/工人/服务人员低于总体，公司/企业领导/管理人员的比例高于总体。

3.1.4　生态农业的经费投入必要性认知度分析

对于生态农业的投入，95.1%的人选择了应该增加。如果国家给予生态农业补贴，最多的人认为应该在生态农业技术的研究、应用和推广方面进行补贴，占 59.9%，其次是农资综合补贴，占 44%，良种补贴 40.5%。

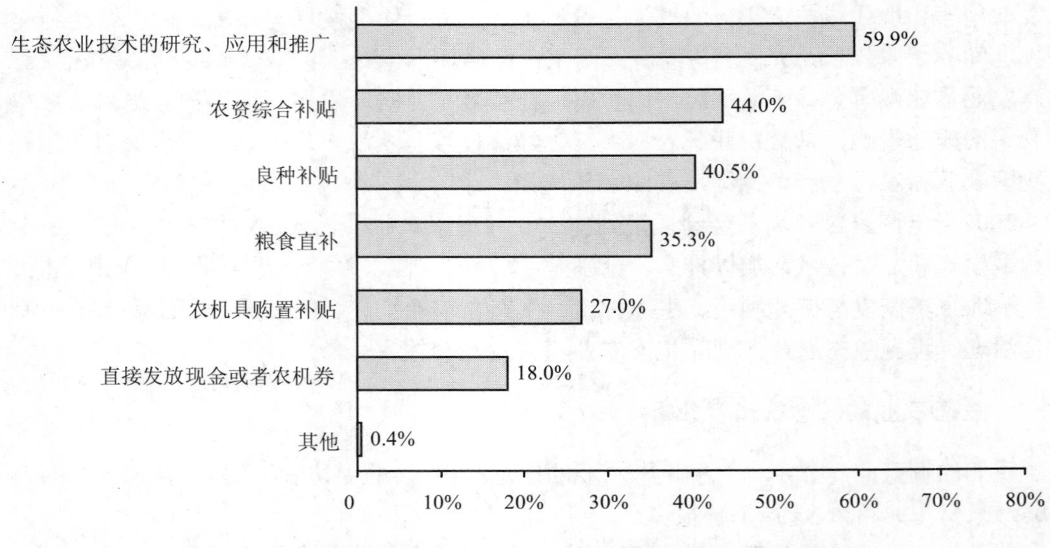

图 1　公众认为对于生态农业，国家应该加大对哪方面的投入

3.2　生态农产品的公众态度分析

3.2.1　生态农产品的公众关注度分析

对于生态农业的文章、政策或新闻的阅读兴趣，73.5%的人选择了有，26.5%的人选择了没有。

3.2.2　生态农产品的公众偏好度分析

在被问到购买食品时，是否会确定该食品无化学残留时，48.3%的人选择了“偶尔会”，其次是 28.3%的人选择了“每次都会”，另外，还有 23.4%的人选择了“不会”。如果加上选择“偶尔会”的人群，关注农药残留问题的人的比例达 76.6%。从所属区域来看，选择“每次都会”的，上海市区的比例（41.5%）明显高于其他区域。

从收入来看，选择“每次都会”的，8 000～10 000 元/月的比例最高，占到 38.5%，除了低于 1 500 元/月的比例明显偏低外，其他收入段选择每次都会的比例差别不大。

从学历上来看，研究生及以上的人选择“每次都会”的比例明显高于其他学历的人群。

3.3　生态农产品和转基因农产品的公众需求和市场潜力分析

3.3.1　生态农产品购买渠道分析

关于农产品的购买渠道，公众认为最主要的渠道是农贸菜市场（82.3%）、超级市场（80.7%），此外，也有 23.2%的人选择便利店/小超市。

与农产品不同，生态农产品的购买渠道，公众选择最多的是超级市场（64.6%），其次是农贸菜市场（41.2%）。

3.3.2　生态农产品购买频次分析

关于农产品的每周购买频次，33.0%的人选择了 3～4 次，其次是 1～2 次，选择比例为 21.4%，选择 7 次及以上、5～6 次、从来没买过的比例分别为 20.7%、18.0%、4.3%。另外有 2.6%的人选择说不清楚。

关于生态农产品每周的购买频次，57.5%的人选择了 1～2 次，其次是从来没买过，选择比例为 20.4%，选择 3～4 次、7 次及以上、5～6 次的比例分别为 14.3%、2.8%、2.8%。另外有 2.1%的人选择说不清楚。

去掉不清楚和从来没买过的人，有意识地购买生态农产品的比例相对于购买农产品总体的购买指数（购买生态农产品的百分比/购买农产品的百分比×100%）为 83.1%。

3.3.3　生态农产品购买考虑因素分析

关于选择农产品时首先考虑的因素，接近一半（48.7%）的人选择了新鲜度，其次是是否绿色安全，选择比例为 25.1%，选择价格、购买便利性、形状外观的比例分别为 17.0%、7.8%、1.3%。

关于影响购买生态农产品的因素，最主要的是价格过高（53.2%），其次是对所谓的生态农业产品的真实性有怀疑（36.2%）、不知道哪些属于生态农业产品（36.1%）。

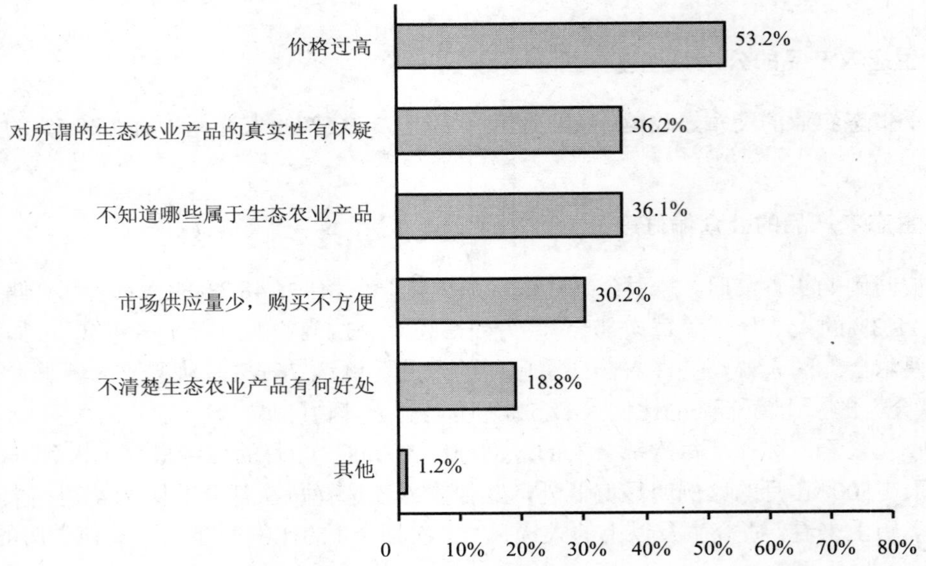

图 2 影响购买生态农产品的因素

3.3.4 生态农产品购买类别偏好分析

人们最愿意购买的生态农产品前三个品类依次是：蔬菜、水果（85.1%），粮食（57.7%），肉类（39.3%）。对于转基因农产品接受度最高的三个品类依次是：蔬菜、水果（49%），其次是粮食（38.6%），食用油（20.8%）。

3.3.5 生态农产品支付意愿分析

被问到购买生态农业产品，愿意多支付多少钱时，46.5%的人选择了比普通农产品高出 0.5 倍以下，其次是不愿意支付任何高出的价格，选择比例为 28.5%，选择 0.5～1 倍、1～2 倍的比例分别为 19.5%、4.8%。另外有 0.7%的人选择 2 倍以上。

也就是说，如果不考虑价格因素，至少有 71.5%的消费者更愿意购买生态农业产品，我们从座谈会上得知，消费者普遍更愿意购买生态农业产品，主要原因是认为生态农产品对自己和家人的健康的有好处，但是当前的市场不能满足这部分需求。

考虑到众多的消费者越来越重视食品的安全问题，但是目前市场上的生态农产品价格一般在市场价的 3～4 倍，并不能为广大消费者接受，因此政府追加投入，给予生态农业农户，尤其是广大中小规模农户一定的补贴，降低生态农产品的生产成本及价格，保证让更多的消费者吃上健康放心的安全食品成为亟待解决的问题。

3.3.6 生态农产品需求潜力分析

关于价格下降是否会增加生态农产品的购买，82.7%的人选择了会，其次是说不好，选择比例为 11.4%。另外有 5.9%的人选择不会。也就是说大部分的公众在价格合理的情况下更愿意选择生态农产品，生态农产品有着很大的需求和市场潜力。

4 公众对于转基因政策的知情及参与情况的补充调查

近几年来，国家大力发展转基因技术并增加进口转基因作物。因此本研究也对转基因作物的研发和审批的公众关注进行了调查，并和过往的民调结果进行了对比。这部分调查的样本来自清华大学媒介调查实验室在线样本库，在全国范围内随机抽取了 500 名公众进行网络问卷调查。结果表明，大部分受访公众希望政府相关部门进一步增加公众对于转基因食品的知情权和审批过程中的参与权。

4.1 超过八成的人希望了解国家对于转基因作物安全证书审批和发放的流程

数据显示，84.5%的受访者希望了解国家对于转基因作物安全证书审批及发放的流程，表示不希望和无所谓的比例分别为 10.2%和 5.3%。同样的问题在 2010 年和 2011 年的调查中，标识希望了解转基因作物审批过程并参与决策的公众分别占受访者的 87%和 95%。可见，大多数受访者希望了解国家对转基因的相关政策。

4.2 超过五成的人认为转基因审批过程不公开透明

调查数据显示，53.8%的人认为目前的转基因审批过程不是公开透明的，认为是公开透明的只占到 31.8%。另外有 14.3%的人目前对此情况不了解。

4.3 认为所有的转基因食品都需要标识的人达到九成

在调查所有的转基因食品是否都需要标识方面，90.4%的受访者表示需要，认为不需要的仅有 8.1%，另外有 1.5%的人持无所谓态度。2010 年和 2011 年的调查结果中分别有 59%和 82%的受访公众认为转基因食品需要标识。对比这三年的调查结果，可以看出越来越多的公众希望“吃的明白”，相关部门需尊重公众意愿和知情权及时更新农业转基因作物标识目录并切实执行。

5 生态农业从业者的认知及面临的困难

经调查，生态农业从业者多数转行自其他行业，也有大学毕业后，因为家里原来是农民，有土地，自己也有兴趣，于是成为“新时代的农民”。他们的土地面积从几十亩到几百亩不等，转行的初衷除了希望保证家人的健康、个人爱好，也有部分原因是出于环保和公益方面的考虑。

虽然多数从业者对生态农业的经济前景普遍持乐观态度，但也有从业者认为“做生态农业事件很高尚的事情，不能为了钱，要有爱心，有环保和公益理念”。

生态农业从业者遇到的难题主要可以归纳为以下 6 个方面。

5.1 缺少政策支持

生态农产品的价格高于市场平均价格，导致不少官员和普通民众都把生态农业当成面向少数有钱人的“贵族农业”，认为生态农业从业者获得了超额利润，所以应该让生态农业从业者完全靠自身实力参与市场竞争，优胜劣汰，国家不应给予政策补助。

但是事实并非如此，目前绝大多数生态农业从业者都是半路出家，缺乏农业方面的基本知识和经验，而农业经验的积累又需要较长的周期，再加上在改善土壤质量、选种、捉虫等方面需要投入更多的人力、资金，很多从业者从事生态农业多年后仍然处在亏损状态。一位被访女性从业者之前在互联网企业工作，辞职从事生态农业，至今已经 7 年，仍然处于亏损状态，亏损总额已经超过了 300 万元。

国家虽然对于农业有很多补助，但是对于生态农业这样的改善型农业基本没有补贴，从业者们希望政府可以给予一定的资金补贴或者政策扶持。

政府对于生态农业的补贴在国外也是有例可循的，例如，奥地利于 1995 年即实施了支持有机农业发展特别项目，国家提供专门资金鼓励和帮助农场主向有机农业转变。法国也于 1997 年制定并实施了“有机农业发展中期计划”。日本农林水产省已推出“环保型农业”发展计划，2000 年 4 月份推出了有机农业标准，于 2001 年 4 月正式执行。美国依阿华州规定，只有生态农场才有资格获得“环境质量激励项目”；明尼苏达州规定，有机农场用于资格认定的费用，州政府可补助 2/3。应该说政府的政策鼓励是促成这些国家的生态农业由单一、分散、自发的民间活动转向政府自觉倡导的全球性生产运动的重要推动因素。

专家指出，我国官员和农民没有共同利益，想靠政府推动生态农业发展，只有：①把环保指标和 GDP 一起，列为官员的政绩考核指标；②细分行政职能，建立专门科室负责推动生态农业的发展；③在《农业法》修订时，将发展生态农业提上立法的日程，使生态农业纳入长效机制发展轨道。

5.2 缺乏融资渠道

缺乏资金是生态农业从业者面临的一大难题。农业生产的特点，如资金回收慢、周期长、利润低，决定了想从银行、投资公司等常规商业渠道获得贷款几乎是不可能的事，因此很多对生态农业感兴趣的人望而却步，资金成为阻碍生态农业扩大的一个壁垒。

目前小规模生态农业从业者多是前期自筹资金，然后通过预收会员费来支持来年的生产，但是这样大大限制了产业的规模。解决融资问题，需要国家和相关机构制定相关金融政策，对于生态农业的融资贷款给予一定倾斜。

5.3 缺少专家的技术支持

很多生态农业的农场主是半路出家，对于农业经验的积累花费了很多精力，他们迫切希望遇到问题时能够请教专家，但是即便是在专家资源比较丰富的北京，农场主们也普遍反映咨询无门。

现在他们根本不奢望专家们到他们的农场亲自给予指导和解答问题，只希望中国农业科学院等单位的专家可以定期开讲座，最好现场可以针对生态农业从业者可能经常遇到的问题发些小册子。但是也有人认为，农科院是政府拨款的事业单位，大部分时间都用于完成政府的科研任务，很少有时间直接为农民服务。

在技术的应用上，也有从业者说“不能只关注前沿技术，已有的技术用好就行了”。

正是因为专家的稀缺，农场主们更加希望有技术能力的农民可以来和自己一起工作，这就是我们接下来要分析的问题。

5.4　缺乏愿意从事生态农业的人才

缺乏具有农业或者生态农业专业知识的人才是目前生态农业从业者面临的又一难题。

目前在北京、上海周边郊区都有以农庄或公司形式存在的生态农业样式，他们需要大量的人手，但是不少负责人反映招不到相关高校相关专业的优秀人才，这些人才具备农业领域的前沿专业知识，对农业具备一定的兴趣和感情，正是生态农业需要的人才。

拥有技术和能力的专业人才不愿意从事生态农业，究其原因，主要是两方面：一方面，农业与其他行业相比，薪水待遇明显偏低；另一方面，工作所在地一般远离城市中心区，选择生态农业作为职业等于选择了一种远离尘世、寂寞、辛苦的生活方式，这对于不少正值青春年华的年轻人来说也是难以习惯和接受的。

5.5　缺少获取信息的渠道

有生态农业从业者反映，想要获取关于生态农业的最新相关新闻或者信息，但苦于没有渠道。

我们在访问中提到东北某地区采用先进技术精确测量每平方米耕地需要的化肥量，从而根据配方精确施肥，在尽量少用化肥的基础上保证农作物产量。从业者马上追问信息来源，并问还有哪里可以得到类似行业内信息。

也有从业者通过不定期的生态农产品集市，和同行业者交流信息，但是并不是每位从业者都有这样的机会。

因此，他们希望能够有一些服务于生态农业从业者的行业报纸或杂志，报道最新行业技术、人物、政策及新闻。

5.6　缺乏销售渠道

生态农业的中小农户普遍没有有机或绿色农产品认证，这和认证的收费制度有关，目前中国的有机认证按照种类、批次收费，每个种类收费上万元，无疑这是中小农户承受不起的。在实践中，中小农户大多采用三种销售模式：CSA（社区支持农业）模式、生态农产品专门集市和专柜三种。

目前发展比较好的模式是 CSA（community support agricultural）模式，它源于美国，就是消费者为了寻找安全的食物，与农民携手合作，建立的经济合作关系。农民按照客户的要求来耕作，人们只需要每年预定一定数量的产品与品种，就可以随时吃到完全按照自己意愿生产出来的粮食和蔬菜。它由此缩短了产品从生产到消费间的距离，强调“越本土、越新鲜并更值得信赖”的消费观念。

中小农户要求社区居民先预付一定的费用（一般是一年的费用），把全年对蔬菜的消费变成直接的投资行为（期菜），并开放农场随时供消费者参观监督，一位农场主甚至每周会接待 3 次私立学校的孩子们到农场学习和游玩。生产者负责每周把社区居民预订的菜送到居民家里。这种模式降低了农户的产品卖不出去的风险，但是同时对生产者的物流配送能力有一定要求，也要求生产者的作物种类要尽量丰富多样。

第二种模式的代表是北京的有机农夫市集。它由一群消费志愿者发起，目的在于帮消费者找到安全放心的食品，也帮农户拓宽销售渠道、增加收入。市集的定位是服务于中小农户。但是这种模式的缺点是时间、地点不固定，只有老顾客才会提前得知举办时间和地点，不利于扩大消费者群体。

作为以上两种模式的补充的第三种专柜模式，就是农户在目标消费者比较集中的地点开辟专柜用于销售生态农产品。

无论哪种模式，都要求生产者开放农场，供消费者随时参观考察，以保证买卖双方的信任，可以说在没有认证和认证制度不被消费者和农户任何的情况下，生态农业得以存在的基础就是买卖双方的信任。

营销方面，有些农场主会利用微博营销自己，农场的会员都是自己的粉丝，潜在消费者也可以通过微博联系到自己，也有农场主会花钱在网络上打广告，不过最多的还是依靠口碑传播。

6 生态农业专家和政府官员的认知及生态农业发展建议

受访专家普遍认为在中国粮食安全是第一位的，因此，中国的生态农业的定义可以不同于国外，即允许适量地施用化肥农药，以保证产量不会减少。

在投入方面，多数专家认为无论是在生态农业的技术研究及推广，还是对认证过程的监测方面，政府都应该增加对生态农业的投入。目前中国已经解决了“吃得饱”的问题，越来越多的人关注“吃得好”，因此政府要从对食品数量安全的重视转移到对食品质量安全的重视上来，保证国民可以吃到安全健康的食品。

关于如何才能同时解决食品安全和应对粮食危机，这是一个长期的艰巨的任务，专家认为发展生态农业是一条比较切实可行的道路，具体的建议大致可以归纳为以下6个方面。

6.1 政府发挥监督指导职能

专家们一方面认为国家应当给予生态农业补助，另一方面也承认这在当前可能性很小。中国作为一个13亿人口的大国，实现了粮食自给率高达95%，以7%的耕地养活了占世界22%的人口。1949年中国粮食人均占有量不到210 kg原粮。如今，我国人均粮食产量已经达到400 kg左右，超过世界平均水平。这些都为生态农业的发展创造了条件，但是不可否认，在目前的技术条件下生态农业产量较低。

我们在调查中了解到中国政府对于生态农业和转基因产品都持较谨慎的态度，尤其对转基因产品如果大规模商业化种植，可能引起的物种方面的不可测风险抱有强烈担心。受访的政府官员对于生态农业是持肯定态度的，但是也表示政府补助的可能性不大，毕竟我国的粮食储备不可与国外某些粮食充足的国家同日而语。

有专家建议，在目前的生产水平下，不一定采取直接补贴的方式，比较现实可行的是政府在生态农业的发展过程中起监督和指导的作用，监督相关部门认真执行相关政策，相关部门为农民测量土地并给出施肥配方，指导农民根据气候土壤情况合理种植作物和精确使用化肥农药，降低成本，增加产出，达到经济价值、食品安全和环境保护的平衡，逐步推动把环保因素列入官员考核指标。

6.2 严格执行认证制度，依法惩处违法行为

生态农业作为一种新型农业形式，从业人员素质普遍较高，尤其要重视规范市场秩序，杜绝投机和违法行为，保证公平公正，为其长期稳定发展创造良好的社会环境。

调查中，接受访问的公众无一例外地表示不相信目前中国市场上所谓的“有机食品”、“绿色食品”、“无公害农产品”认证。信任是生态农业得以存在的重要前提，否则高价格得不到认可，产品销售不出去，整个产业无以为继。认证本应是降低买卖双方信任成本的有力手段，为何却得不到买卖双方的认可？

据专家介绍，国外生态农业发展较早的国家，此类认证都由专门的第三方机构负责执行，而在中国这是由政府包办的，在执行的过程中，这种权利又被下放给很多市场化的公司。一个批次的一个品种的认证费就要上万元，对于多数中小农户来说，这都是无法承受的，“有钱就能弄到认证”已经成为业内共识。负责认证的公司为了发展业务，就给一些根本不符合条件的农户颁发了认证，这些农户的产品打着“有机”的旗号，高价售卖给企业或消费者，导致消费者对整个生态农业的不信任。

中小农户为了得到消费者的信任，只好采用会员制，让会员随时参观考察农场，并负责物流配送，但是这样严重限制了生态农业的规模及发展，正是因为信任机制导致的销售渠道受限，使得生态农业成为少数富人的特权，被视为“贵族农业”。

中小农户普遍表示“如果是免费或收费很少，我们肯定是愿意去认证的”，消费者也表示，这么高的费用都由农户承担明显是不合理的。看来，高昂的认证费成为认证机制无法发挥应有作用的一个门槛，降低认证费用是市场主体的要求，是保证认证制度切实可行的一个重要前提。

有专家表示，农产品认证之所以不被消费者信任，和整个社会的信任机制有关。在我国违法成本低，守法成本高，只有加大对违法行为的惩罚力度，才能保证认证的真实可信，农户和认证机构不敢随意造假。为了保证认证的质量，专家建议：必须严格执行认证的各个环节，对此过程中出现的违法行为依照《食品安全法》《反不正当竞争法》《刑法》等予以严惩，为生态农业提供一个公正公平的社会环境，保证生态农业的健康稳定发展。

6.3 成立生态农业协会

接受我们访问专家，都一致认为应该成立农业协会。农业协会是致力于实现靠个体会员难以达到的共同目标，其首要目标不是获得最大利润，而是在于为全体会员服务和维护与增进全体会员的共同利益的自发的、非营利性的行业组织。

农业行业协会在治理农业经济活动中的重要作用主要有以下几个方面：①通过游说，联合或影响政府管理者制定和实施政策，有时还要协助政府管制经济和实施援助计划；②为农民提供包括信息在内的各种服务，如价格协调，在国内协调本行业产品价格，在国外市场上保护本国产品的合理价格，以减少国际贸易摩擦；③介绍和引进国际上先进的技术，发展深加工项目，进行技术推广和促进行业技术进步；④加强与国外有关组织的联系，协调国际纠纷，如帮助农民开拓国外市场，参加国外反倾销应诉等；⑤创立许可程序并设安全、质量和竞争的标准，建立治理市场行为规范。

从这些功能可以看出，如果有类似的生态农业行业协会出现，那么农户面临的无法获

取需要的信息，没有技术支持，缺乏融资渠道，产品缺乏销售渠道等种种问题皆可得到解决。

其实，中国已经存在一些农业协会的雏形，例如，由一群关注生态农业和“三农”问题的消费者志愿发起的北京有机农夫市集（Country Fair），搭建了一个平台，让从事有机农业的农户能够和消费者直接沟通、交流，既帮助消费者找到安全、放心的产品，也帮助农户拓宽市场渠道，鼓励更多农户从事有机农业，从而减少化肥和农药带来的环境污染、维护食品安全、实践公平贸易。目前每个周末举办市集，时间和地点不固定。通过豆瓣小站和新浪微博@北京有机农夫市集发布开集消息。

6.4 民间环保组织的积极推动

民间环保组织在环境保护工作中，发挥了积极作用，已经成为推动中国和世界环境事业发展不可缺少、不可替代和不可忽视的重要力量。它的作用主要是起到了政府职能所不易做、不便做的拾遗补缺的补充作用；起到政府与社会之间的沟通、交流和融合作用；起到监督政府、保护百姓环境权益的作用；起到宣传群众、引导群众、组织群众参与各种与环境有关的活动以及咨询和服务等作用。

我国环保民间组织的起步晚、发育慢。1978 年 5 月，由政府部门发起成立的环保民间组织即中国环境科学学会第一家成立。直到 1991 年、1994 年，民间自发组成的环保民间组织辽宁省盘锦市黑嘴鸥保护协会和“自然之友”才先后成立。因此，我国民间环保组织在专业性、组织民众参与活动以及国际民间环境交流与合作等方面的能力均有待提高，需要向国际环保组织学习借鉴相关经验。

环保方面的专家建议：民间环保组织应当成为推动生态农业的动力之一。让中国政府和公众认识到：既要发展生产，又要保护环境，维持农业生态系统的良性循环，走可持续发展的道路，农业的发展目标应该由单纯追求规模扩大和效率提高转变为重视农业的多功能性和自然循环功能的维持与促进，将是一项十分艰巨而重要的任务。

6.5 提高全社会的食品安全意识

发展生态农业，食品安全是比保护环境更为有效的切入角度。解决食品安全问题不应该仅仅看监管环节，而更应该从农产品生产环节找问题，只有发展生态农业才能确保从源头上解决食品安全问题。有专家指出：“解决食品安全问题，发展有机农业、全面农业才是出路。”

提高全社会对食品安全问题的重视，这是生态农业发展的最主要动力。而从我们对北京、上海两地的居民所做的市场调查结果显示，有相当部分居民还未充分意识到传统农业的生产方式带来的食品安全问题，或者虽然意识到了，但是并不愿意为此支付额外费用，这反映了食品安全还未引起人们足够的重视。

提高全社会对食品安全问题的重视，依赖于媒体的宣传报道，政府和执法部门对危害食品安全的行为予以打击和惩处，当然也要进一步降低生态农产品的价格，使更多人“吃得起”。

6.6　零售商给予销售支持

前面介绍过，目前生态农业农户的 3 种主要销售模式，但是需要指出的是，只要生态农产品的销售仍然依靠农户个人，而无法进入超市、农贸市场等渠道，就只能是少数人的特权产品，无法摆脱“贵族农业”的标签。

国外的主要零售商在促进本国生态农业推广方面做出了卓有成效的尝试。2000 年春季，英国最大的销售连锁商冰岛公司宣布：该公司将把货架上的所有食品都换成生态农产品，而且价格和原来一样。这个举动随即在整个市场引起连锁反应。使得生态食品在英国已不再只是一种时髦。

专家建议：中国的零售商可以在有条件的门店率先引进“生态食品”专区。这既可以成为零售商的营销亮点，也可以吸引部分高端消费者，同时树立了关注商品质量和消费者健康、热衷环保的公益形象，一举多得。相信中国主要零售商的努力可以帮助生态食品从少数人的专利走入寻常百姓家。

参考文献

[1]　林祥金. 世界生态农业的发展趋势. 中国农村经济. 2003，7：76-80.

[2]　农业部推出十大生态农业模式. 农民日报.

专题 3　转基因生物的赔偿责任与补救

Session 3　Liability and Redress for Damage from GMOs

The Nagoya–Kuala Lumpur Supplementary Protocol on Liability and Redress to the Cartagena Protocol on Biosafety：An analysis and implementation challenges

《卡塔赫纳生物安全议定书关于赔偿责任和补救的名古屋-吉隆坡补充协议》：分析与实施挑战

Lim Li Lin and Lim Li Ching
Third World Network

Abstract: The fifth meeting of the Parties to the Cartagena Protocol on Biosafety ended in Nagoya，Japan on 15 October 2010 with the adoption of a new environmental treaty，the Nagoya – Kuala Lumpur Supplementary Protocol on Liability and Redress to the Cartagena Protocol on Biosafety. Developing countries wanted to have substantive rules on international liability and redress to address damage resulting from living modified organisms（LMOs）included in the Cartagena Protocol itself，but this proved too contentious to resolve at the time. The compromise，contained in Article 27，was to negotiate liability and redress rules at a later stage，after the entry into force of the Cartagena Protocol. This began in earnest in 2005. The negotiations of the Supplementary Protocol were difficult and were opposed by those with an interest in the production and export of LMOs. Developing countries and some developed countries，on the other hand，maintained that an international regime to deal with damage caused by LMOs was necessary because of the unique risks of LMOs and their transboundary nature，and in order to ensure that those responsible would be held liable. Most developing countries wanted a binding international regime that would establish substantive rules on civil liability，whereby victims of damage from LMOs can turn to national courts for redress. Instead，because of compromises made in the negotiations，the Supplementary Protocol takes an 'administrative approach'，whereby response measures are required of the operator（person or entity in control of the LMO）or the competent authority，if the operator is unable to take response measures. This would cover situations where damage to biodiversity has already occurred，or when there is a sufficient likelihood that damage will result if timely response measures are not taken. The provision of response measures in the event of damage，or a sufficient likelihood of damage，resulting from LMOs is the central obligation of Parties to the Supplementary Protocol. However，Parties can still provide for civil liability in their domestic law and the first review of the Supplementary Protocol（five years after its entry into force）will assess the effectiveness of its provision on civil liability. This could trigger further work on an international civil liability regime. Other contentious issues include that of financial security. In cases where damage does occur，a liability regime should ensure that financial resources are available to enable or compensate for necessary measures to redress the damage. The Supplementary Protocol merely retains Parties' right to provide, in domestic law, for financial security. However, the provision on financial security mandates the first meeting of the Parties to the Supplementary Protocol to request the Secretariat to undertake further work on financial security. In addition, the first review of the Supplementary Protocol includes a review of the effectiveness of the provision on financial security. The Supplementary Protocol sets the minimum international standards on liability and redress for damage resulting from LMOs. Much has actually been left to Parties to determine and implement at the national level. Parties to the Supplementary Protocol are required to provide, in their domestic law, for rules and procedures that address damage. Thus, national legislation on liability and redress should implement and augment the Supplementary Protocol.

Keywords: Liability and redress, Damage, Living modified organisms, Transboundary movements, Administrative approach, Civil liability

Introduction

Liability is an obligation of a (natural or legal) person to provide compensation or take redress measures for damage resulting from an action or a situation for which that person is deemed to be responsible under the applicable law.[①]

In relation to living modified organisms (LMOs), the issue of liability will only arise when there has been damage caused by LMOs. (The Cartagena Protocol on Biosafety uses the term living modified organisms, for what are commonly known as genetically modified organisms, GMOs.) It must be established – in fact and in law – that the harm is directly attributed to the LMO (in particular its properties, their reproduction or modification) or the activity in relation to it. Further, it must be established that there is a person who can be identified as being responsible. Only then will the issue of compensating for the harm done arise[②].

1 History and introduction to the Nagoya-KL Supplementary Protocol

During the negotiations of the Cartagena Protocol on Biosafety, the issue of liability and redress rules being developed and included in the Protocol was raised. This however proved too controversial to be resolved at the time, and as a compromise, it was agreed that such rules would be negotiated at a later date.

---

① Secretariat of the Convention on Biological Diversity (2011). Liability and redress: Basic concepts. Workshop material no. 1. SCBD, Montreal.

② Nijar, G.S. (2007). Liability and redress for damage arising from genetically modified organisms: Law and policy options for developing countries. *In* Lim L.C. and Traavik, T. (eds). (2007). Biosafety First: Holistic approaches to risk and uncertainty in genetic engineering and genetically modified organisms. Tapir Academic Press, Trondheim.

As such，Article 27 of the Cartagena Protocol committed the Conference of the Parties serving as the meeting of the Parties to the Protocol（COP-MOP，the governing body of the Protocol）to adopt，at its first meeting，a process with respect to the appropriate elaboration of international rules and procedures in the field of liability and redress for damage resulting from transboundary movements of LMOs. The COP-MOP was to endeavour to complete this process within four years.

Accordingly，at the first meeting of the COP-MOP in 2004，an Open-ended Ad Hoc Working Group of Legal and Technical Experts on Liability and Redress was established，and Terms of Reference and an indicative work plan agreed. In accordance with Article 27，the work was scheduled to end four years later，in 2008.

Five meetings of the Working Group were held in May 2005，February 2006，February 2007，October 2007 and March 2008. However，the negotiations were difficult and were opposed by those with an interest in the production and export of LMOs. Parties failed to complete the process in 2008，as mandated，despite meeting in a small group setting just before COP-MOP 4 and in a contact group during COP-MOP 4 in Bonn，Germany.

Consequently，COP-MOP 4 renewed the mandate for further work and established a new format for the negotiations，in the form of a Group of the Friends of the Co-Chairs of the former Working Group，to continue the work on liability and redress.

This Group met four times，in February 2009，February 2010，June 2010 and October 2010. It finally agreed to the text of the Nagoya – Kuala Lumpur Supplementary Protocol on Liability and Redress to the Cartagena Protocol on Biosafety，which was adopted on 15 October 2010 at COP-MOP 5 in Nagoya，Japan.

The Supplementary Protocol is an additional international legal treaty to the Cartagena Protocol on Biosafety. It deals specifically with liability and redress arising from damage caused by GMOs that have been subject to transboudary movement. It will enter into force after 40 Parties to the Cartagena Protocol ratify it.

2　Key elements of the Nagoya – Kuala Lumpur Supplementary Protocol on Liability and Redress

2.1　Nature and approach of the instrument

Most developing countries wanted a binding international regime that would set substantive rules on civil liability，whereby victims of damage from GMOs can turn to national courts for redress，by establishing the responsibility of another person for damage suffered. In the case of civil liability，remedies are monetary compensation and/or injunctive relief.

However，the issue of the nature of the regime，and whether the instrument or parts of it should be legally binding，became so contentious that negotiations almost collapsed in Bonn in 2008. The emergence of a group of 'Like-Minded Friends'（around 80 developing country

Parties and Norway，and including all of the Africa Group）however，pushed through a compromise agreement that the international instrument on liability and redress should have binding elements on civil liability.

The package agreed to included an agreement that the international liability and redress regime would be legally binding and would comprise administrative approaches；and that while there would be no legally binding international civil liability rules and procedures，there would be a clause that would preserve the right of Parties to put in place domestic laws and policies on civil liability and redress which should include elements as stipulated in guidelines to be negotiated. The Bonn agreement was a heavily compromised proposal，given that most developing countries and Norway have been firmly behind a comprehensive and binding international civil liability regime，and had argued for this throughout the years of negotiations.

Therefore，because of the compromises made，the Supplementary Protocol takes an 'administrative approach'，whereby liability would be a matter to be resolved administratively between the liable entity and the executive arm of a government. Response measures are required of the operator（person or entity in control of the LMO）or the competent authority，if the operator is unable to take response measures. The provision of response measures in the event of damage，or a sufficient likelihood of damage，resulting from LMOs，is thus the central obligation of Parties to the Supplementary Protocol.

2.2 Definition of damage，scope，liable entity

Damage is defined in the Supplementary Protocol as an adverse effect on the conservation and sustainable use of biological diversity，taking into account risks to human health. A threshold applies，in that the damage has to be measurable or otherwise observable，and significant. An indicative list of factors that should be used to determine the significance of an adverse effect is included. Once the threshold of significant damage has been met，the need for response measures arises.

The Supplementary Protocol applies to damage resulting from LMOs that find their origin in a transboundary movement. The LMOs referred to are those（i）intended for direct use as food or feed，or for processing；（ii）destined for contained use；and（iii）intended for intentional introduction into the environment. It also applies to damage resulting from unintentional transboundary movements and illegal transboundary movements.

Furthermore，domestic law implementing the Supplementary Protocol shall also apply to damage resulting from transboundary movements of LMOs from non-Parties；this is a critical issue as some of the major producers and developers of LMOs are non-Parties to the Cartagena Protocol and hence cannot be Parties to the Supplementary Protocol，unless they first ratify the Cartagena Protocol.

The liable entity is identified in the Supplementary Protocol as the 'operator'. This is defined as any person in direct or indirect control of the LMO. As appropriate，and as determined by national law，this could include，*inter alia*，the permit holder，person who placed the LMO on the market，developer，producer，notifier，exporter，importer，carrier or supplier.

2.3 Response measures

The central obligation that Parties to the Supplementary Protocol assume is to provide for response measures in the event of damage，or a sufficient likelihood of damage，resulting from LMOs. Response measures are defined as reasonable actions to (i) prevent，minimize，contain，mitigate or otherwise avoid damage，as appropriate；and (ii) restore biological diversity. Response measures must be implemented by，and in accordance with，domestic law.

Response measures are required in both situations where damage to biodiversity has already occurred，and when there is a sufficient likelihood that damage will result if timely response measures are not taken. If there is damage，the operator has obligations to immediately inform the competent authority，evaluate the damage and take response measures. Meanwhile，the competent authority has to identify the operator responsible，evaluate the damage and determine which measures should be taken by the operator. The competent authority may itself implement appropriate response measures，including when the operator has failed to do so.

It is understood that the operator is responsible for paying for the costs incurred in the exercise of its obligations under the Supplementary Protocol. In addition，the competent authority has the right to recover from the operator the cost and expenses of，and incidental to，the evaluation of the damage and the implementation of response measures.

3 Some key issues for developing countries

3.1 Civil liability

The issue of civil liability proved to be one of the most contentious in the negotiations. As indicated earlier，most developing countries had actually envisioned that the international liability regime under the Cartagena Protocol would encompass substantive civil liability rules. Due to the compromises made during the course of the negotiations，the Supplementary Protocol contains only one legally binding clause on civil liability，which preserves Parties' rights to put in place domestic civil liability rules and procedures.

In the obligation to provide，in domestic law，for rules and procedures that address damage，Parties shall provide for response measures. Parties may also assess whether response measures are already addressed by their domestic civil liability law.

For damage to biodiversity，taking into account risks to human health，Parties may apply general existing civil liability rules and procedures，apply or develop specific civil liability rules and procedures，or apply or develop a combination of both. This is an additional tier that Parties may turn to in order to seek redress，since the Supplementary Protocol already provides for an administrative approach to such damage.

While the Supplementary Protocol *per se* does not cover traditional damage(which includes personal injury, loss or damage to property or economic interests), Parties are obliged to continue to apply existing general civil liability law, develop and apply or continue to apply specific civil liability law or develop and apply or continue to apply a combination of both, with regard to material or personal damage. This extends the reach of the Supplementary Protocol to traditional damage associated with damage to biodiversity, taking into account risks to human health. However, questions of interpretation still remain.

In all these damage scenarios, Parties are implicitly obliged to review their domestic laws to assess whether or not they have in place adequate rules and procedures on civil liability, and act accordingly. Civil liability is entrenched as an obligatory requirement in the national laws of countries, as countries must either apply and/or develop existing or specific civil liability laws to address damage caused by LMOs.

When developing new specific civil liability laws, rules and procedures to address damage caused by LMOs, Parties shall, *inter alia*, address damage, the standard of liability (including strict or fault-based liability), channelling of liability where appropriate and the right to bring claims. In the course of negotiations after the Bonn session, the civil liability guidelines that were initially on the table were dropped.

Arguably, the Supplementary Protocol does not establish an international legal obligation to implement a civil liability system. However, the first review of the Supplementary Protocol (five years after its entry into force) will assess the effectiveness of the provision on civil liability. This may lead to further work on an international civil liability regime.

3.2 Financial security and additional compensation measures

In cases where damage does occur, a liability regime should ensure that financial resources are available to enable or compensate for necessary measures to redress the damage. This issue was also very contentious, with some Parties opposing any mention of financial security, while others argued that without provision for financial security, the Supplementary Protocol would have placed obligations on importing Parties (which may have cost implications) without ensuring that the costs are borne by the person or entity concerned.

In the end, the Supplementary Protocol merely retains Parties' right to provide, in domestic law, for financial security. This right has to be exercised in a manner consistent with the rights and obligations under international law.

However, the provision on financial security requires the first meeting of the Parties to the Supplementary Protocol to request the Secretariat to undertake a comprehensive study on financial security. This study should address, *inter alia*, (i) the modalities of financial security mechanisms; (ii) an assessment of environmental, economic and social impacts of such mechanisms, in particular on developing countries; and (iii) an identification of the appropriate entities to provide financial security.

The first review of the Supplementary Protocol, five years after entry into force, also includes a review of the effectiveness of the provision on financial security.

Furthermore，the Decision adopted at COP-MOP 5 on liability and redress states that where the costs of response measures have not been covered，such a situation may be addressed by additional and supplementary compensation measures. These may include arrangements to be addressed by the COP-MOP in the future.

3.3 'Products thereof'

Whether or not 'products thereof' should be included in the scope of the Supplementary Protocol was very contentious. Some Parties argued that the Cartagena Protocol itself does not include 'products thereof' in its scope.（The scope of the Cartagena Protocol is limited only to *living* modified organisms，however the issue of 'products thereof' is covered by the Cartagena Protocol with regard to information sharing，and in information required in notifications for the advance informed agreement and simplified procedure，as well as in the principles for risk assessment.）

Others argued that the rationale for the inclusion of 'products thereof' lies in their mention in the Cartagena Protocol in relation to risk assessment，which places them squarely within liability considerations as mandated by Article 27 of the Cartagena Protocol，for if that risk materializes then there must be provisions for liability and redress. Therefore the inclusion of 'products thereof' would provide for a comprehensive and adequate instrument dealing with damage from LMOs.

While mention of 'products thereof' was eventually removed from the operative text of the Supplementary Protocol，the report of the fourth meeting of the Group of the Friends of the Co-Chairs records an understanding that Parties may apply the Supplementary Protocol to damage caused by processed materials that are of LMO-origin，provided that a causal link is established between the damage and the LMO in question.

This understanding is significant as it clarifies that the Supplementary Protocol may apply to damage caused，not only by LMOs，but also by their products，which may be non-living material. In any case，if a Party so wishes，it can include 'products thereof' in its domestic law implementing the Supplementary Protocol.

4 Conclusions and way forward

The Supplementary Protocol sets international minimum standards on liability and redress for damage resulting from LMOs，and specifically covers one approach to liability（the administrative approach）. As such，domestic law should implement and augment the Supplementary Protocol，and much is left to countries at the national level to determine and implement. Parties to the Supplementary Protocol should put in place substantive laws with regard to liability and redress to address damage from LMOs，and this could include specific civil liability rules and procedures.

Domestic law is specifically mentioned in relation to：Defining the 'operator'，applying to damage from import of LMOs from non-Parties, establishing the causal link, situations in which the operator may not be required to bear costs, recourse to remedies, implementation of response measures, exemptions or mitigations, time limits, financial limits, provision of financial security, rules and procedures on civil liability.

The Supplementary Protocol is therefore just the beginning in the journey for liability and redress. Remaining gaps will need to be addressed at the national level in domestic law, or through future work under the Supplementary Protocol, such as that mandated for civil liability and financial security. This could be aided by complementary capacity-building measures, which are already envisaged for the development and/or strengthening of human resources and institutional capacities relating to the implementation of the Supplementary Protocol.

References

[1] Nijar, G.S. (2012). The Nagoya - Kuala Lumpur Supplementary Protocol on Liability and Redress to the Cartagena Protocol on Biosafety: An analysis and implementation challenges. Int Environ Agreements, Springer Science+Business Media B.V..

[2] Nijar, G.S. (2007). Liability and redress for damage arising from genetically modified organisms: Law and policy options for developing countries. In Lim L.C. and Traavik, T. (eds). (2007). Biosafety First: Holistic approaches to risk and uncertainty in genetic engineering and genetically modified organisms. Tapir Academic Press, Trondheim.

[3] Secretariat of the Convention on Biological Diversity (2011). Liability and redress: Basic concepts. Workshop material no. 1. SCBD, Montreal.

[4] Lim, L.C and Lim, L.L (2011). The Nagoya - Kuala Lumpur Supplementary Protocol on Liability and Redress: Process, provisions and key issues for developing countries. Third World Network, Penang.

[5] Third World Network (2012). Liability and Redress for Damage Resulting from GMOs: The Negotiations under the Cartagena Protocol on Biosafety. TWN, Penang.

转基因生物安全及其损害赔偿机制研究

On Compensation mechanism for Damage from GMOs

刘燕，张振华
（环境保护部南京环境科学研究所生物多样性中心副主任，南京 210042）
LIU Yan，ZHANG Zhen-Hua
（*Nanjing Institute of Environmental Sciences*，*Ministry of Environmental Protection*，*Nanjing* 210042）

摘　要: 转基因生物自诞生之日起，有关其安全性的争论就从未停止过。为确保在改性活生物体（LMOs）的越境转移、过境、处理和使用过程中采取充分必要的保护措施，国际社会就此形成了《卡塔赫纳生物安全议定书》和《卡塔赫纳生物安全议定书关于赔偿责任和补救的名古屋——吉隆坡补充议定书》等国际公约。本文尝试对"生物安全"、"环境损害"、"生物安全损害赔偿"等概念进行界定，在目前尚无"生物安全损害赔偿"案例的情况下，对有关环境损害的国际立法和实践进行梳理，归纳其特征，在借鉴美国、欧盟有关生物安全损害及赔偿立法的基础上，分析我国在相关领域的立法现状并提出建议。
关键词: 转基因生物；损害；赔偿；生物安全议定书

Abstract: Science the appearance of genetically modified organismsn（GMOs），the argument about its safety has never stopped. To make sure that living modified organisms（LMOs）transboundary movement，transit，handling and use of the full process to take the necessary protective measures，the international community promulgated "the Cartagena Protocol on Biosafety" and "Cartagena Protocol on Biosafety Protocol on Liability and Redress to the Nagoya - Kuala Lumpur supplementary Protocol "and other international conventions. The objective of this research was to give the definition such as "bio-security"，"environmental damage"，"Biosafety damages" and summarize the international legislation and practice of environmental damage as well as their characteristic. Drawing from the laws of the compensation for biosafety damages in United States and European Union，the current situation of related regulations and rules in China could be evaluated and the recommendations could be given.
Keywords: Genetically Modified Organisms，Damage，Compensation，Biosafety Protocal

1　引言：转基因生物技术与转基因生物安全

从 1983 年世界第一例转基因植物到 1993 年世界第一种转基因食品投放市场，转基因作物显现出传统作物所难以企及的优势，市场化推进速度惊人。从 1996 年开始转基因作物的商业化种植，此后的十余年时间里，全球转基因作物的种植面积增长了 100 倍（ISAAA，2012）。

转基因生物技术在很大程度上能对传统农业和相关产业产生很大的促动：第一，转基因生物技术可以改良品种。第二，它还可以延长食品保存时间及增加营养成分。第三，通过此技术给作物植入防虫防毒防菌基因，使作物本身产生抵抗病虫害侵袭的能力，因而可减少农药的使用量，有利于环境保护。第四，转基因技术及基因食物正向医疗药用方面扩展。

然而，从第一例转基因生物诞生之日起，人类有关转基因技术和转基因食品安全性的争论就从未停止过。如含有抗虫害基因的食品是否会威胁人类健康；转基因产品对环境的影响；转基因产品是否会破坏生物多样性；转基因产品带来的伦理问题等。综合起来说，目前对转基因食品的担心主要集中在对人体健康和对环境的影响两个方面。当然，人们对转基因技术的最大担忧还是来自对环境的影响方面。例如，外源基因的导入可能会造就某种强势生物，从而破坏原有生物种群的动态平衡和生物多样性。

第一次提到生物安全（biosafety）概念的是美国国立卫生研究院（NIH）专门针对生物安全制定的规范性文件——《NIH 实验室操作规则》。此规则对生物安全概念的界定，主要针对的是转基因生物，所调控的是一种狭义的生物安全，即将病源性微生物控制在实验室内的安全使用。生物安全与生物技术紧密相连，没有现代分子生物技术的发展，没有对遗传物质在物种间进行转移的科技能力，就不会有生物安全问题的出现。生物安全是基于转基因生物及其产品而导致的确定或不确定的潜在风险。

2 “生物安全”、“环境损害”、“生物安全损害赔偿”之界定

转基因生物安全，不同领域的专家都从自己的专业角度对这个问题进行了很多研究。这些研究主要集中在生态学、科技伦理、环境科学等方面，主要通过技术层面来分析转基因生物的安全性问题。随着研究的深入，人们发现生物安全问题对人类社会的深远影响已经超过人们的预期，原来那种希望仅仅依科学技术的完善来解决人们对生物安全的疑虑和不安的想法已经显得过于简单。于是，法律手段作为环境管理的重要工具之一得到重视。作为转基因生物安全法律问题研究的逻辑起点，转基因“生物安全”及“环境损害”的概念界定是一个核心和基础性问题[1]。

2.1 “生物安全”之界定

按照《国际科学技术发展报告》的解释，生物安全有狭义和广义之分。狭义生物安全是指防范由现代生物技术（主要指转基因技术）的开发和应用所产生的负面影响，即对生物多样性、生态环境及人体健康可能构成的威胁或潜在风险。广义生物安全则不仅针对现代生物技术的开发和应用，还包括了更广泛的内容，大致分为三个方面：一是指人类的健康安全；二是指人类赖以生存的农业生物安全；三是指与人类生存有关的环境生物安全[2]。

在我国法学界，对于生物安全的理解不一。[3]王灿发教授认为，“生物安全应当是指生物的正常生存和发展以及人类生命和健康不受人类生物技术活动和其他开发利用活动侵害和损害的状态”。[4]蔡守秋教授认为，“生物安全是指生物种群的生存发展不受人类不当

活动干扰、侵害、损害、威胁的正常状态……所谓人类不当活动是指违背自然生态规律的人类活动，包括开发、利用生物资源的生产活动、交易活动和技术活动”。[5]还有学者将转基因生物安全定义为“是指为使转基因生物及其产品在研究、开发、生产、运输、销售、消费等过程中受到安全控制、防范其对生态和人类健康产生危害以及救济转基因生物所造成的危害、损害而采取的一系列措施的总和”。

由此可见，在法律的视野中，转基因生物安全通常包括两个层面的内容：其一是生态和健康上的风险防范。因为“科学不确定性”的存在，使得法律对转基因生物的风险规制努力只能是“防范”而非“根除”隐患；其二是指一旦转基因生物造成了损害，法律是否加以救济，提供怎么样的法律框架予以救济。

2.2 “环境损害”之界定

环境损害可以从广义和狭义两方面来理解。其中，[6]狭义的环境损害局限于对自然资源如空气、水体、土壤、动植物等的损害；[7]广义的环境损害范围除自然资源外，还包括物质财产（主要指人类文化遗产和考古学），风景和环境舒适以及上述各项因素的相互联系。

从立法和学术研究的角度看，对因环境污染和破坏而侵害他人权益这一现象的称谓并不统一。英美法文献根据环境遭受污染或破坏的事实，通常称其为“环境污染”。[8]在大陆法系国家，“环境污染”虽为常用之术语，但在立法上，德国法采用“干扰侵害”，而法国法则称“近邻妨害”和“公态损害”。[9]日本的立法和学术研究均采“公害”一说。[10]

在我国的法学研究中，“环境损害”、“环境侵害”、“环境侵权”等概念均有使用。从描述的状态上看，“环境损害”强调受到损害的结果，“环境侵害”则侧重权益遭受不利的状态，“环境侵权”则包括环境污染和环境破坏两方面的内容，而且包含环境侵害状态和环境损害后果两个方面。有学者认为，从赔偿责任角度来看，“环境损害”更能“统称对环境造成污染和对资源进行破坏的所有不法行为。”[11]

2.3 “生物安全损害赔偿”之界定

传统民法理论上的损害赔偿，是指行为人因故意或过失侵害了他人的财产权利和人身权利并造成了损失而应当承担的民事法律责任。“环境损害”属于广义的民事赔偿救济，而“生物安全损害赔偿”则是属于环境损害的一种形式。生物安全损害赔偿属于特殊的侵权损害赔偿，是环境责任的承担方式之一，它是一种基于特殊侵权的侵权之债。

生物安全损害赔偿制度（the Damage Compensating System of Biosecurity），是指在生物安全风险转变为现实的损害之后，有关主体根据相关生物安全法律法规的规定，向被损害方给予赔偿的一整套措施。其中损害，是指伤害（Injury），或者所蒙受的损失（Loss）；生物安全的损害，亦应包括这两个方面：即一是有形的伤害，主要是对人体的伤害；二是有形的或者无形的损失，主要是因生物安全风险转化为现实之后而造成的经济上的损失。[12]

3 生物安全损害及赔偿的国际与国内立法

国际上制定有关生物技术安全的制度、规定和指南等最早可以追溯到 80 年代中期。

面对生物技术对生物安全可能带来的风险，国际上近年来开始了生物安全方面的立法，采取法律措施来防止这种风险的现实发生，为此已经制定了一些法律文件，使得环境立法领域进一步得到加强。其中主要有以下这些文件：《世界自然宪章》（也叫《大自然宪章》）、《生物多样性公约》、《国际生物技术安全技术准则》、《21 世纪议程》、《世界贸易组织条约》、《卡塔赫纳生物安全议定书》等，最后是联合国粮农组织和世界卫生组织制定的有关国际法等。

从出现时间方面看，生物安全的国家立法要早于国际立法。然而，生物安全国际法一经出现，即在很大程度上推动了各国的生物安全立法的发展。迄今为止，已有数十个国家在生物安全方面开展研究，并陆续制定了有关生物技术实验研究、工业化生产和环境释放等一系列安全准则、条例、法规或法律。各国生物安全立法的管理模式主要可分为两类：一类是以美国、加拿大等国为代表的基于产品的管理模式；另一类是欧盟等基于技术的管理模式。基于产品的管理模式认为，以基因工程为代表的现代生物技术与传统生物技术没有本质区别，因而应针对生物技术产品而不是生物技术本身进行管理。而基于技术的管理模式认为，现代生物技术本身具有潜在的危险性，因此，只要与重组 DNA 相关的活动都应进行安全性评价并接受管理。

4 我国相关立法现状分析与建议

作为生物多样性保护国际法发展的积极推动者和实践者，中国不仅积极参与了 1992 年《联合国生物多样性公约》和 2000 年《卡塔赫纳生物安全议定书》的起草和谈判，并签署和批准了该《公约》和《议定书》，同时，中国还积极履行相关的国际生物安全保护义务，加强生物安全管理，并已取得初步成效。在国内，中国已经制定了一些生物安全专门立法，但存在诸多问题，这些问题有待于通过健全生物安全立法体系、制定专门的综合性生物安全立法予以解决。

4.1 我国生物安全立法的现状

中国目前有关生物安全的专门立法主要包括基因工程安全、农业转基因生物安全、转基因食品安全、转基因药品安全和转基因微生物安全五个方面。

其一，基因工程安全专门立法。迄今为止，中国关于基因工程安全的专门立法只有《基因工程安全管理办法》一部。该《办法》就适用范围和主管机关、安全评价制度、安全控制制度、许可制度、法律责任制度等方面做出了规定。遗憾的是，该法未能得到有效执行。

其二，农业转基因生物安全专门立法。在中国目前转基因生物安全立法体系中，有关农业转基因生物安全的专门立法的规定最完备。这方面的立法包括：2001 年《农业转基因生物安全管理条例》；2002 年颁布的《条例》的三个配套规章，即《农业转基因生物安全评价管理办法》、《农业转基因生物标识管理办法》和《农业转基因生物进口安全管理办法》；2002 年颁布的三个管理程序，即《农业转基因生物安全评价管理程序》、《农业转基因生物进口安全管理程序》和《农业转基因生物标识审查认可程序》；2002 年 3 月、2002 年 10 月和 2003 年 3 月三次决定延长转基因农产品安全管理临时措施实施期限的决定。这些立法从适用范围、主管机关和协调机制、安全评价制度、安全措施制度、标识制度、报告制度、许可制度、资料存档制度、监督检查制度、应急处理制度、法律责任制度等方面，就农业转基因生物安全管理做出了规定。

其三，转基因食品安全专门立法。1995 年，中国颁布了《食品卫生法》。该法是中国食品安全管理的法律依据，对利用新资源生产的食品、食品添加剂的新品种的管理做出了相应的规定，其中内在地包括了利用转基因生物生产的食品和食品添加剂。2001 年，中国又颁布了《农业转基因生物安全管理条例》。该《条例》从农业转基因生物安全管理的角度，就转基因食品安全做出了一些规定，其中主要内容包括两个方面，即部分转基因食品形态；卫生行政管理部门对转基因食品安全的管理职责。2002 年，卫生部颁布《转基因食品卫生管理办法》。该《办法》从适用范围、基本要求、生产者义务、评价制度、许可制度、标识制度、监督管理制度等方面做出了规定。

其四，转基因药品安全专门立法。中国关于转基因药品安全的专门立法主要是《新生物制品审批办法》。该《办法》从适用范围、命名及分类、研制要求、许可制度等方面做出了规定。

其五，转基因微生物实验室安全专门立法。2004 年，中国颁布了《病原微生物实验室生物安全管理条例》。该《条例》包括适用范围和主管机关、病原微生物管理、实验室管理、实验室感染控制、实验室监督管理、法律责任等内容。

4.2 我国生物安全立法体系的不足

尽管这些上述立法为中国生物安全管理的某些方面提供了重要的法律依据，但与中国生物技术的发展水平和生物安全管理的实际需要相比，中国的生物安全管理的立法却相对滞后，无法满足生物安全管理的需要。这主要表现为五个方面：

其一，缺少综合性的专门立法。综观中国现行生物安全立法，一些专门法规主要是有关部门从本部门管理的角度制定的屈指可数的法规和规章，并没有一部从整个生物安全的角度对生物安全管理做出全面、系统规定的综合性立法。在现有的专门立法中，《基因工程安全管理办法》曾一度是其他转基因生物安全立法的范本和依据，但该《办法》的基本出发点更多的是技术管理，而不是法律规制。同时，由于其立法层次较低，也无法担当起中国转基因生物安全牵头法规的重要角色。其他立法均从转基因生物安全管理的某一方面作出规定，也不具有牵头法律的综合性。农业转基因生物安全立法以《农业转基因生物安全管理条例》为核心，虽然形成了一个较为完整的法律体系，但其涉及的只是农业领域。缺乏综合性专门立法导致了多头管理、重复管理、管理规范缺失等诸多方面的问题。

其二，法规体系不健全。要对生物安全进行全面而有效的管理和监督，就应具备一套健全完善的法规体系。目前中国的生物安全立法尚没有一个健全的法规体系，当然更谈不上完善的体系。其后果是，各部门只能根据自身的情况和需要制定部门规章，而较少考虑其他部门的同类规定。在这种情况下，针对有些问题的规定是重复的，但针对另一些重要的问题却未作规定，从而极大地影响了执法效果。由于法规体系不健全，在处理某些案件过程中出现了中国生物安全立法目前存在的主要问题是：缺乏一部牵头的生物安全管理法律，缺乏一套健全的法规体系，缺乏有效的管理制度，尚未建立起协调统一的管理程序。中国生物安全立法存在的主要问题是：缺乏一部牵头的生物安全管理法律或法规；缺乏一个健全的法规体系；缺乏有效的管理制度；缺乏统一监督管理机构和管理程序的规定。

其三，管理制度不完善。中国目前生物安全管理制度仍然有待完善，这主要体现为三个方面：首先，调整阶段不连贯。转基因生物安全立法应当实行研究、试验、保管、运输、进口、出口、应用、商品化、废弃等全过程管理制度，但中国目前的立法大都忽视了其中的某几个重要阶段和环节（如对废弃物的处置和回收等），使得在某些阶段内出现“管理真空”。其次，调整手段不全面。在转基因生物安全管理过程中，一些重要的制度是不可或缺的，如转基因生物的研制、应用推广和进口的环境影响评价制度、损害赔偿保险制度、越境转移的事先知情同意制度、公众参与制度、奖励制度、应急处理制度、听证制度、损害赔偿与补救制度等。然而，中国目前的有些转基因生物安全立法在这些重要制度上或者相关规定存在问题，或者重视程度不够，有的制度甚至基本上未作规定。最后，调整领域不完整。中国十年以来出台的转基因生物安全立法主要集中于基因工程安全、农业转基因生物安全、转基因食品安全、转基因药品安全和转基因微生物实验室安全等领域。随着中国生物技术的迅速发展，在林业、野生动植物、微生物等方面的转基因生物安全问题也日益暴露出来，但中国至今没有出台关于这些领域的专门立法。

其四，管理机构和管理程序不统一。这主要体现为两个方面：第一，管理机构不统一。在转基因生物安全管理领域，中国各主管部门分散立法。即使是国务院发布的有关行政法规，也是委托某一主管部门进行牵头起草，之后以国务院的名义发布。这样，每一部立法均规定了相应的管理机构；但由于缺乏综合性的转基因生物安全立法，目前没有一个对转基因生物安全进行统一监督管理的机构。相应地，任何一个部门的机构都不能掌握全面的转基因生物安全管理信息，更无法基于此进行协调和统一。这样就导致以下后果：其一，在国内管理方面，管理机构的复杂性使得行政相对人有时在接受管理时无所适从，甚至造成重复登记、重复审批；其二，在对外协调和合作方面，国外的有关机构和其他主体有时不知道应当与中国的哪一个部门进行联系、办理手续，非常不利于中国履行对有关国际条约的承诺。第二，管理程序不统一。由于各部门在有关立法中均作出自己的规定，其中有些程序性规定与其他部门相关立法的程序性规定没能很好地衔接和协调，由此出现相关立法在管理程序上不统一的现象。其后果是，不仅行政相对人的办事难度大大提高，而且很容易使不同行政机关在对同一事务进行管理的过程中产生潜在冲突，难以适应高效行政的要求。

其五，立法技术不成熟。中国有关生物安全保护专门立法在立法技术上也有待改进，主要体现为：首先，必要内容欠缺。例如，《农业转基因生物安全管理条例》规定的转基因食品仅包括转基因农产品的直接加工食品，但不包括使用转基因生物原料加工的食品。该《条例》中没有规定对这类食品应当如何管理。此外，法律责任规定的缺失也屡见不鲜，这在《转基因食品卫生管理办法》和《新生物制品审批办法》中均有体现；在《基因工程安全管理办法》中，民事赔偿的范围也没有明确界定。其次，立法结构不合理。在国内立法中，术语解释可以置于总则中，也可以置于附则中，但一般置于附则中。但在中国的一些转基因生物安全专门立法中，名词解释有时在总则和附则部分同时出现。例如，《基因工程安全管理办法》将“基因工程”的定义置于总则部分规定，同时在“附则”部分也进行了规定。再如，在《基因工程安全管理办法》中，将审批机关工作人员为申报者保守技术秘密在法律责任部分规定。事实上，这应当属于审批机关工作人员的“义务”；而法律责任则应规定违反此种“义务”所应承担的否定性法律后果。从立法技术的角度讲，这些情况是不应当发生的。再次，法律术语运用不当。例如，《基因工程安全管理办法》规定，负有责任的违法主体应承担“停止损害行为”的民事责任。事实上，“损害行为”提法本身是不准确的。因为在法律上，“损害”通常指一种后果，很少与“行为”连用。在此，正确的提法应当是“停止侵害行为”；如果一定要用“损害”，那么也应改为“造成损害的行为”，而不应为“损害行为”。

4.3 我国生物安全立法体系的完善

要从根本上解决前述问题，首先需要健全生物安全立法体系。根据生物安全管理内容和范围的要求以及中国的法律体制，中国生物安全法规体系应当由不同级别、层次所组成。在国家立法中，可以分为三个层次。第一个层次，应当有一部由全国人民代表大会常务委员会通过的生物多样性保护法，其中包括关于生物安全的原则规定，也可以是制定一部《生物安全法》。第二个层次，应当由国务院颁布一项生物安全管理条例或者生物安全法的实施细则，对生物安全管理做出较为全面、具体的规定。第三个层次，由国务院授权主管生物安全工作的部门就具体管理问题发布行政规章。

生物安全立法体系应当由七个方面的立法构成：其一，综合性的管理立法，即对生物安全进行全面统一管理的立法，这主要是指环境保护法和生物多样性保护法有关生物安全管理的规定、国务院发布的管理条例等。其二，专门生物安全管理立法，即针对生物工程所涉及的各方面进行的专门立法，包括人体基因管理立法、野生动物基因管理立法、野生植物基因管理立法、农业生物基因工程管理立法、家禽和家畜基因管理立法等。其三，生物工程环境影响和安全评价立法，包括评价机构管理立法、评价程序立法、评价技术规则立法等。其四，生物技术成果越境转移管理立法。其五，生物安全标准立法，此种标准可以根据不同生物技术的特点和生物安全要求进行制定，也可以根据每一种生物技术不同阶段的生物安全风险程度制定相应的标准。其六，生物技术损害赔偿责任承担和纠纷处理程序立法。其七，其他有关立法中关于生物安全管理的规定，包括动植物检疫立法中关于转基因生物检疫的规定、农业法中关于生物安全的规定、森林法中关于转基因林木管理的规定、草原法中关于生物安全的规定、野生动植物保护法中关于转基因动植物管理的规定、种子法中关于转基因种子管理的规定、渔业法中关于转基因生物管理的规定、自然保护法中关于生物安全的规定、药品管理法中关于转基因药品管理的规定、食品卫生法中关于转基因食品管理的规定、进出口商品检验法中关于进出口转基因商品的规定等。

上述各个层次和方面的法律、法规、规章以及技术性规范相互配合、相互协调，共同构成了中国生物安全法的完整体系。

参考文献

[1] 胡敏飞. 跨国环境侵权的国际私法问题研究. 上海：复旦大学出版社，2009，2.

[2] 亦云. 国际科学技术发展报告. 中国生物技术信息网，2006.

[3] 王灿发. 国际贸易. 北京：中国商务出版社，2000.

[4] 蔡守秋. 论生物安全法. 河南省政法管理干部学院学报，2002，2.

[5] 王明远. “转基因生物安全”法律概念辨析. 法学杂志，2008，1.

[6] Thomas Schoenbaux. Environmental Damage in the Comman Law，Environmental Damage in International and Comparative Law，England：Oxford University Press，2002，213.

[7] Philippe Sands. Principles of international environmental law，England，Cambridge University Press，2003，876.

[8] 邱聪智. 公害法原理. 台湾：三民书局，1984，5.

[9] 王明远. 环境侵权救济法律制度. 北京：中国法制出版社，2001，205.

[10] 乔世明. 环境损害与法律责任. 北京：中国经济出版社，1999，37.

[11] 1993 年欧洲理事会《卢加诺公约》第 8 条 d 款。

[12] 夏少敏，陈真亮. 生物安全损害赔偿问题初探，2009. 10.

生物安全损害赔偿法律问题研究

On Legal Liability for Damage from GMOs

于文轩*
（中国政法大学，北京，100088）
YU Wen-Xuan
（China University of Political Science and Law，Beijing，100088）

摘　要：生物安全法上的损害赔偿，是指在生物安全风险转变为现实损害后，致害方根据生物安全法的规定，对受害方给予赔偿。在国际法层面，生物安全损害赔偿法律机制应关注赔偿主体、索赔主体、索赔程序、赔偿金额、责任限制、金融保障等方面。我国现行与生物安全损害赔偿相关的立法主要包括 1993 年《基因工程安全管理办法》、2001 年《农业转基因生物安全管理条例》、2004 年《病原微生物实验室生物安全管理条例》中的相关规定。在进一步立法中，我国应主要从归责原则、共同致损、免责条件、限额赔偿、环境损害赔偿金等方面入手，完善生物安全损害赔偿法律机制。

关键词：生物安全；损害赔偿；法律责任；立法

Abstract: Damage from GMOs includes the negative impacts of GMOs related activities to human health and biodiversity. In international law, the redress to the GMOs related damages should focus more on such aspects as subjects of compensation and claim, procedures, amounts, liability limits and financial supports. In China, there are some regulations and rules relevant to GMOs related damages and redress, while facing some challenges. Doctrine of liability fixation, joint liability, liability limit, compensation and exemptions should be put emphasis on in the further legislation.

Keywords: Biosafety, Damage, Legal Liability, Legislation

1　引言

法律规则由行为模式和法律后果组成。[1]行为模式，是指“法律规则中规定人们如何具体行为的方式或范式”，一般认为包括应为模式、可为模式和勿为模式三种；法律后果是指“人们遵守或违反法律规则中规定的行为模式所产生的后果”。[2]生物安全法律规则亦

* 于文轩，男，满族，法学博士，中国政法大学副教授，主要研究方向为环境与资源保护法。

循此一般原理。在生物安全法中，如果法律主体的行为违反了应为模式或者勿为模式的规定，即在事实上和法律上造成了相应的后果。其中，事实层面的后果包括因转基因生物研究开发、生产、运输、环境释放、贮存和销售等活动而对人类健康和生物多样性保护和可持续利用所产生的负面影响，即损害；法律方面的后果是赔偿，亦即使受损的法益获得弥补。生物安全法上的损害赔偿，是指在生物安全风险转变为现实损害后，致害方根据生物安全法的规定，对受害方给予赔偿。损害赔偿制度是生物安全法的其他制度得以顺利贯彻和实施的重要保障。

2 生物安全损害赔偿国际法的制度设计

2010 年《名古屋-吉隆坡补充议定书》是迄今为止国际社会达成的第一个也是唯一一个专门针对生物安全损害赔偿的国际条约，具有重要意义。然而，由于生物安全损害赔偿问题涉及利益相关方的重大利益，该议定书是多年谈判过程中利益相关方不断妥协的成果，非常明显地体现出国际环境法的“软法”特征。这一特征，也决定了该条约在一些方面并非尽如人意。因此，有必要借鉴国际环境法其他领域解决损害赔偿问题的经验，为生物安全损害赔偿国际法律制度的进一步完善提供参考。

2.1 国际环境损害赔偿制度的借鉴意义

迄今为止，国际社会制定了数十部关于环境损害赔偿的重要法律文件，所涉领域包括核损害、油污损害、由空间实体造成的损害、与生物安全相关的环境损害、危险废物转移造成的损害、与海洋和南极相关的环境损害以及国际法未加禁止的行为所引起的损害等。按照时间顺序，这些国际法文件主要包括：1963 年《关于核损害的民事责任的维也纳公约》、1969 年《对公海上发生油污事故进行干预的国际公约》、《国际油污损害民事责任公约》、1971 年《设立国际油污损害赔偿基金公约》、《油轮油污责任暂行补充协定》、1972 年《空间实体造成损失的国际责任公约》、1978 年《油轮所有人自愿承担油污责任协定》、1982 年《联合国海洋法公约》、1992 年《生物多样性公约》、1996 年《国际法未加禁止之行为引起有害后果之国际责任条款草案》、1999 年《关于危险废弃物越境转移及其处置所造成损害的责任和赔偿问题议定书》、《南极条约》协商会议的有关共识、2000 年《卡塔赫纳生物安全议定书》等。

根据这些国际法文件，国际环境损害赔偿立法的主要原则包括非歧视原则、无过错责任与过错责任相结合原则、责任分担原则和国家责任原则四个方面。其中，非歧视原则是指国际环境法主体在处理环境损害赔偿争端时，不应因各主体之间经济发展水平、社会制度、意识形态、宗教信仰等方面的差异而适用不同的条件，而应平等对待、一视同仁；无过错责任与过错责任相结合原则，是指在国际环境损害赔偿法中不是采用单一的归责原则，在有些情况下采用无过错责任原则，而在另一些情况下则采用过错责任原则；责任分担原则，是指如果同一环境事故中存在多个致害方，或者受害人同时也对事故负有相应责任，则损害赔偿责任应由各相关主体共同承担；国家责任原则，是指国家作为国际法主体需承担损害赔偿责任，而不得基于主权理由而豁免。

根据前述国际法文件，国际环境损害赔偿立法的要点包括赔偿主体、金额限制、责任限制和资金支持等方面。在赔偿主体方面，最应注意的是国家在某些情况下也应对其国际不当行为承担损害赔偿责任；金额限制，是指考虑致害方的承受能力以及其他相关因素而对致害方的最高赔偿金额进行的限制；责任限制，是指在一定条件下，国际法主体对其不当行为不承担损害赔偿责任；资金支持，是指为确保受害方能够及时充分地获得赔偿而建立基金、保险和保证等制度。[3]

2.2 生物安全损害赔偿国际法的基本原则

根据国际环境法在损害赔偿方面的主要原则，以及生物安全损害赔偿自身的特点，在生物安全损害赔偿国际立法中，应确立的主要原则应包括无过错责任原则、责任分担原则和国家责任原则。

2.2.1 无过错责任原则

尽管在国际环境法的其他方面采用无过错责任与过错责任相结合的原则，但在生物安全损害赔偿领域应采用无过错责任原则。无过错责任原则是生物安全国际法风险预防原则在归责原则上的体现。由于生物安全与生物最本质的性质、物种的生存和繁衍甚至与人类的生命健康密切相关，故其风险性之大为国际社会所共同关注；同时，一旦生物安全风险转化为现实，其后果往往又是非常严重的，甚至是不可逆转的。因此，生物安全问题本身存在着巨大的风险性。这样，风险预防原则便成为生物安全国际法的基本原则之一。与风险预防原则相适应，在生物安全损害赔偿国际立法中确立无过错责任原则，有利于督促国际环境法主体更好地约束自己的行为，尽最大的注意义务，从而最大限度地避免环境损害的发生，而这正是生物安全损害赔偿国际立法的重要目标之一。

2.2.2 责任分担原则

现代生物技术给人类社会带来了正反两方面的影响。“如能在开发和利用现代生物技术的同时亦采取旨在确保环境和人类健康的妥善安全措施，则此种技术可使人类受益无穷”；但“现代生物技术扩展迅速，公众亦日益关切此种技术可能会对生物多样性产生不利影响，同时还需顾及对人类健康构成的风险”。正是基于这一考虑，国际社会对生物技术的普遍态度是，在推动生物技术进步的同时，通过各种措施尽量避免可能由此发生的消极后果，也即“谨慎发展”。在生物安全损害赔偿国际立法中规定责任分担原则，有利于生物安全国际法谨慎发展原则在归责原则方面的实现。

在生物安全领域，责任分担原则要求有关各方均应尽到足够的注意义务；在同一环境事故中，如果存在多个致害方，或者受害方同时对事故负有相应责任，共同致害方或者负有责任的受害方也应当分担相应的损害赔偿责任。对于前一种情况而言，共同致害方应对受害方分别或者连带地承担责任——至于是选择分别承担责任还是连带承担责任，则应根据致害行为是否在事实上（而不是在主观上）具有关联性而确定：如具有关联性，则应承担连带责任，反之则应分别承担责任。对于后一种情况而言，致害方在受害方自身应当承担责任的范围内免除相应的损害赔偿责任，但在此同样有一点需要注意，即在此免除的仅是相应的损害赔偿责任，而不包括其他形式的国际环境责任，如恢复原状的责任等；而且对于承担相应责任的受害方而言，其分担相应责任的事实，并不妨碍其承担其他应当承担的国际环境责任，如国际环境刑事责任等。

2.2.3　国家责任原则

在生物安全国际法领域，国家基于主权等理由的豁免不应适用。“国家责任是一国实施国际不当行为所引起的法律后果，具体体现为行为国与受害国之间的一种新的权利义务关系。……一国一经确认为实施了国际不当行为，即应负担停止其不当行为和补偿该行为造成的损害后果的义务，受害国则相应地享有要求或者迫使行为国履行其上述义务以及在特定情况下对行为国进行制裁的权利”。而“赔偿是国家责任的逻辑后果”，而“将非法行为的一切后果消除掉”常常被用来作为赔偿的标准。[7]

在生物安全领域，国家责任更有其存在和适用的合理性。一方面，在国际公法领域，国家是最主要的国际环境法主体。如果允许国家基于其主权等原因而得到豁免，则在事实上在很大程度上“架空”了国际环境法责任，使之有名无实。另一方面，在国际私法领域，私法主体所从事的与生物安全相关的行为很大一部分有国家或者政府在政策或者法律上给予支持。对于这部分损害他方生物安全权益的行为，国家至少应承担监管不力的责任——更何况其中不乏在国家的明示或者暗示授意下所从事的行为。

2.3　生物安全损害赔偿法律制度的要点设计

2010 年《名古屋-吉隆坡补充议定书》为形成健全的生物安全损害赔偿法律制度体系奠定了重要的基础。进一步国际努力应从赔偿主体、索赔程序、责任限制、金融保障等方面展开。

2.3.1　赔偿主体

国际环境损害赔偿立法主要将赔偿主体确定为设备或者装置的所有人、管理人、共同管理人等与环境事故有密切关系的主体。在生物安全领域，这些设备或者装置的所有人、管理人和共同管理人等主体如果因其不当行为而造成了损害，无疑应承担损害赔偿责任。与此同时，国家在某些情况下也应对其国际不当行为承担损害赔偿责任。一方面，这是前述生物安全损害赔偿国际立法“国家责任原则”的必然要求；另一方面，国际环境法作为国际法的一个新的分支，目前其发展已经越来越多地体现出公法与私法相融合的特征，这一点在生物安全国际法领域体现得尤为明显，赔偿主体就不应仅限定为设备或者装置的所有人、管理人、共同管理人等传统的“私”的主体，而应进一步将国家涵盖其中。

具体而言，赔偿主体应当包括损害行为人、国家和有责任的第三方等。其中，损害行为人（生产商、出口商、进口商、承运商、销售商等）承担主要责任；国家承担补充性的民事赔偿责任；如果损害全部或者部分地由第三方引起，则在其承担损害赔偿责任之后，还享有对该第三方的追偿权。索赔主体应当包括受害者和国家。其中，如果生态系统（生态环境）受到生物技术研发应用活动的损害，国家可主动行使索赔权；如果受害者没有能力或者丧失能力提出损害赔偿请求，受害者可以申请国家代其索赔。

2.3.2　索赔程序

生物安全损害赔偿的索赔程序，可以依照《生物多样性公约》的相关规定和国际法的相关规则处理。《生物多样性公约》第 27 条规定了国家间的协商程序：缔约国之间在就公约的解释或适用方面发生争端时，有关缔约国应通过谈判方式寻求解决；如果有关缔约国无法以谈判方式达成协议，可以联合要求第三方进行斡旋或要求第三方出面调停。国家间的仲裁和调解可以适用《生物多样性公约》附件二的规定。同时，民事程序中的大多数问题，如受理的法院或者仲裁庭、适用的法律等，可适用国际私法的一般程序。

2.3.3 责任限制

在责任限制方面，应主要关注赔偿金额、责任范围、责任豁免和责任时限。在赔偿金额方面，对人身伤亡和财产损失实行实额（全额）赔偿制度，而对生态损害的赔偿则应实行限额赔偿制度。在责任范围方面，国家和政府（行政）因素不应成为责任限制的原因。在责任豁免方面，不可抗力应设定为免责条件。在此处的“不可抗力”情形下，损害完全由于不可抗力引起，同时经从事现代生物技术研发应用活动的一方及时采取补救措施，仍然不能避免造成该损害。在责任时限方面，相对时限可以设定为 3 年，绝对时限可以设定为 20 年。

2.3.4 金融保障

在金融保障方面，应特别关注基金机制的完善。首先应明确生物安全损害赔偿基金的强制性。在资金来源方面，现代生物技术研发应用者在从事有关活动前，应当支付一定数额的资金作为生物安全损害赔偿基金的组成部分，支付的数额应当与风险评估结论相衔接。在基金使用方面，应根据致害方向生物安全损害赔偿基金实际支付的资金数额与法定缴资额的比例、法定赔付限额以及基金使用规则，由基金赔付其造成的损害；如果依此赔付的金额无法完全弥补受害方的损害，则由致害方承担其余的赔偿责任；如果致害方仍然无力赔偿全部损失，则根据国家赔偿责任，由国家承担补充性赔偿责任。

3 我国生物安全损害赔偿立法现状与完善

我国目前尚未制定专门的生物安全损害赔偿立法，相关立法的内容亦不完善。应根据加强生物安全管理、保护生物多样性和人体健康的现实需求，完善我国生物安全损害赔偿立法。

3.1 我国生物安全损害赔偿立法的现状

我国现行立法中涉及生物安全损害赔偿的内容并不多见，主要分散在 1993 年《基因工程安全管理办法》、2001 年《农业转基因生物安全管理条例》和 2004 年《病原微生物实验室生物安全管理条例》之中。

1993 年《基因工程安全管理办法》第 28 条：“违反本办法的规定，造成下列情况之一的，负有责任的单位必须立即停止损害行为，并负责治理污染、赔偿有关损失；情节严重，构成犯罪的，依法追究直接责任人员的刑事责任：①严重污染环境的；②损害或者影响公众健康的；③严重破坏生态资源、影响生态平衡的。”这一规定对何为“严重”污染环境、何为“影响”公众健康、何为“严重破坏”生态资源等未作出明确的规定。

2001年《农业转基因生物安全管理条例》第54条规定："违反本条例规定，在研究、试验、生产、加工、贮存、运输、销售或者进口、出口农业转基因生物过程中发生基因安全事故，造成损害的，依法承担赔偿责任。"这一条款对赔偿主体、赔偿数额、免责条件等问题均未作明确的规定。

2004年《病原微生物实验室生物安全管理条例》第57条第2款规定："因违法作出批准决定给当事人的合法权益造成损害的，作出批准决定的卫生主管部门或者兽医主管部门应当依法承担赔偿责任。"这一条款失于笼统，缺乏应有的可操作性。

3.2 我国生物安全损害赔偿立法的完善

完善我国生物安全损害赔偿立法，应主要从归责原则、共同致损、免责条件、环境损害赔偿金、限额赔偿等方面入手。

其一，归责原则。我国生物安全损害赔偿立法应采取无过错责任原则。该原则是指生物安全法主体就其所从事的现代生物技术的开发利用行为造成的财产损失、人身损害或者生态损害后果依法承担损害赔偿责任时，不考虑其是否存在主观过错的归责原则。具体法律条文可以表述为："造成转基因生物安全危害的单位和个人，有责任排除危害，并对直接遭受损失的单位或者个人赔偿损失。"

其二，共同致损。共同致损是指损害结果由两个以上法律主体共同造成。此时，确定各加害人责任大小的主要依据应是其在客观上造成损失或者损害的大小（客观归责），而不是其是否有主观过错或者各自过错的大小。具体法律条文可以表述为："两个或者两个以上的单位和个人共同造成转基因生物安全危害的，共同承担排除危害、赔偿损失的责任。"在实践中，如果无法区分责任，则应实行连带责任。

其三，免责条件。完全由于不可抗拒的自然灾害，并经及时采取合理措施，仍然不能避免造成环境危害的，或者损害完全由受害人自己的故意行为造成的，造成损害的单位和个人免予承担责任。在此，免责条件的要点有二：一是发生灾害；二是采取措施仍无法避免。二者缺一不可。

其四，环境损害赔偿金。从事转基因生物体封闭研究、中间试验、环境释放和商品化、运输、贮存、越境转移、废弃物处理和处置的单位和个人造成环境污染或者生态破坏的，应当支付环境损害赔偿金。赔偿金额应以使生态环境恢复至受损前的状态（生态功能）所需花费为标准。评价生态环境的状态应以环境质量标准为依据。所谓"环境质量标准"，是指为保护人体健康、社会物质财富安全和维护生态平衡，对环境中有害物质或者因素含量的最高限额和有利环境要素的最低要求所作的规定。

其五，限额赔偿。赔偿额度的确定，一般应考虑社会经济发展政策和水平、立法目标、其他相关机制的发展程度等方面的因素。在一个力图最大限度地推动经济发展的政策背景下，环境损害赔偿数额一般以不对经济发展构成实质影响为限度；而一国经济发展水平则在总体上决定了社会经济主体对损害赔偿数额的承受能力，一般与赔偿数额成正比。如果立法旨在补偿受害人的人身伤害和财产损失，即属补偿性赔偿；如果同时意图对加害方施加更大的威慑，即属惩罚性赔偿，体现在环境损害赔偿的数额方面，后者的损害赔偿额上限一般高于前者。如果一国资金支持机制较为发达，则在确定环境损害赔偿数额的上限时一般不会仅考虑加害方的实际承受能力因素，同时还会考虑到资金支持机制对环境风险的分担功能。

参考文献

[1]　刘金国，舒国滢. 法理学教科书. 北京：中国政法大学出版社，1999，50.
[2]　刘金国，舒国滢. 法理学教科书. 北京：中国政法大学出版社，1999，51-52.
[3]　王灿发，于文轩. 生物安全国际法导论. 北京：中国政法大学出版社，2006，174-175.
[4]　王灿发，于文轩. 生物安全的国际法原则[J]，现代法学，2003（4）：130.
[5]　2000 年《生物多样性公约卡塔赫纳生物安全议定书》，序言.
[6]　梁淑英. 国际公法. 北京：中国政法大学出版社，1993，77.
[7]　王铁崖. 国际法. 北京：法律出版社，1995，152.

会议闭幕
Closing Remarks

Lim Li Ching 女士的闭幕讲话

Closing Remarks of Lim Li Ching

Lin Li Ching
（The Third World Network）

谢谢薛达元教授，谢谢所有的参会者，这两天讨论得非常好，内容很多，现在我稍微小结一下。很多内容不一定小结得非常准确，因为我们谈得非常宽泛，我个人在这两天学到的几个要点，对我来说可能翻译方面的原因会落掉一些内容，包括一些政治、社会、经济方面的层面，包括评估。

我就一个一个来说，第一天我印象很深的一个内容，就是中国的一个独特性，也就是说在这一方面的独立研发工作，关于生物安全方面工作中国已经开展了，这一方面研发工作要尽最大的努力，看看风险在哪里，GMO 对人类健康和环境有什么样的影响，我们应该使用哪种研发的手段，进行风险方面的评估。

第二，风险评估。也是我印象非常深刻的一点，我们同时要看一下风险评估的时间限制，我们应该考虑从中期角度和长期角度来规划我们应该做的工作，比如说 *Bt* 棉花的风险评估如何在中国开展，一开始 *Bt* 棉会出现一些有利影响，体现在害虫减少方面，但是，经过一段时间又发出其他警告，出现了一些抗性以及其他相关次生害虫的出现。我们应该意识到这些对科学家都是一种挑战，对法规的建设者和相关监管部门都是一种考验。如果有一些技术方面的问题能够解决，我们能不能在问题发生之前，通过相关技术预测一些新基因的特征来预防一些问题。去年有一些论文发表出来了，结果和最初的结论不太一样，证明最初的一些想法是错误的，市场上的确出现了一些新的挑战，我们应该采取灵活的方法来应对这些挑战。

第三，生物技术研究。其实最后说来说去都是要开展生物学方面的研究，我们应该继续这一方面的工作，同时我们应该承认我们的知识还是有缺口的，还有不确定性，还有很多未知的问题没有解决。而且，这些方面的问题，甚至我们一开始都没有设想到以及如何解决。有一些法律文书，比如说《卡塔赫纳生物安全议定书》的确意识到这些方面的问题，而且充分考虑到了这些不确定性因素，包括风险评估问题，此外，我们手头上已经有了一些指南性的文件考虑了这些问题。我们看到今天早上的一个讨论，有些事情发生不是在真空当中发生，是在事实当中发生的，有各种各样的影响因素，还有一些国际层面的争论，所以我们在社会经济层面、政治层面以及机构设置等方面面临挑战。比如说我们进行一个新的风险评估时，应该充分考虑这几个方面的问题，特别是公众的参与和协商，这样的话可以把我们的范围扩大到公众，他们的想法是应该包括进来的，这一点我想我们应该有一个共识。《卡塔赫纳生物安全议定书》中的一些条款也是这样说的，其他国家应该尽量考虑公众的意见。目前我们碰到的挑战，在国家层面有非常热烈的讨论，我也希望发展中国家能够多多参与进来，因为许多发展中国家从一开始议定书的协商阶段就意识到了这个问题是非常重要的，今天上午和下午我们也讨论了这一方面的问题。我们的确需要广泛的辩论，这种辩论可能包括相关的一些问题：科学之路怎么走，相关公共研究、公共意识，什么样的技术应该被采用，什么样的方法最适合我们的需要，同作为发展中国家，我们优先考虑的问题是不一样的，所以需要从各自的情况出发，然后看是否有其他可选的方法，哪些问题适合我们解决才是最重要的。

第四，损害赔偿的问题。我们缺少损害赔偿方面的讨论，在《卡塔赫纳生物安全议定书》从来没有深入讨论过，《名古屋-吉隆坡补充议定书》有相关规定，这又开辟了一个新的讨论机会。从原则角度来说，假如带来了损害我们怎么去赔偿，伤害方如何去赔偿受害方，如果环境受到了伤害我们怎么去救济等。我们要看看自己的法律是否已经跟上，是否有一些程序已经到位，是否有一些相关做法已经实施。实际上，有一些国际文件及相关补充文件，对这方面是没有规定的，就是说各国应该承担一些各自的义务，包括如何把我们的工作做得更好。损害赔偿是一个整体问题，需要一些法律建设及其他方面的措施解决。另外，我们在国家层面讨论了损害的问题，实际上涉及赔偿，只是在社会方面、环境方面等问题面临更大的困难，但是不管怎么说，这些问题我们应该放在桌面上去谈。其实刚才魏伟教授也谈到了这个问题，这个问题大家争论还挺热烈，我们认为议定书已经做了一些规定，而且，议定书也要求保持信息畅通和交流。如果我们有一些反对的意见，如果我们在现实当中发现一些 GMO 带来的损害，我们一定要采取一些措施进行赔偿。另外关于损害的起因问题，我们应该采取正确的方法和手段审视这些的损害，预测其在长时间内会产生什么样的损害。可能短期没法确定其损害风险，但是我们要有一个时间框架。

此外，还有财务方面的安全性等问题，所以我们讨论的方面的确是非常多，不可能一个一个都讨论得那么具体，但是不管怎么说，我们讨论的质量和成熟度还是非常不错，参与者非常积极。对于各自的国家来说，我们应该非常公开地将这种讨论和辩论继续下去，把相关的问题带进来，用不同的立场去看这个问题，使知识不断更新，让其他的经验得到分享。像研讨会这种讨论会也是一种平台，可以促使我们继续做这种工作。这个问题的解决的确需要在国家层面开展生物安全的讨论，农业方面一定要有政策方面的可选手段，哪一些对公众最有利，以及我们如何整合相关问题，如何将不同人对不同问题的意见综合等等。

最后，非常感谢在过去 10 年安排的五次研讨会，特别感谢在过去 10 年当中我们能够和薛达元教授共同合作举办这些活动，这些由环保部以及中央民族大学等发起的项目活动。我们在北京有办事处，我们非常荣幸能够与来自不同部门的代表开展合作，我们会继续为这些活动提供支持，再次衷心感谢薛达元教授，感谢在座每一位朋友的参与，并与我们分享了丰富的经验，我也从中学到了很多。

谢谢！

薛达元教授闭幕发言

Closing Remarks of Xue Da-Yuan

薛达元[1,2]
（1. 环境保护部南京环境科学研究所；2. 中央民族大学生命与环境学院）
Xue Da-Yuan[1,2]
（1. Nanjing Institute of Environmental Science，MEP；
2. College of Life and Environmental Sciences，Minzu University of China）

非常感谢 Lin Li Ching 的精彩总结，刚才她做了这两天的总结，但我想做一个 10 年的总结。从 2004 年“生物安全国际论坛第一届研讨会”开始，到这次举办的 2013 年“生物安全国际论坛第五届研讨会”，已经跨越了 10 年的时间，共举办了 5 次国际研讨会。为什么做 10 年的总结？因为我过两年就到退休年龄了，不知道以后还能不能继续主办这个论坛的研讨会，不肯定。所以，10 年当中我有一点感受，我举办这个系列论坛会议，完全是靠自己的兴趣以及对社会的责任。

2004 年、2005 年、2008 年和 2011 年的四届会议都得到了德国技术合作公司（GTZ/GIA）项目的支持，这是德国政府支持中国生物安全能力建设的项目。我们和第三世界网络（TWN，The Third World Net）也有很多合作，自 2005 年以来一直得到 TWN 的合作项目支持，已有七八年时间了。“第三世界网络”是名符其实的“第三世界的网络”，就是他们始终站在发展中国家的立场上，在国际舞台上非常活跃，包括《生物多样性公约》及其《卡塔赫纳生物安全议定书》、《关于获取遗传资源和公正公平分享其利用惠益的名古屋议定书》、《关于赔偿责任和补救的名古屋-吉隆坡补充议定书》的谈判和后续履行活动，他们都是积极参与，并发挥很大的作用。例如，下周（2013 年 6 月初）《生物多样性公约》秘书处将在加拿大蒙特利尔召开有关《名古屋议定书》能力建设的专家组会议，全球五大洲只遴选了 15 个专家代表和 3 个国际机构观察员，我发现 TWN 就是其中的一个观察员，他们的总干事 CHEE Yoke Ling 女士将出席会议，我本人也将作为亚洲专家代表参加会议。对此，我第一感觉是 TWN 的影响很大，能够作为全球非政府环保组织参加国际事务；第二感觉是 TWN 很敬业，能够始终如一地参与国际公约的谈判进程，不管大会小会都有身影；第三感觉是 TWN 很专业，有专业水准，能够对生全球物多样性保护和生物安全管理做出贡献。

因此，我特别感谢能和第三世界网络合作多年，我们合作得非常愉快，也富有成效。我们合办了一份内部刊物——《生物多样性与传统知识简报》。这份刊物的原身是《国外生物安全信息简报》，专门讲国外转基因生物安全方面的动态进展，特别是风险方面的研究进展和新闻报道，多为原文直接翻译。由于经常报道转基因生物的负面影响，很多人对此都不满意，就连美国大使馆都有官员找我谈过，说你能不能不做这个，这对我们孟山都等生物技术公司具有不利影响。由于多种原因，2006 年始，我就将杂志改名为《生物多样性与传统知识简报》，并从一开始就与 TWN 合作，由我和 CHEE Yoke Ling 做主编，编辑部设在中央民族大学。这份杂志有四个栏目，包括遗传资源获取与惠益分享（揭露国际上的“生物剽窃”）；生物多样性相关传统知识的保护与传承；转基因生物安全；气候变化与农业可持续发展。虽然转基因生物安全只是一个栏目，但相关的内容仍然很多。我们的合作很成功，我们的杂志为双月刊，自 2006 年年底开始至今，已出版 40 期。每期印刷 1 000 册，免费寄送给相关政府官员（包括中央部委和地方省厅）、还有从事转基因研究的高校和科研院所的专业技术人员、还有感兴趣的公众，他们很多人反映这个东西还是有帮助的。

本次会议共有三个主办方，环境保护部南京环境科学研究所、中央民族大学和第三世界网络，我同时在环境保护部南京环境科学研究所和中央民族大学工作，在此我代表这两个单位以及合作方 TWN 对所有参会者表示特别的感谢。我还要特别感谢原环保部国际合作司的程伟雪副司长，他作为司局级的干部，10 年内的五次研讨会都参加了，而且他都是坚持每次会议从头到尾地认真听会和积极参与讨论，有时也帮助我主持会议。还有国家林业局科技发展中心的黄发强主任，他主管国家林业局的转基因林木生物安全及遗传资源获取与惠益分享方面的工作，他也总是从头至尾参加了每次研讨会。此外，外交部、农业部、国家质检总局等相关部门也非常重视转基因生物安全问题，每次会议都会派人参加，并致辞或做报告，我衷心感谢各部门及朋友们的关照和支持。环境保护部作为《卡塔赫纳生物安全议定书》以及《生物多样性公约》的牵头部门，成立了国家生物安全管理办公室，他们对主办“生物安全国际论坛”一直给予坚定的支持，而且这 10 年中举办的五次研讨会都是由环保部官方正式批准的。最近两三届会议还得到国家民委的官方批准。因此，我们的会议不是自己开的，而是在相关政府主管部门的领导和支持下召开的。

还有很多其他的部门给予我们很多支持。这次会议比以前会议缩小了规模，邀请的部门不多，但以前的几次会议邀请了很多部门的官员参与，包括卫生部、知识产权局、食品和药品管理局以及科技部等。这次会议的邀请了很多专家，主要是来自高等院校和科研院所的专家，也邀请了非政府组织和消费者公众代表，有些公众做了很多自愿性的研究工作，如尹帅军先生不仅是一个消费者，也是一名研究者，他手头上拥有许多转基因生物的数据，他从一个公众的角度来研究这个事，还做的这么专业，花了不少时间和精力，令人佩服。还有汪文蓉女士作为私营小公司的业主，私人出资资助转基因棉花的生物安全调查研究，这是非常不简单的。还有很多专家都是常年支持我们，使我很感动，假如我不能主办第六次研讨会的话，希望年轻人把这个论坛继续办下去。

感谢参加人员做的这些报告，我们这个平台特别体现了一个民主性，政府、非政府、科学家、生物技术公司、消费者公众，甚至还有媒体等的广泛参与，形成这么一个平台，这在中国其他地方很难见到的。中国官方研讨会全部都是政府官员与研究人员，基本上不邀请非政府组织，最多邀请国际组织，而不邀请草根组织，但我们连公众都邀请。第四次会议还特别邀请生物技术公司，全世界五大公司：孟山都、杜邦（先锋）、先正达、拜耳、巴斯夫等公司的代表都参加了会议。加拿大的一位教授给我发邮件说：在中国能做成这样，政府和非政府能够对话，能够坐下来谈，这一点是很重要的，他说这一点在加拿大都很难做到，他感到很惊讶，我们这种研讨会是把所有的利益相关方聚集在一起。我们的这种形式将来要继续保持下去。

虽然一次研讨会并解决了不中国的转基因生物安全问题，甚至没有什么结果，但是有这个论坛平台跟没有这个论坛平台是不一样的。事情总要有人来做，尽管我们解决不了什么问题，但是这个平台是重要的，如果大家同意这一点就鼓掌。

最后，对所有做报告的，参加讨论的，以及没有发言但在这里认真听报告的，我们都表示衷心感谢，非常感谢大家参加我们这个会议，并且对会议做出贡献。感谢大家，感谢所有的远程而来的外地专家和外国专家。感谢中央民族大学的研究生担负了会务支持，同时我们也感谢同传译员的辛勤工作。再一次感谢大家！

会议总结报告

Summary Report

王艳杰[1]，薛达元[1,2]
（1.中央民族大学生命与环境科学学院；2.环境保护部南京环境科学研究所）
Wang Yan-Jie[1]，Xue Da-Yuan[1,2]
（1.College of Life and Environmental Sciences，Minzu University of China；
2.Nanjing Institute of Environmental Sciences，MEP）

1 会议背景

转基因生物对生态环境、人类健康和社会经济的影响在许多国家包括中国仍然是重要的问题。在国际层面，2000 年在《生物多样性公约》下产生了专门针对转基因生物的《卡塔赫纳生物安全议定书》，进而在 2010 年，又在《卡塔赫纳生物安全议定书》产生了一项新的针对转基因生物损害的国际法——《有关赔偿责任与补救的名古屋-吉隆坡补充议定书》。另外，2012 年，《卡塔赫纳生物安全议定书》第 6 次缔约方大会针对风险评估和社会经济影响等作出相关决定，也推动了履行议定书相关议题的工作。

为了研讨国际和国内有关转基因生物安全的最新进展，促进国家之间和专家之间以及管理人员与研究人员之间的信息交流，经环境保护部和国家民族事务委员会正式批准，由环境保护部南京环境科学研究所与中央民族大学以及第三世界网络（TWN）主办，中国生态学会民族生态学专业委员会、中国环境科学学会生态与自然保护分会以及中国民族地区环境资源保护研究所协办，于 2013 年 5 月 27～28 日在北京湖北大厦召开了“生物安全国际论坛——第五次研讨会”。研讨会议程主要包括两个主题报告和三个专题下的 16 个报告以及开放式讨论。

2 会议开幕

会议由中央民族大学教授、环境保护部南京环境科学研究所研究员薛达元主持。他介绍说，此次研讨会共有 65 名代表参加，分别来自中国、加拿大、马来西亚、美国、南非、荷兰、奥地利、拉脱维亚等多个国家和相关国际机构。国内参会者包括来自中国环境保护部、农业部、外交部、国家林业局、国家质检总局等政府部门的官员和代表，高校和科研院所的专家与研究人员，TWN 及绿色和平等非政府组织代表，以及对转基因生物安全感兴趣的公司、媒体和公众代表。

在研讨会开幕式上，环境保护部国家生物安全管理办公室王捷处长、外交部条法司付长华处长（一秘）、环境保护部南京环境科学研究所科技处李维新处长、中央民族大学国际合作处何克勇处长以及第三世界网络代表 Lim Li Ching 女士分别致辞。

3 主题报告

主题报告主要针对生物安全独立研究的最新科学进展，以及当前生物安全研究和风险评估面临的挑战进行讨论。复旦大学生命科学学院生态与进化生物学系主任卢宝荣教授做了题为“转基因生物技术的利益及其生物安全挑战”的报告，报告以转基因水稻例分析了转基因逃逸及其生态环境影响，并对环境风险评价的管理策略进行阐述。报告指出由于栽培稻与野生近缘种共同分布，转基因会从栽培稻向野生稻逃逸，产生的生态后果受基因本身、环境、种植状态的多重影响。奥地利环境保护局生物安全主管官员 Andreas Heissenberger 博士做了题为“转基因风险评估的原则、挑战及方法”的报告。报告介绍了《议定书》关于风险的定义和指导原则，风险评估过程、面临的问题及解决的措施。报告指出尽管有了这些指南性文件，但是需要研究更具体的方法、并获得更多的数据以解决不同国家和地区生物安全的风险评估问题。

4 专题会议

会议共有 3 个专题。

专题 1：转基因生物的风险评估进展。该专题主要根据《卡塔赫纳生物安全议定书》（以下简称《议定书》）的“风险评估与风险管理指南”而进行的风险评估国际标准的讨论，包括中国转基因生物安全的管理进展，《议定书》下风险评估标准的进展，转基因生物风险评价中的主要考虑，以及转基因水稻、转基因油菜、转基因棉花等对环境和非靶标害虫以及植物病害等方面影响的研究。

专题 2：转基因生物相关的社会经济影响。该专题主要讨论了转基因生物决策过程中考虑的社会经济因素。《议定书》允许各国在决策过程中考虑转基因生物的社会经济因素，制定相关国家生物安全法规。但是《议定书》并没有规定具体的评价标准和方法，而且调查数据不足，导致各国实施困难。该专题讨论了中国民众对转基因棉花的认知程度调查，世界各国应对转基因生物社会经济影响而采取的政策制度，南非转基因棉花的种植情况，全球和区域条件下转基因生物的社会经济影响，包括粮食安全、人类健康及经济效益等。

专题 3：转基因生物的赔偿责任与补救。该专题针对《有关赔偿责任与补救的名古屋-吉隆坡补充议定书》进行了讨论，阐述了生物安全损害、环境损害、生物安全损害赔偿、补救等概念，针对目前尚无“生物安全损害赔偿”案例的现状，分析各国有关生物安全损害及赔偿的立法、实践及面临的挑战，并提出相关建议。

开放式讨论：研讨会主题报告和各个分主题报告讨论之后，开放式讨论为参会者提供了一个自由发表意见和观点的空间。在开放讨论中，发言集中在转基因生物的损害赔偿机制、案例及转基因生物对环境、人类健康的影响，转基因生物安全的立法、监管方面等。此外，还针对转基因对传统育种和传统作物种质资源保护的影响，转基因生物风险评估时间的长短问题，如何调动公众参与与协商，如何利用法规政策及生物技术手段解决新出现的问题等进行了讨论。

5 会议闭幕

在会议总结和闭幕式上，第三世界网络（TWN）的 Lim Li Ching 女士和大会主席薛达元教授对此次研讨会进行了总结。Lim Li Ching 认为中国开展生物安全方面工作的态度和方法值得学习和借鉴，而且中国也正在和将继续努力进行转基因生物的环境风险及社会经济影响评估。薛达元教授对生物安全国际论坛（2004—2013 年）的 10 年工作进行了简要总结。在过去的 10 年中，本论坛共召开 5 次国际研讨会，有 500 多中外人员参加了论坛。他指出，该论坛的显明特点是为所有利益相关方提供了公平和自由交流的平台，包括研究人员、政府管理人员、生物技术公司、媒体、非政府社会团体及公众等，本论坛促进了中国专家与国外专家、政府与非政府机构、消费者与生物技术公司、媒体与公众之间在转基因生物安全问题上的交流。他希望这个平台能够继续下去，并希望更多的年轻人能够勇于担当责任，将此平台传承下去并发扬光大。

附 录
Appendix

附录一　会议日程

Appendix 1　Workshop Program

2013 年 5 月 26 日（周日）　26 May 2013（Sunday）

会议报到：2013 年 5 月 26 日 15：00 开始在湖北大厦注册报道

Registration：3 pm，26 May 2013 at Beijing Hubei Hotel

2013 年 5 月 27 日（周一）27 May 2013（Monday）

09：30—9：30　Registration　继续报到注册

09：30—10：00　开幕式（主持人：薛达元）

Opening Ceremony（Chair：XUE Dayuan）

致辞　Remarks

- 环境保护部国家生物安全管理办公室王捷处长致辞
  Mr. WANG Jie，Division Director，National Biosafety Office，Ministry of Environmental Protection（MEP）
- 外交部条法司付长华处长致辞
  Mr FU Changhua，Division Director，Ministry of Foreign Affairs
- 环境保护部南京环境科学研究所科技处李维新处长致辞
  Mr LI Weixin，Division Director of Nanjing Institute of Environmental Science，MEP
- 中央民族大学国际合作处处长何克勇教授致辞
  Prof. HE Keyong，Director，International Cooperation Office，Minzu University of China
- 第三世界网络代表 Lim Li Ching 女士致辞
  Ms Lim Li Ching，representative of TWN

Group photo（合影）

10：00—12：00　主题报告　Keynote Presentations

主持人：薛达元（中央民族大学教授，环境保护部南京环境科学研究所研究员）

Chair：Pro. XUE Dayuan（Professor，Minzu University of China and Nanjing Institute of

Environmental Science，MEP）

10：00—10：50： 转基因生物技术的利益及其生物安全挑战

Advantages of transgenic biotechnology and its challenges of biosafety
卢宝荣，复旦大学（上海）教授
- Prof. LU Baorong，Fudan University，Shanghai，China
Questions and Discussion（提问与讨论）

10：50—11：10 Break（茶歇）

11：10—12：00 转基因生物的风险评估：主要原则、面临挑战和应对方法

Risk assessment of GMOs：Key principles，challenges and ways forward
Dr. Andreas Heissenberger，Environment Agency Austria（奥地利环境保护局）
Questions and Discussion（提问与讨论）

12：00—13：30 Lunch（午餐）

13：30—17：30 专题 1：转基因生物的风险评估

Session 1：Risk Assessment of GMOs
主持人（Chair）：Ms. LIM Li Lin

13：30—15：20 Presentations（20 minutes each）（报告，每人 20 分钟）

- 中国农业转基因生物安全管理进展
  The Progress on Biosafety Regulation of Agri-GMOs in China
- 付仲文，中国农业部科技发展中心 副处长、高级农艺师
  Mr. FU Zhongwen，Deputy Division Director，Center of Science and Technology Development，Ministry of Agriculture of China

- 推进《卡塔赫纳生物安全议定书》下的风险评估
  Advancing Risk Assessment under the Cartagena Protocol on Biosafety
  Ms. Lim Li Ching，Experts of Third World Network（第三世界网络生物安全专家）

- 遗传修饰生物体风险评价中的主要考虑
  Main Considerations for Risk Assessment of GMOs
  魏伟，中国科学院植物研究所研究员
  Dr Wei Wei，Institute of Botany，Chinese Academy Sciences

- 转 *Bt* 基因水稻及常规 SY63 水稻叶片结构对臭氧浓度升高的响应
  Leaf Morphology and Ultrastructure Responses to Elevated O_3 in Transgenic Bt

（Cry1Ab/Cry1Ac）Rice and Conventional Rice under Fully Open-air Field Conditions

刘标，环境保护部南京环境科学研究所研究员

Dr Liu Biao，Nanjing Institute of Environmental Sciences，MEP

Discussion（讨论）

15：20—15：40 Break（茶歇）

15：40—17：30 Presentations（20 minutes each）（报告，每人 20 分钟）

- 转基因油菜的风险评估：油菜 3 与其近缘野生种之间的基因流

  Risk Assessment of GM Oilseed Rape：Gene Flow between Oilseed Rape and its Wild Relatives

  刘勇波，中国环境科学研究院生态所副研究员

  Dr. LIU Yongbo，China Research Academy of Environmental Science
- 转基因大豆风险评估：转基因耐除草剂大豆饲喂鹌鹑 90 天的安全评估

  Risk Assessment of GM Soybean：A 90-day Safety Assessment of Genetically Modified Glyphosate-Tolerant Soybean in Japanese Quails

  王长永，环境保护部南京环境科学研究所研究员

  Dr. WANG Changyong，Nanjing Institute of Environmental Sciences，MEP
- 转 *Bt* 基因作物对植物-害虫-天敌食物链影响研究进展

  Influence of Transgenic Crops on the Food Chain：Plant-Pests-Predators

  赵彩云，中国环境科学研究院 副研究员

  Dr ZHAO Caiyun，Chinese Reasearch Academy of Environmental Science
- 中国六省转基因抗虫棉种植对棉花害虫的影响调查

  Investigation on Impacts of GM Cotton's Plantation on the Non-target Pests in the Six Provinces of China

  陈晨/汪文蓉，中央民族大学生命与环境学院 研究生/上海有机道文化发展有限公司

  CHEN Chen/WANG Wenrong，College of Life and Environmental Science，Minzu University of China/Shanghai Organic Food Culture Development Company

Discussion（30 minutes）（30 分钟讨论）

18：00—20：00 Dinner（晚餐）

2013 年 5 月 28 日（周二）28 May 2013（Tuesday）

9：30—12：00 专题 2：转基因生物相关的社会经济影响

Session 2：Socio-Economic Considerations related to GMOs

主持人：薛达元

Chair：Pro. XUE Dayuan

Presentations（each 20 minutes plus 5 minutes for questions）

报告（每人 20 分钟+5 分钟提问）

- 中国转基因棉花的社会经济影响
  Socio-economic Impacts of GM Cotton in China
  何佩生/赵珩，荷兰莱顿大学教授
  Prof. HO Peter/Prof. ZHAO Heng，Leiden University，The Netherlands
- 转基因生物的社会经济影响评估
  Socio-economic Aspects in the Assessment of GMOs
  Dr. Andreas Heissenberger，Environment Agency Austria
  奥地利环境保护局
- 南非转 *Bt* 基因棉花的种植经验
  Experiences of Bt Cotton from South Africa
  Mariam Mayet，African Centre for Biosafety
  非洲生物安全中心（南非）
- 接受转基因生物的社会经济影响：比较分析不同国家的选择
  Socio-Economic Impacts of the Adoption of GMOs：Comparative Analysis of Various National Choices
  Dr. Yves Tiberghien，Director，Institute of Asian Research，Associate Professor，Department of Political Science，University of British Columbia（UBC），Canada
  加拿大不列颠哥伦比亚大学亚洲研究院院长
- 转基因生物的社会经济影响：全球和区域水平
  The Social and Economic Impacts of Genetically-modified Organisms：Global and Local
  Dr. Jerry McBeath，Professor and Chair，Department of Political Science，University of Alaska Fairbanks，USA
  美国阿拉加斯大学教授，政治系主任

Discussion in 25 minutes（讨论 25 分钟）

12：00—13：30 Lunch（午餐）

13：30—15：30 专题 3：转基因生物的赔偿责任与补救

Session 3：Liability and Redress for Damage from GMOs

Chair（主持人）：Ms. LIN Li Lim

Presentations（20 minutes each）（报告，每人 20 分钟）

- 《卡塔赫纳生物安全议定书关于赔偿责任与补救的名古屋-吉隆坡补充议定书》的分析与实施挑战
  The Nagoya–Kuala Lumpur Supplementary Protocol on Liability and Redress to the Cartagena Protocol on Biosafety：An Analysis and Implementation Challenges

Ms. LIM Li Lin，第三世界网络专家

- 转基因生物安全及其损害赔偿机制研究

  On mechanism of liability and redress for genetic modified organisms

  刘燕/张振华，环境保护部南京环境科学研究所生物多样性中心副主任

  Ms. LIU Yan，Nanjing Institute of Environmental Sciences，MEP

- 生物安全损害赔偿法律问题研究

  Legal Liability And Redress For Damage from GMOs

  于文轩，中国政法大学副教授

  Dr. YU Wenxuan，Associate Professor，China University of Political Science and Law，Beijing

Discussion（60 minutes）（60 分钟讨论）

15：30—16：00 会议成果总结与闭幕

Summary and Closing Ceremony

Chair（主持人）：XUE Dayuan/LIM Li Ching

附录二　转基因生物安全国际论坛第五次研讨会参会人员名单

Appendix 2　List of Participants for International Biosafety Forum-5th

序号	姓名（Name）	单位（Organization of Affiliation）	职务（Title）	电子邮箱（Email）	地址（Address）	电话（Phone）	传真（Fax）	移动电话（Mobile phone）
国外专家　（Foreign Experts）								
1	Yves TIBERGHIEN	University of British Columbia Institute of Asian Research and Dept of Political Science	Director/Prof.	Yves.tiberghien@ubc.ca	Vancouver，CANADA			131 475 09805（China）
2	Lim Li Ching	Third World Network，Malaysia	Senior Researcher	ching@twnetwork.org	Petaling Jaya，Selangor，Malaysia	+603-23002585		+6012-2079744
3	Lim Li Lin	Third World Network，Malaysia	Senior Researcher	lin@twnetwork.org	Petaling Jaya，Selangor，Malaysia			
4	Andreas Heissenberger	Umweltbundesamt - Environment Agency Austria	Unit Director	andreas.heissenberger@umweltbundesamt.at	Vienna，Austria			
5	Jerry McBeath	University of Alaska Fairbanks，US	Professor，Chair of Department	gamcbeath@alaska.edu	Fairbanks，AK 99709，USA			

序号	姓名（Name）	单位（Organization of Affiliation）	职务（Title）	电子邮箱（Email）	地址（Address）	电话（Phone）	传真（Fax）	移动电话（Mobile phone）
6	Jerry Huang McBeath	Agricultural and Forestry Experiment Station，University of Alaska Fairbanks，US	Professor	jhmcbeath@alaska.edu	Fairbanks，AK 99709，USA	907-474-7431	907-474-6099	
7	Mariam Mayet	African Centre for Biosafety	Director	mariammayet@mweb.co.za	South African	011-6460699	011-4861156	832694309
8	Jim Harkness（郝克明）	Institute for Agriculture and Trade Policy	President	jharkness@iatp.org	Minneapolis MN55404，USA			
9	Shefali Sharma	Institute for Agriculture and Trade Policy		ssharma@iatp.org	Minneapolis MN55404，USA			
10	Chad Futrell	IES BEIJING	Visiting Scholar	wchadfutrell@gmail.com	Latovia			+86-18610542012
11	Karlis Rokpelnis	Minzu University of China	Ph.D Student	Karlis@yahoo.com	Latovia			+86-15201594730
12	Solvita Poseiko		Visiting Scholar	solvita.poseiko@gmail.com	Latovia			
国 内 报 告 专 家 （Domestic Experts）								
13	卢宝荣 Lu Baorong	复旦大学生命科学学院 College of Life Science，Fudan University	教授 Professor	brlu@fudan.edu.cn	上海邯郸路 220 号，200433	+86-21-65643668	+86-21-65643668	+86-13681959109
14	付仲文 Fu Zhongwen	农业部科技发展中心 Development Center for Science and Technology，Ministry of Agriculture，P. R. China	副处长 Deputy Division Director	fuzhongwen@agri.gov.cn	北京市朝阳区东三环南路 96 号农丰大厦 717 房间，100122	+86-10-59199389	+86-10-59199391	+86-18210095388

序号	姓名（Name）	单位（Organization of Affiliation）	职务（Title）	电子邮箱（Email）	地址（Address）	电话（Phone）	传真（Fax）	移动电话（Mobile phone）
15	刘标 Liu Biao	环境保护部南京环境科学研究所 Nanjing Institute of Environmental Sciences，China Ministry of Environmental Protection（MEP）	研究员 Research Fellow	85287064@163.com	南京蒋王庙街8号，210042	+86-25-85287064	+86-25-86411611	+86-13814065155
16	赵彩云 Zhao Caiyun	中国环境科学研究院 Chinese Research Academy of Environmental Sciences	副研究员 Associate Research Fellow	zhaocy@craes. org.cn	北京朝阳区安外大羊坊8号院	+86-10-84910906	+86-10-84910906	+86-13811149270
17	魏伟 Wei wei	中国科学院植物研究所 Institute of Botany，The Chinese Academy of Sciences	副研究员 Associate Research Fellow	weiwei@ibcas.ac.cn	北京香山南辛村20号，100093	+86-10-62836275	+86-10-82596146	+86-15801348973
18	王长永 Wang Changyong	环境保护部南京环境科学研究所 Nanjing Institute of Environmental Sciences，MEP	研究员 Research Fellow	wcy@nies.org	南京蒋王庙街8号，210042	+86-25-85287223	+86-25-85287131	+86-13405806268
19	于文轩 Yu Wenxuan	中国政法大学 China University of Political Science and Law	副教授 Associate Professor	wenxuanyu@mail.tsinghua.edu.cn	北京市海淀区西土城路25号中国政法大学环境资源法研究和服务中心，100088			+86-13693605826

序号	姓名（Name）	单位（Organization of Affiliation）	职务（Title）	电子邮箱（Email）	地址（Address）	电话（Phone）	传真（Fax）	移动电话（Mobile phone）
20	刘勇波 Liu Yongbo	中国环境科学研究院 Chinese Research Academy of Environmental Sciences	副研究员 Associate Research Fellow	liuyb@craes.org.cn	北京香山南辛村 20 号			+86-13718660809
21	张振华 Zhang Zhenhua	环境保护部南京环境科学研究所 Nanjing Institute of Environmental Sciences，MEP	助理研究员 Assistant Research Fellow	zhang1984922@hotmail.com	南京蒋王庙街 8 号，210042	+86-25-85287251	+86-25-85287251	+86-15951869925
22	赵珩 Zhao Heng	中央民族大学生命与环境科学学院 College of Life and Environmental Sciences，Minzu University of China（MUC）	教授 Professor	hengzhao2000@gmail.com	北京市海淀区中关村南大街 27 号，100081	+86-315-03637224	+86-315-03637225	+86-13718369720
23	陈晨 Chen Chen	中央民族大学生命与环境科学学院 College of Life and Environmental Sciences，MUC	研究生 Graduate Student	chenchenxf@126.com	北京市海淀区中关村南大街 27 号，100081			+86-18611700619

序号	姓名（Name）	单位（Organization of Affiliation）	职务（Title）	电子邮箱（Email）	地址（Address）	电话（Phone）	传真（Fax）	移动电话（Mobile phone）
				参会人员（Participants）				
24	黄发强 Huang Faqiang	国家林业局科技发展中心 Center of Science and Technology Development，State Forestry Administration	副主任 Deputy Director General	hfq8833@sina.com	北京市东城区和平里东街 18 号	+86-10-84238703	+86-10-84238703	+86-13701240158
25	付长华 Fu Changhua	外交部 Ministry of Foreign Affairs	一秘/处长 First Secretary	fu_changhua@mfa.gov.cn	北京市朝阳区朝阳门南大街 2 号			+86-18701398409
26	孙卓婧 Sun Zhuoqian	农业部科技发展中心 Development Center of Science and Technology，Ministry of Agriculture，China	项目官员 Program Officer	szjman@163.com	北京市朝阳区农展南里 11 号	+86-10-59199390		+86-15210845201
27	付伟 Fu Wei	中国检验检疫科学研究院 Chinese Academy of Inspection and Quarantine（CAIQ）	副主任 Deputy Division Director	fuwei0212@163.com	北京市朝阳区惠新西街惠新里 241 号 1003 室	+86-10-64913887		+86-15001098698
28	王捷 Wang Jie	环境保护部 国家生物安全管理办公室 National Biosafety Management Office，China Ministry of Environmental Protection（MEP）	处长 Director	wang.jie@mep.gov.cn	北京市西直门内南小街 115 号，100035	+86-10-66556325		+86-13701278069

序号	姓名（Name）	单位（Organization of Affiliation）	职务（Title）	电子邮箱（Email）	地址（Address）	电话（Phone）	传真（Fax）	移动电话（Mobile phone）
29	关潇 Guan Xiao	环境保护部国家生物安全管理办公室 National Biosafety Management Office，MEP	项目官员 Program Officer	guanxiao815@yahoo.com.cn	北京市西直门内南小街 115 号，100035	+86-10-66556327	86-10-66556327	+86-18601278718
30	于之的 Yu Zhidi	环境保护部对外合作中心 Foreign Economic Cooperation Office，MEP	处长 Division Director	yu.zhidi@mepfeco.org.cn	北京市西城区后英房胡同 5 号，100035			+86-13810467838
31	程伟雪 Cheng Weixue	环境保护部国际合作司 Department of International Cooperation，MEP	高级顾问 Senior Consultant	cheng.weixue@mep.gov.cn	北京市西直门内南小街 115 号，100035			+86-13522937445
32	李维新 Li Weixin	环境保护部南京环境科学研究所 Science and Technology Department，Nanjing Institute of Environmental Sciences，Ministry of Environmental Protection，China	处长 Division Director	lwx@nies.org	南京蒋王庙街 8 号，210042			+86-15366090936
33	何克勇 He Keyong	中央民族大学国际交流处 International Cooperation Office，MUC	处长 Director	keyonghe@163.com	北京市海淀区中关村南大街 27 号，100081			+86-15810105931

序号	姓名（Name）	单位（Organization of Affiliation）	职务（Title）	电子邮箱（Email）	地址（Address）	电话（Phone）	传真（Fax）	移动电话（Mobile phone）
34	韦贵红 Wei Guihong	北京林业大学法学院 Department of Law, Beijing Forestry University	副教授 Associate Professor	weiguihong@gmail.com	北京市海淀区清华东路 35 号			+86-13051050877
35	康定明 Kang Dingming	中国农业大学 China Agricultural University	教授 Professor	kdm@pku.edu.cn	北京市海淀区北京大学生命科学院生物技术楼	+86-10-62756091	+86-10-62751847	+86-13651119146
36	杨君 Yang Jun	大连理工大学 Dalian University of Technology	副教授 Associate Professor	junyang@dlut.edu.cn	大连市凌工路 2 号大连理工大学生命科学与技术学院	+86-411-84707245		+86-15541150567
37	汪文蓉 Wang Wenrong	上海有机道文化发展有限公司 Shanghai Organic Food Culture Development Company	总经理 Director	wwenrong@msn.com	上海市徐汇区裕德路 168 号徐汇商务大厦 1813 室，200030	+86-021－54651119	+86-021－54651119	+86-13012836996
38	周玖璇 Zhou Jiuxuan	云南思力生态替代技术中心 Yunnan Pesticide Eco-Alternatives Center	主任 Director	zhou.jiuxuan@gmail.com	中国云南昆明市	+86-871-5656769	+86-871-5656373	+86-15925129216

序号	姓名（Name）	单位（Organization of Affiliation）	职务（Title）	电子邮箱（Email）	地址（Address）	电话（Phone）	传真（Fax）	移动电话（Mobile phone）
39	俞江丽 Yu Jiangli	绿色和平东亚办公室（Greenpeace East Asia）	部门副经理 Assistant Manager	yu.jiangli@greenpeace.org jy2381@gmail.com	东城区新中街68号聚龙花园7号楼聚龙商务楼3层	+86-10-65546931-107	+86-10-65546932	+86-13910742099
40	张菁 Zhang Jing	绿色和平东亚办公室（Greenpeace East Asia）	项目主任 Campainer	zhang.jing@greenpeace.org	东城区新中街68号聚龙花园7号楼聚龙商务楼3层	+86-10-65546931-141	+86-10-65546932	+86-113540052920
41	尹帅军 Yin Shuaijun	海南海洋安全与合作发展研究院（representative of public）	特约研究员 Visiting Researcher	antintheheaven@163.com	北京市海淀区北安河村5街30号			+86-13693302550
42	武建勇 Wu Jianyong	环境保护部南京环境科学研究所 Nanjing Institute of Environmental Sciences，MEP	副研究员 Associate Researcher	wujy10@hotmail.cn	南京蒋王庙街8号，210042	+86-25-85287230	+86-25-8528723	+86-15062205261
43	降蕴彰 Jiang Yunzhang	经济观察报 Economic Observer（News paper）	记者 Journalist	Jiangnan72@126.com	北京市东城区安德路47号院甲11号楼	+86-10-64208996	+86-10-64208996	+86-18910079373
44	白田田 Bai Tiantian	经济参考报 Economic Information Daily	记者 Journalist	tiantiannews@yeah.net	北京市宣武门西大街57号			+86-15811212695

序号	姓名（Name）	单位（Organization of Affiliation）	职务（Title）	电子邮箱（Email）	地址（Address）	电话（Phone）	传真（Fax）	移动电话（Mobile phone）
45	杨红军 Yang Hongjun	云南省生态农业研究所 Yunnan Institute of Ecological Agriculture	副所长 Deputy Director	327842781@qq.com	云南省昆明市高新区海源北路丽阳星城 99 号	+86-871-68323744	+86-871-68322277	+86-13608710018
46	陈惜平 Chen Xiping	第三世界网络（TWN Beijing Office）	高级顾问 Senior Consultant	kchen@twnetwork.org	北京市建国门外外交公寓 4-1-132	+86-10-85324730		+86-13801237269
47	朱贞艳 Zhu Zhenyan	第三世界网络（TWN Beijing Office）	法律顾问 Legal Counsultant	zhenyan@twnetwork.org	北京市建国门外外交公寓 4-1-132	+86-10-85324730		+86-18862390953
48	关正君 Guan Zhengjun	中国科学院植物研究所 Institute of Botany，The Chinese Academy of Sciences	博士后 Postdoctorate	zhengjunguan@126.com	北京市海淀区香山南辛村 20 号，100093			+86-15300203284
49	方伟伟 Fang Weiwei	中国科学院植物研究所 Institute of Botany，The Chinese Academy of Sciences	研究生 Graduate Student	fangww@ibcas.ae.cn	北京市海淀区香山南辛村 20 号，100094			+86-18810274791
50	曹迪 Cao Di	中国科学院植物研究所 Institute of Botany，The Chinese Academy of Sciences	研究生 Graduate Student	caodi0227@sohu.com	北京市海淀区香山南辛村 20 号，100095			+86-15101009459
51	薛达元 Xue Dayuan	中央民族大学 Minzu University of China 环境保护部南京环境科学研究所 Nanjing Institute of Envi. Sci.，MEP	教授 Professor	xuedayuan@hotmail.com	北京市海淀区中关村南大街 27 号，100081	+86-10 6893 1632	+86-10 6893 1632	+86-13910176361

序号	姓名（Name）	单位（Organization of Affiliation）	职务（Title）	电子邮箱（Email）	地址（Address）	电话（Phone）	传真（Fax）	移动电话（Mobile phone）
52	成功 Cheng Gong	中央民族大学生命与环境科学学院 College of Life and Environmental Sciences，MUC	讲师 Lecturer	vitorchenggong@gmail.com	北京市海淀区中关村南大街27号，100081	+86-10 6893 1632	+86-10 6893 1632	+86-13488664474
53	吴力 Wu Li	中央民族大学生命与环境科学学院 College of Life and Environmental Sciences，MUC	行政助理 Administrator	bfsuwl@yahoo.com.cn	北京市海淀区中关村南大街27号，100081	+86-10 6893 1632	+86-10 6893 1632	+86-13811269362
54	尹仑 Yin Lun	中央民族大学生命与环境科学学院 College of Life and Environmental Sciences，MUC	博士生 PhD Candidate	yinlun@cbik.ac.cn	北京市海淀区中关村南大街27号，100081	+86-10 6893 1632	+86-10 6893 1632	+86-13888267735
55	王艳杰 Wang Yanjie	中央民族大学生命与环境科学学院 College of Life and Environmental Sciences，MUC	博士生 PhD Candidate	yanjie2005@126.com	北京市海淀区中关村南大街27号，100081	+86-10 6893 1632	+86-10 6893 1632	+86-18911075284
56	杜玉欢 Du Yuhuan	中央民族大学生命与环境科学学院 College of Life and Environmental Sciences，MUC	博士生 PhD Candidate	yuhuan0306@163.com	北京市海淀区中关村南大街27号，100081	+86-10 6893 1632	+86-10 6893 1632	+86-15210633226

序号	姓名（Name）	单位（Organization of Affiliation）	职务（Title）	电子邮箱（Email）	地址（Address）	电话（Phone）	传真（Fax）	移动电话（Mobile phone）
57	刘春晖 Liu Chunhui	中央民族大学生命与环境科学学院 College of Life and Environmental Sciences，MUC	研究生 Graduate Student	liuchunhui1022@hotmail.com	北京市海淀区中关村南大街27号，100081	+86-10 6893 1632	+86-10 6893 1632	+86-15116995520
58	郑燕燕 Zheng Yanyan	中央民族大学生命与环境科学学院 College of Life and Environmental Sciences，MUC	研究生 Graduate Student	yan3168586@126.com	北京市海淀区中关村南大街27号，100081	+86-10 6893 1632	+86-10 6893 1632	+86-15501067502
59	张家楠 Zhang Jianan	中央民族大学生命与环境科学学院 College of Life and Environmental Sciences，MUC	研究生 Graduate Student	venus_nan@126.com	北京市海淀区中关村南大街27号，100081	+86-10 6893 1632	+86-10 6893 1632	+86-13426036038
60	张爽 Zhang Shuang	中央民族大学生命与环境科学学院 College of Life and Environmental Sciences，MUC	研究生 Graduate Student	xzhangshuang@sina.com	北京市海淀区中关村南大街27号，100081	+86-10 6893 1632	+86-10 6893 1632	+86-18311096318
61	黎平 Li Ping	中央民族大学生命与环境科学学院 College of Life and Environmental Sciences，MUC	研究生 Graduate Student	Liping2005hot@126.com	北京市海淀区中关村南大街27号，100081	+86-10 6893 1632	+86-10 6893 1632	+86-13466644698

序号	姓名（Name）	单位（Organization of Affiliation）	职务（Title）	电子邮箱（Email）	地址（Address）	电话（Phone）	传真（Fax）	移动电话（Mobile phone）
62	葛小芳 Ge Xiaofang	中央民族大学生命与环境科学学院 College of Life and Environmental Sciences，MUC	研究生 Graduate Student	gexiaof@163.com	北京市海淀区中关村南大街27号，100081	+86-10 6893 1632	+86-10 6893 1632	+86-15652663526
63	麦尔哈巴·阿布拉 Merhaba·Abula	中央民族大学生命与环境科学学院 College of Life and Environmental Sciences，MUC	研究生 Graduate Student	41691878@qq.com	北京市海淀区中关村南大街27号，100081	+86-10 6893 1632	+86-10 6893 1632	+86-18810626254
64	王新 Wang Xin	中央民族大学生命与环境科学学院 College of Life and Environmental Sciences，MUC	研究生 Graduate Student	wangxin51007@hotmail.com	北京市海淀区中关村南大街27号，100081	+86-10 6893 1632	+86-10 6893 1632	+86-13426045181
65	梁晨 Liang Chen	中央民族大学生命与环境科学学院 College of Life and Environmental Sciences，MUC	研究生 Graduate Student	chen.liang1122@gmail.com	北京市海淀区中关村南大街27号，100081	+86-10 6893 1632	+86-10 6893 1632	+86-13811121391